Building Codes ILLUSTRATED
FOR HEALTHCARE FACILITIES

Building Codes
ILLUSTRATED
FOR HEALTHCARE
FACILITIES

A Guide to Understanding the 2006 International Building Code® for Healthcare Facilities

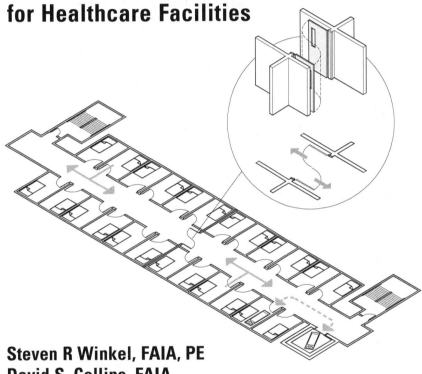

Steven R Winkel, FAIA, PE
David S. Collins, FAIA
Steven P. Juroszek, AIA

Building Codes Illustrated Series Advisor
Francis D.K. Ching

BICENTENNIAL
1807
WILEY
2007
BICENTENNIAL

John Wiley & Sons, Inc.

Published by John Wiley & Sons, Inc., Hoboken, New Jersey
Published simultaneously in Canada

For general information about our other products and services, please contact our Customer Care Department within the United States at (800) 762-2974, outside the United States at (317) 572-3993 or fax (317) 572-4002.

Wiley also publishes its books in a variety of electronic formats. Some content that appears in print may not be available in electronic books. For more information about Wiley products, visit our web site at www.wiley.com.

Winkel, Steven R.
  Building codes illustrated for healthcare facilities / Steven R. Winkel, David Collins, Steven P. Juroszek.
    p. cm. -- (Building codes illustrated series)
  Includes bibliographical references and index.
  ISBN 978-0-470-04847-4 (pbk.)
  1. Health facilities--Design and construction--Standards. 2. Building--Standards. 3. International building code. I. Collins, David S. (David Smith), 1945- II. Juroszek, Steven P. III. Title. IV. Title: Healthcare facilities.
  RA967.W56 2007
  725'.51--dc22

                                        2007002360

10 9 8 7 6 5 4 3 2

Disclaimer

The book contains the authors' analyses and illustrations of the intent and potential interpretations of the *2006 International Building Code®* (IBC) as it applies to healthcare facilities. The illustrations and examples are general in nature and not intended to apply to any specific project without a detailed analysis of the unique nature of the project. As with any code document, the IBC is subject to interpretation by the Authorities Having Jurisdiction (AHJ) for their application to a specific project. Designers should consult the local Building Official early in project design if there are questions or concerns about the meaning or application of code sections in relation to specific design projects.

The interpretations and illustrations in the book are those of the authors. The authors do not represent that the illustrations, analyses, or interpretations in this book are definitive. They are not intended to take the place of detailed code analyses of a project, the exercise of professional judgment by the reader, or interpretive application of the code to any project by permitting authorities. While this publication is designed to provide accurate and authoritative information regarding the subject matter covered, it is sold with the understanding that neither the publisher nor the authors are engaged in rendering professional services. If professional advice or other expert assistance is required, the services of a competent professional person should be sought.

The authors and John Wiley & Sons would like to thank the International Code Council for their thorough review of the manuscript. Their review does not reflect in any way the official position of the International Code Council. Any errors in the interpretations or illustrations in the book are solely those of the authors and are in no way the responsibility of the International Code Council.

Acknowledgments

The authors would like to acknowledge the contributions of Francis D.K. Ching, whose drawings in *Building Codes Illustrated* provided the foundation and standard for the illustrations in this book. They would also like to acknowledge the contributions of Nathan Crane, Trevor Lunde, Gerard Robinson, and Raluca Vandergrift, who assisted with some of the illustrations in this book.

Contents

Preface

The primary purpose of this book is to familiarize code users with the use of the *2006 International Building Code®* (IBC) as it applies to healthcare facilities. It is intended as an instructional text on how the Code was developed and how it is organized, as well as a reference document on how to use the Code for the design of healthcare facilities. It is intended to be a companion to the IBC, not a substitute for it. This book must be read in concert with the IBC. Note that the design of many healthcare facilities will also involve the use of the National Fire Protection Association "Life Safety Code," also known as "NFPA 101." Designers must often consult both the International Building Code and NFPA 101 in concert with each other in the design of healthcare facilities. The IBC and NFPA 101 come from separate and distinct code development processes and are not necessarily coordinated. This book focuses on how the International Building Code applies to healthcare occupancies. It does not address the relationship between the requirements of NFPA 101 and the IBC. Designers must review with the regulatory authorities having jurisdiction over their project how the regulators will apply the often overlapping code requirements contained in these two codes.

This book is designed to give an understanding of how the International Building Code was developed, and how it is likely to be interpreted when applied to the design and construction of healthcare facilities. The intent of this book is to give a fundamental understanding of the relationship of codes to practice for healthcare design professionals, especially those licensed or desiring to become licensed as architects, engineers, or other related design professionals. Code knowledge is among the fundamental reasons for licensing design professionals, for the protection of public health, safety and welfare. It is our goal to make the acquisition and use of code knowledge easier and clearer for code users.

Many designers feel intimidated by building codes. They can seem daunting and complex at first glance. It is important to know that they are a product of years of accretion and evolution. Sections start simply and are modified, and new material is added to address additional concerns or to address interpretation issues from previous code editions. The complexity of a building code often comes from this layering of new information upon old without regard to overall continuity. Building codes are living documents, constantly under review and modification. It is vital to an understanding of codes to keep in mind that they are a human institution, written by ordinary people with specific issues in mind or specific agendas they wish to advance.

BUILDING CODE

Webster's Third New International Dictionary defines a building code as: "A set of rules of procedure and standards of materials designed to secure uniformity and protect the public interest in such matters as building construction and public health, established usually by a public agency and commonly having the force of law in a particular jurisdiction."

How This Book Is Organized

The first two chapters of this book give background and context regarding the development, organization and use of the IBC. Chapters 3 through 16 are organized and numbered the same as the corresponding subject-matter chapters in the IBC. Chapter 16 summarizes the structural provisions of IBC chapters 16 through 23. Book chapters 17 through 25, cover the remaining IBC chapters addressed by the book in the same order as the chapters in the code. Corresponding IBC chapter numbers are cross-referenced in the book index.

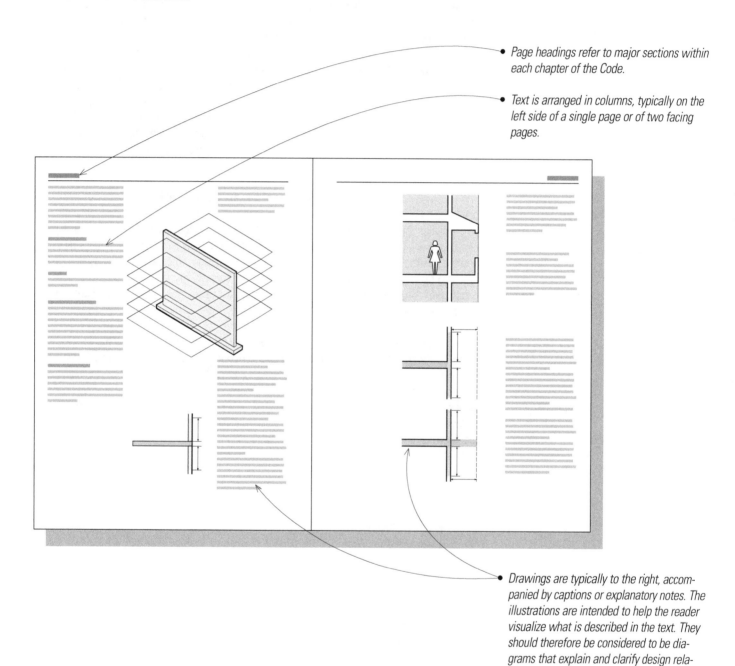

• *Page headings refer to major sections within each chapter of the Code.*

• *Text is arranged in columns, typically on the left side of a single page or of two facing pages.*

• *Drawings are typically to the right, accompanied by captions or explanatory notes. The illustrations are intended to help the reader visualize what is described in the text. They should therefore be considered to be diagrams that explain and clarify design relationships rather than representing specific design solutions.*

Target Audiences

This book addresses code issues specific to the design and construction of healthcare facilities. It accompanies and expands upon the basic principals addressed in the more general Building Codes Illustrated, which is intended for a general audience. This volume assumes some basic knowledge of building code applications and goes into greater detail regarding specific code requirements for healthcare facilities.

For Emerging Professionals

Whether encountered during the design, production, management or construction administration phases for healthcare facilities, codes and standards are an integral and inescapable part of the practice of architecture and engineering. New practitioners need to refine their skills and knowledge of codes to make their projects safe and buildable with few costly changes. The more practitioners know about the code the more it can become a tool for design rather than an impediment. The better the underlying criteria for code development and the reasons for code provisions are understood the easier it is to create code-compliant designs. Early understanding and incorporation of code-compliant design provisions in a project reduces the necessity for costly and time-consuming rework or awkward rationalizations to justify dubious code decisions late in project documentation, or even during construction. Code use and understanding should be part of accepted knowledge for professionals, so that it becomes a part of the vocabulary of design.

For Experienced Practitioners

The greatest value of this book is that it is based upon the newly adopted International Building Code. This is a code that is similar but by no means identical to the three model codes, the Uniform Building Code, the National Building Code, and the Southern Building Code, that most experienced practitioners have used in the past. New state and federal standards have been developed using this new code and the new requirements, while similar, are by no means identical to those in prior codes. This book will guide experienced practitioners out of the old grooves of code use they may have fallen into with the old codes. The code-analysis methods and outcomes will vary from the old codes to the new IBC. While there are seemingly familiar aspects from each code interspersed throughout the new code, the actual allowable criteria and how they are determined are often quite different. It is likely that the illustrations and the underlying reasons for the development of each code section will look familiar to experienced practitioners. The experienced practitioner must not rely on memory or old habits of picking construction types or assemblies based on prior practice. Each building must be looked at anew until the similarities and sometimes-critical differences between the new code and old habits are understood and acknowledged.

It is also worth remembering that building officials and plan checkers are not as familiar with these new codes as well. We are now in a period of transition during which dialogue between designers and plan reviewers will be essential. The precedents that people on each side of the plan-review counter in the building department are most familiar with may no longer apply. Designers and building officials must arrive at new consensus interpretations together, as they use the new code for specific projects.

How to Use This Book

This book focuses on the use and interpretation of primarily the nonstructural provisions of the International Building Code as they specifically apply to healthcare facilities. There are references to basic structural requirements, but this book does not attempt to go into structural requirements in depth. That is the subject for another volume.

The organization of this book presumes that the reader has a copy of the 2006 version of the IBC itself as a companion document to this book. The book is intended to expand upon, interpret and illustrate various provisions of the Code. The IBC has been adopted in many jurisdictions. As it is now being extensively applied, there is an evolving body of precedent in application and interpretation. It is our hope that the analysis and illustrations in the book will aid the designer and the Authorities Having Jurisdiction (AHJ) in clarifying their own interpretations of the application of code sections to projects.

The book is not intended to take the place of the *2006 International Building Code®* in any way. The many detailed tables and criteria contained in the IBC are partially restated in the book for illustrative purposes only. For example, we show how various tables are meant to be used and how we presume certain parts will be interpreted. When performing a code analysis for a specific project, we anticipate the reader will use our book to understand the intent of the applicable code section and then use the Code itself to find the detailed criteria to apply. One can, however, start with either the IBC or this book in researching a specific topic.

Beginning with the *2006 International Building Code®*:
• Search Contents or Index.
• Read relevant section(s).
• For further explanation and/or clarification, refer to this book.

Beginning with *Building Codes ILLUSTRATED*:
• Search Code Index for section number or Subject Index for topic.
• Refer back to specific text of *2006 International Building Code®*.

The text is based upon the language of the Code and interprets it to enhance the understanding of the user. The interpretations are those of the authors and may not correspond to those rendered by the AHJ or by the International Code Council (ICC). This book, while based upon a publication of the ICC, does not in any way represent official policies, interpretations, or positions of the ICC. We would encourage the users of the book to confer with the AHJ, using the illustrations from this book to validate interpretations. Reconciling text with construction drawings often benefits from additional illustrations. We trust that this will be the case with the explanations and graphics in this book.

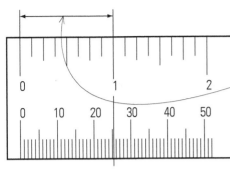

Metric Equivalencies

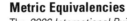

The *2006 International Building Code®* uses the following SI units.

Length
- 1 inch = 25.4 mm
- 1 foot = 304.8 mm
- All whole numbers in parentheses are millimeters unless otherwise noted.

Area
- 1 square inch = 645.2 mm^2
- 1 square foot (sf) = 0.0929 m^2

Volume
- 1 cubic foot (cf) = 0.028 m^3
- 1 gallon (gal) = 3.785 L

Angle
- 1 radian = 360/2π = 57.3°; 1 degree = 0.01745 radian (rad)

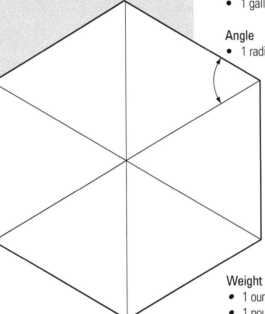

Weight
- 1 ounce = 28.35 g
- 1 pound = 0.454 kg = 0.004448 kN

Force
- 1 pound per square inch (psi) = 6.9 kPa
- 1 pound per linear foot (plf) = 1.4882 kg/m = 0.01459 kN/m
- 1 pound per square foot (psf) = 4.882 kg/m^2 = 0.0479 kN/m^2 = 0.0479 kPa
- 1 pound per cubic foot (pcf) = 16.02 kg/m^3

Light
- 1 foot-candle = 10.76 lux

Speed
- 1 mile per hour (mph) = 0.44 m/s = 1.609 km/h

Heat
- 1 British thermal unit (Btu) = 0.293 watts (w)
- °C=[(°F)-32]/1.8

1
Building Codes

The existence of building regulations goes back almost 4,000 years. The Babylonian Code of Hammurabi decreed the death penalty for a builder if a house he constructed collapsed and killed the owner. If the collapse killed the owner's son, then the son of the builder would be put to death, if goods were damaged then the contractor must repay the owner, and so on. This precedent is worth keeping in mind as you contemplate the potential legal ramifications of your actions in designing and constructing a building in accordance with the code. The protection of the health, safety, and welfare of the public is the basis for professional licensure and the reasons that building regulations exist.

Various civilizations over the centuries have developed building codes. The origins of the codes we use today lie in the great fires that swept American cities regularly in the 1800s, Chicago developed a building code in 1875 to placate the National Board of Fire Underwriters, who threatened to cut off insurance for businesses after the fire of 1871. It is essential to keep the fire-based origins of the codes in mind when trying to understand the reasoning behind many code requirements.

"If a builder build a house for some one, and does not construct it properly, and the house which he built fall in and kill its owner, then that builder shall be put to death.

If it kill the son of the owner, the son of that builder shall be put to death.

If it kill a slave of the owner, then he shall pay slave for slave to the owner of the house.

If it ruin goods, he shall make compensation for all that has been ruined, and inasmuch as he did not construct properly this house which he built and it fell, he shall re-erect the house from his own means.

If a builder build a house for some one, even though he has not yet completed it; if then the walls seem toppling, the builder must make the walls solid from his own means."

Laws 229-233
Hammurabi's Code of Laws
(ca.1780 BC)

From a stone slab discovered in 1901 and preserved in the Louvre, Paris.

The various city codes and often conflicting codes were refined over the years and began to be brought together by regional nongovernmental organizations to develop so-called model codes. The first model codes were written from the point of view of insurance companies to reduce fire risks. Model codes are developed by private code groups for subsequent adoption by local and state government agencies as legally enforceable regulations. The first major model-code group was the Building Officials and Code Administrators (BOCA), founded in 1915. They published the BOCA National Building Code. Next was the International Conference of Building Officials (ICBO), formed in 1922. The first edition of their Uniform Building Code was published in 1927. The Southern Building Code Congress, founded in 1940, published the Southern Building Code.

These three model-code groups published the three different building codes previously in widespread use in the United States. These codes were developed by regional organizations of building officials, building materials experts, design professionals, and life safety experts to provide communities and governments with standard construction criteria for uniform application and enforcement. The ICBO Uniform Building Code was used primarily west of the Mississippi River and was the most widely applied of the model codes. The BOCA National Building Code was used primarily in the north-central and northeastern states. The SBCCI Standard Building Code was used primarily in the Southeast. The model-code groups have merged together to form the International Code Council and BOCA, ICBO and SBCCI have ceased maintaining and publishing their legacy codes.

The International Building Code

Over the past few years a real revolution has taken place in the development of model codes. There was recognition in the early 1990s that the nation would be best served by a comprehensive, coordinated national model building code developed through a general consensus of code writers. There was also recognition that it would take time to reconcile the differences between the existing codes. To begin the reconciliation process, the three model codes were reformatted into a common format. The International Code Council, made up of representatives from the three model-code groups, was formed in 1994 to develop a single model code using the information contained in the three current model codes. While detailed requirements still varied from code to code, the organization of each code became essentially the same during the mid-1990s. This allowed direct comparison of requirements in each code for similar design situations. Numerous drafts of the new International Building Code were reviewed by the model-code agencies along with code users. From that multiyear review grew the original edition of the International Building Code (IBC), first published in 2000. There is now a single national model code, maintained by a group comprised of representatives of the prior three model-code agencies, the International Code Council, headquartered in Falls Church, Virginia. The three organizations have now accomplished a full merger of the three model-code groups into a single agency to update and maintain the IBC.

Note that most local jurisdictions make modifications to the codes in use in their communities. For example, many jurisdictions make amendments to require fire sprinkler systems where they may be optional in the model codes. In such cases mandatory sprinkler requirements may change the design options offered in the model code for inclusion of sprinklers where not otherwise required by the code. It is imperative that the designer determines what local adoptions and amendments have been made to be certain which codes apply to a specific project.

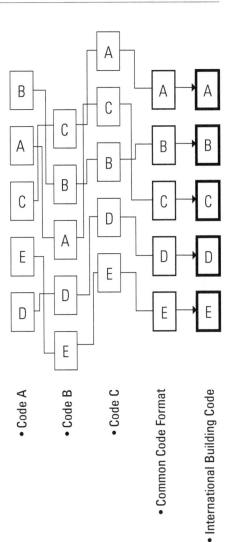

- Code A
- Code B
- Code C
- Common Code Format
- International Building Code

There are also specific federal requirements that must be considered in design and construction in addition to the locally adopted version of the model codes. Foremost among these for designers is the Americans with Disabilities Act of 1990.

Americans with Disabilities Act

The Americans with Disabilities Act (ADA) of 1990 is federal civil-rights legislation requiring that buildings be made accessible to persons with physical and certain defined mental disabilities. The ADA Accessibility Guidelines (ADAAG) are administered by the Architectural and Transportation Barriers Compliance Board (ATBCB), and the regulations are administered by the U.S. Department of Justice. Enforcement of the law is through legal actions brought by individuals or groups asserting violations of their rights of access, as civil rights.

The ADA is not subject to interpretation by local building officials; it is enforced by legal action, through the courts. Access is to be provided for all disabilities, not just for people with mobility impairments. These include hearing, vision, speech, and cognitive impairments, as well as persons of short stature and with limited mobility not necessarily requiring the use of a wheelchair. The ADA effectively applies to all new construction. The ADA also requires that barriers to access be removed from existing buildings where such work is readily achievable. The definition of readily achievable is an economic one and should be addressed by the building owner, not the architect.

The ADA is one of the few building regulations—in this case a law, not a code—that requires retrofitting of projects apart from upgrading facilities during remodeling or renovation. Most codes apply to existing buildings only when renovation is undertaken. Under the ADA those access improvements that are readily achievable should be undertaken by the owner whether or not any other remodeling work is to be done. The owner, not the architect, must make this determination.

The ADA is not enforced by local building officials unless the ADAAG guidelines are adopted as the local access provisions. We will concentrate here only on those accessibility codes that are enforced locally and are subject to review and interpretation as part of the permit process. Designers must first concentrate on complying with codes and standards adopted locally, but must also keep national statutory requirements such as the ADA in mind. It is prudent to review design work against ADAAG at the same time as the model-code review. It is often a judgment call as to which is the most stringent requirement where requirements between codes and legislation differ. In these situations, it is essential and prudent to make the client aware of these discrepancies and have them actively participate in any decisions as to which part of which requirements will govern the design of project components.

Space requirements for accessibility are related to ergonomics. Bigger is not automatically better. For example, specifying an 18" (457) dimension between a toilet and adjacent grab bars is based on reach ranges and leverage for movement using one's arms. A longer reach reduces leverage and thus may be worse than too little space.

State Building Codes

Each state has a separate and distinct code adoption process. In the past many states adopted one of the three previous model codes, and some states even had their own building codes. The geographic areas for state model-code adoptions corresponded roughly to the areas of influence of the three previous model codes as noted on page 3. The BOCA National Building Code predominated in the northeastern United States. The Southern Building Code was adopted throughout the southeastern United States. The Uniform Building Code was adopted in most states west of the Mississippi River. Many states allowed local adoption of codes so that in some states, such as Texas, adjacent jurisdictions in the same state had different building codes based on different model codes. Now, the advent of the International Codes has altered this landscape drastically. The "I Codes" are now the basic model codes in essentially every state. However, be aware that most state processes still allow amendments to the IBC, which means that there will likely be state adopted amendments to the IBC. Make certain you know what code you are working with at the permitting level.

Local Building Codes

Many localities adopt the model-code documents with little modification except for the administrative chapters that relate to local operations of the building department. Larger cities such as Los Angeles, New York City, Chicago and San Francisco typically adopt much more sweeping revisions to the model codes. The codes for such cities often bear little resemblance to the underlying model codes, and in some cases have no basis in them at all. Interpretations, even of the unaltered model code made by big-city building departments, often tend to be very idiosyncratic and non-uniform when compared to smaller jurisdictions that use less modified versions of the model codes. The adoption of the IBC at the state level has generated a review of big-city building codes so that these city codes are moving toward greater conformity with the model codes. For example, San Francisco and Los Angeles are currently using a UBC based state code, which will soon be converted to an IBC based state code. This will require a careful analysis of the city code amendments for conformance with the new model code. This re-development of codes has also been occurring in other large cities such as Dallas

and New York as their states adopt the IBC. Be aware of local modifications and be prepared for varying interpretations of the same code sections among various jurisdictions. Do not proceed too far in the design process based on review of similar designs in another jurisdiction without verification of the code interpretation in the jurisdiction where the project is located. Similarly, although this book offers opinions of what code sections mean, all such opinions are subject to interpretation by local authorities as they are applied to specific projects.

Occupancy Specific Codes for Healthcare

Many healthcare uses and especially hospitals, are regulated by state regulations which overlay model code requirements. These are often tied to state licensing of facilities and are related to the level of care provided at the facility in question. It is essential that the designer determine which agency, or agencies in some cases, will be the Authority Having Jurisdiction (AHJ) to be able to determine which set of rules will be

applicable to your project. It is also essential to determine at an early stage in the design into which the category the facility will be placed by the AHJ.

State and federal agencies often review and certify various health care facilities which involve application of regulations that may not directly parallel the model building codes. Often they do so in conjunction with either a licensing requirement or following a program funding mandate. Local and state licenses are usually focused on the types of services and the qualifications and numbers of personnel that perform the services in the care facility. Other programs mandate a type of evaluation which may create conflicts between the program mandates and the criteria in the IBC.

An example of one such program is the Medicare and Medicaid programs administered through the Department of Health and Human Services. Federal law mandates that facilities receiving such funding must be certified in compliance with NFPA 101's Life Safety Code. Because this is often administered by state agencies, designers should be aware that they often enforce versions of the code that are not based on the federal law. 42 CFR Ch. IV (10–1–06 Edition) states:

The hospital must meet the applicable provisions of the 2000 edition of the Life Safety Code of the National Fire Protection Association. The Director of the Office of the Federal Register has approved the NFPA 101® 2000 edition of the Life Safety Code, issued January 14, 2000, for incorporation by reference in accordance with 5 U.S.C.552(a) and 1 CFR part 51.

The National Fire Protection Association also publishes various other documents that are adopted to accompany the other model codes. Primary examples are NFPA-13, which governs the installation of fire sprinklers, and NFPA-70, which is the National Electric Code.

Where fire codes in the past were seen as maintenance documents, the International Fire Code is a companion document to the IBC. It serves as a document regulating building design as well as building operations. They are intended to provide for public health and safety in the day-to-day operation of a structure. They are also meant to assure that building life-safety systems remain operational in case of emergency. The model-code agencies have developed model fire codes for these purposes. They are developed with primary input from the fire services and less input from design professionals. Fire codes can have an impact on building design. They have requirements for fire-truck access, locations and spacing of fire extinguishers, as well as requirements for sprinklers and wet or dry standpipes. The Fire Code also may contain requirements for added fire protection related to the ease or difficulty of fire equipment access to structures.

Plumbing Codes often dictate the number of plumbing fixtures required in various occupancies. Some codes place this information in the Building Code, some in the Plumbing Code and some in appendices that allow local determination of where these requirements may occur in the codes. The designer must determine which of these courses of action the local adopting authority has chosen. The determination of the required number of plumbing fixtures is an important design consideration. Which set of plumbing fixture criteria is to be used is often not obvious and must be confirmed with the Authorities Having Jurisdiction early in the design process.

Code Interactions

The Authorities Having Jurisdiction (AHJ)—a catch-all phrase for all planning, zoning, fire and building officials having something to say about building—may not inform the designer of overlapping jurisdictions or duplicity of regulations. Fire departments often do not thoroughly check plan drawings at the time building permit documents are reviewed by the building department. Fire-department deficiencies are often discovered at the time of field inspections by fire officials, usually at a time when additional cost and time is required to fix these deficiencies. The costs of tearing out noncomplying work and replacing it may be considered a designer's error. Whenever starting a project, it is therefore incumbent upon the designer to determine exactly which codes and standards are to be enforced for the project and by which agency. It is also imperative to obtain copies of any revisions or modifications made to model codes by local or state agencies. This must be assured for all AHJs.

The model codes have no force of law unto themselves. Only after adoption by a governmental agency are they enforceable under the police powers of the state. Enforcement powers are delegated by statute to officials in various levels of government. Designers must verify local amendments to model codes to be certain which code provisions apply to specific projects.

There are many different codes that may apply to various aspects of construction projects. Typically the first question to be asked is whether the project requires a permit. There are cost thresholds where permits are required, usually relatively low, often as little as $100. Certain projects, such as interior work for movable furniture or finishes, are usually exempt. Carpeting may be replaced and walls painted without a permit, but moving walls, relocating doors, or doing plumbing and electrical work will require a permit in most jurisdictions.

Traditionally, codes have been written with new construction in mind. In recent years more and more provisions have been made applicable to alteration, repair and renovation of existing facilities. One of the emerging trends in code development is the creation of an International Existing Building Code (IEBC). As the importance of preservation of historic structures and the sustainable design implications of reusing exiting buildings become more important the IEBC will take on greater impact. The reuse of existing buildings is also of concern for accessibility issues. One of the most crucial aspects of remodeling work is to determine to what extent and in what specific parts of your project do building codes and access regulations apply. Most codes are not retroactive. They do not require remedial work apart from remodeling or renovation of a building. A notable exception to this is the Americans with Disabilities Act (ADA), which requires that renovation be undertaken retroactively and provide access for persons with disabilities if it can. However, this is a civil-rights law and not a code. It is typically not enforced by building officials, but note that some jurisdictions have adopted the ADA and the accompanying ADA Accessibility Guidelines as their access code. This does not relieve the building owner from obligations under the ADA. In existing buildings it is critical for the designer to determine with the AHJ what the boundaries of the project are to be and to make certain that the AHJ, the designer and the client understand and agree upon the requirements for remedial work to be undertaken in the project area.

Standard of Care

The designer should always remember that codes are legally and ethically considered to be minimum criteria that must be met by the design and construction community. The protection of health, safety and welfare is the goal of these minimum standards. Registered design professionals will be held by legal and ethical precedents to a much higher standard than the code minimum.

The so-called standard of care is a legal concept defining the level that a practitioner is expected to meet. This is higher than the minimum standard defined by the code. The code is the level that a practitioner must never go below. Because professional work involves judgment, perfection is not expected of a design profes-

sional. The standard of care is defined for an individual designer as being those actions that any other well-informed practitioner would have taken given the same level of knowledge in the same situation. It is a relative measure, not an absolute one.

Life Safety vs. Property Protection

The basis for building-code development is to safeguard the health, safety and welfare of the public. The first and foremost goal of building codes is the protection of human life from the failure of building life safety provisions or from structural collapse. There is also a strong component of property protection contained in code requirements. Sprinklers can serve both purposes. When buildings are occupied, sprinklers can contain or extinguish a fire, allowing the building occupants to escape. The same sprinkler system can protect a structure from loss if a fire occurs when the structure is not occupied. While many systems may perform both life safety and property protection functions, it is essential that code developers keep the issue of life safety versus property protection in mind. Security measures to prevent intrusion into a structure may become hazards to life safety. A prime example of this is burglar bars on the exterior of ground-floor windows that can trap inhabitants of the building in an emergency if there is not an interior release to allow occupants to escape while still maintaining the desired security. In no case should property-protection considerations have primacy over life safety.

Code Development

As described above the three previously existing model-code agencies merged into one organization. The three agencies modified their code development processes into a unified national format. This new format has been modified slightly over the past few years as it had been developed and now seems well settled.

As in the past, any person may propose a code revision. Any designer, material supplier, code official or interested member of the public that feels they have a better way to describe code requirements or to accommodate new life safety developments or new technology may prepare revised code language for consideration. Proposed code changes are published for review by all interested parties. They are then categorized based on what section of the code

is being revised and assigned to a committee of people experienced in those matters for review and consideration. Committees are typically organized around specific issues such as means of egress, fire safety, structural, general and so forth. Anyone may testify at these committee hearings regarding the merits or demerits of the code changes. The committee then votes to make its recommendation to the annual business meeting. At the final hearing testimony will be heard from interested parties, both from non-voting industry representatives and building officials who are given voting privileges. Only governmental members of the organization, typically public employees serving as building, mechanical, electrical, plumbing and fire officials, are those allowed to vote on the proposed changes. This is described as a "governmental consensus process" by the ICC.

The International Building Code is a living document. It is subject to regular, ongoing review and comment cycles. A new code is published at regular intervals, usually every three years. This publication cycle gives some measure of certainty for building designers that the code will remain constant during the design-and-construction process. The code development cycle allows the code to respond to new information, growing by accretion and adaptation.

Performance vs. Prescriptive Codes

There is now an ICC International Performance Code. It presents regulations based on outcomes rather than prescriptions. It encourages new design methods by allowing a broader parameter for meeting the intent of the International Codes. Where adopted locally it may be used in place of the regular IBC provisions. We will discuss briefly the distinctions between prescriptive and performance codes.

The International Building Code, as the codes that preceded it, is predominately prescriptive in nature, but it does have some performance based criteria as well. It is developed to mitigate concerns by creating mostly specific and prescribed responses to problems that have been identified. Designers identify the problem to be addressed, such as the width of corridors, and then they look up the prescribed response in the applicable code section. For example, guardrail heights in most commercial applications are prescribed to be 42" (1067) high and are required when adjacent changes in grade exceed 30" (762). The designer follows the prescribed requirements to avoid the problem the code has identified—that is, preventing falls over an edge higher than 30" (762). The code provides a defined solution to an identified problem.

Performance codes, such as the ICC International Performance Code, define the problem and allow the designer to devise the solution. The word *performance* in this context refers to the problem definition and to the setting of parameters for deciding if the proposed solution solves the problem adequately. These standards define the problem, but do not define, describe or predetermine the solution.

The use of performance codes has been increasing in the past few years, due in large part to the development of new modeling techniques for predicting how a building will react under certain fire, earthquake or other stimuli. Performance codes are used in many countries around the world. Their requirements may be as broad as "the building shall allow all of its prospective occupants to safely leave the building in the event of a fire." Most performance codes have tightly defined requirements, but the exiting requirement stated above is a good example of the essence of what performance-code requirements can be.

The basic form of modem performance-code language can be described as objective-based. Each code requirement is broken into three sections. We will use fall prevention as our example. Note that provision of guardrails is only one example of many solutions to the performance objective, not the only solution:

Objective: What is to be accomplished? In this case the prevention of falls from heights of more than 30" (762).

Functional Statement: Why do we want to accomplish this? We wish to safeguard building occupants by preventing them from accidentally falling from a height great enough to result in an injury.

Performance Requirement: How is this to be accomplished? Performance codes could become prescriptive at this juncture, mandating a guardrail. More likely such a performance standard would require that the barrier be high enough, strong enough and continuous enough to prevent falls under the objective circumstances. Note that a guardrail meeting current code standards would be deemed to satisfy those requirements, but alternate means and methods could also achieve the same ends. For example, landscaping could prevent access to the grade change, or innovative railing substitutes could be designed to function like automobile air bags to catch falling persons without having a visible rail present in most conditions. Let your imagination provide other alternatives.

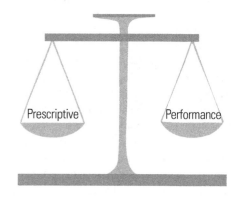

Prescriptive   Performance

Performance codes give designers more freedom to comply with the stated goals. They also require the designer to take on more responsibility for knowing the consequences of their design actions. We anticipate that performance codes will be used in limited ways for innovative projects, but that many typical, repetitive designs will continue to use prescriptive code for speed, clarity and assurance of compliance during design review. Also, given the legal climate designers are often reluctant to take on the responsibility for long-term code compliance for innovative systems.

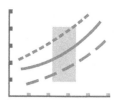

2
Navigating
the Code

The key word to remember about how all building codes are developed and how they all work is *intent*. The intent of the author of a building-code section is to solve a specific design problem. Designers are usually trying to measure visual and spatial expressions against the language of the code. During this process ask what problem, or performance criteria, the code section is addressing. The language may start to make more sense as one tries to go beyond the specific language to determine why the words say what they say.

Designers also have intent. They are trying to achieve certain functional or formal goals in the design of the building. Designers should measure their own intent for the design against their interpretations of the intent of the code. When examined together, the intent of the code and that of the design solution should be concurrent.

Each section of the code was developed to solve a certain problem. The code is typically written in relatively short sections. Sections are organized into chapters based upon common themes, but usually are developed in isolation from one another with little attention to continuity of the entire document. As you look at the code try and visualize the intent of the writer of that section and try to understand the problem they were addressing. Code language usually arises from a specific issue the code writer wishes to address based on experience, or upon a construction or life safety issue. The writer then makes the requirements general so that they apply to more typical conditions than the specific instance that generated the concern.

The intent of the code is a crucial concept to understand. *Why* is a much more important question than *what* when you are puzzled by the actual language of a code passage. The code is a general document that must then be interpreted for its specific application to a specific project. If you know the code in general and think about its intent, you will be in a better position to formulate your own interpretation of code sections as they apply to your specific project. You will thus be in a position to help building officials see the validity of your opinion when interpretation of the code is required for a specific design condition. Confidence will come with experience in use of the code. Learning the code is vital to your success as a well-rounded designer.

Learn to use the table of contents and use the index. It can be very useful to get a copy of the CD-ROM of the Code for use in your practice. This allows key word searches. Don't try and memorize passages of the code, as these may change over time as the code is amended. Learn the organization of the code and learn where to find things that way. Use the index if the table of contents doesn't get you where you want to be. Think of synonyms for the topic you are researching to facilitate key word or index searches. You may have to scan large portions of the index to locate potential items.

Remember that the model code may be amended during adoption by state and local agencies. Be certain to know what local code amendments to the code apply to your projects. Also determine if the local Authority Having Jurisdiction (AHJ) has published written opinions regarding their interpretation of the code in their jurisdiction.

intent
+
interpretation
=

- intent
- intent
- *intent*
- intent
- intent

Alternate Means and Methods

§104.11 states that the provisions of this code are not intended to prevent the installation of any material or to prohibit any design or method of construction not specifically prescribed by this code. While written around prescriptive descriptions of tested assemblies and rated construction, the code recognizes that there may be many different ways of solving the same design problems. It recognizes that there will be innovations in building types, such as mixed-use buildings and atrium buildings that do not fit neatly into prescribed occupancy classifications. There are also innovations in patient care that require new layouts for hospital wards. New treatment technologies may dictate new construction requirements, such as shielding for x-rays or radiation. The code recognizes that there will be innovations in materials and construction technology that may happen faster than code revisions are made. Thus the code sets up a method for the building official to approve proposed alternative designs. Deviations from prescribed standards must be submitted for review and approval of the building official. The criteria they are to use are spelled out in the code. We have highlighted some of the key provisions of the approval in *bold italics.* The alternative is to be approved when "the proposed design is satisfactory and complies with the *intent* of the provisions of this code, and that the material, method or work offered is, for the purpose intended, at least the *equivalent* of that *prescribed in the code in quality, strength, effectiveness, fire resistance, durability and safety"* (emphasis added). These words are also the fundamental criteria for why each and every code section is included in the basic code.

Evaluation of Innovative Products

Innovations in construction materials and methods need to be evaluated for code-compliance. Testing agencies often perform standardized tests on new products. These tests and data about the product must then be evaluated for code-compliance. One popular way of demonstrating compliance to the AHJ for products or construction methods is through the use of ICC Evaluation Service reports.

The ICC-Evaluation Service (ICC-ES) is a non-profit, public benefit corporation affiliated with ICC that does technical evaluations of building products, components, methods, and materials. Reports are prepared at the request of companies wishing their products to be evaluated by ICC-ES. Supporting data such as product information and test reports is reviewed by the ICC-ES technical staff for code-compliance. The evaluation process culminates with the issuance of a report on code compliance. The reports are public documents, readily available on the Internet. They may be used by designers in determining whether an innovative or unusual construction material or process is code compliant. The designer may then use the ES report to demonstrate code compliance by submitting it for review by the AHJ. Designers are very rarely involved in the test and reporting process itself. The evaluation reports are of use for designers in demonstrating code compliance for materials or manufactured products that have already obtained ICC-ES reports.

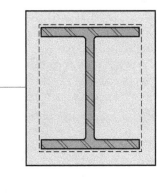

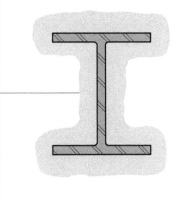

• *Concrete, spray-on fireproofing and gypsum board provide alternate means of fireproofing a structural steel member.*

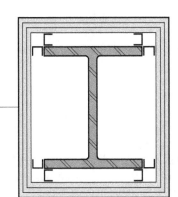

Code Interpretations

Designers and code officials approach interpretations from quite different perspectives. The designer is trying to make a functional or formal design code compliant while satisfying project requirements in an aesthetic, economical and practical way. The AHJ examines completed drawings for compliance with code requirements. While the AHJ is not unaware of the practical requirements contained in the building design, they are charged first and foremost with protecting the health, safety and welfare of the public by verifying code compliance. It is the responsibility of the designer to demonstrate code compliance and to modify noncompliant areas while continuing to meet the project requirements.

Both the designer and the AHJ are working to apply generalized code provisions to a specific project. It is differences in opinion about the application of the general to the specific that most often give rise to differences in interpretation. Code officials also see many more similar examples of the relationship of code sections to various designs. Thus they may generalize interpretations from one project to another even though the projects may be different in significant ways. On the other hand, designers may find that similar designs receive quite different interpretations by the AHJ in different jurisdictions. When differences of opinion about interpretation occur, the designer must work with the AHJ to reconcile the intent of the design to the interpretations of the intent of the code. If reconciliation cannot be reached the designer must decide whether to revise the project to obtain approval or appeal the ruling of the AHJ to some civic body prescribed in the jurisdiction for hearing appeals. In addition, the AHJ can be requested to ask for an opinion from the ICC regarding the intent of a code section in question. In ascending order of processing time the opinions can be: a staff opinion, a written staff opinion, or a formal opinion. Such appeals to the ICC are allowed to be made by any ICC member. It is thus prudent for design professionals to be ICC members to be able to access this service. In addition, members receive discounts on ICC codes and have access to other interpretive and educational materials. Members may also participate in the code development process and gain deeper insights into code interpretations.

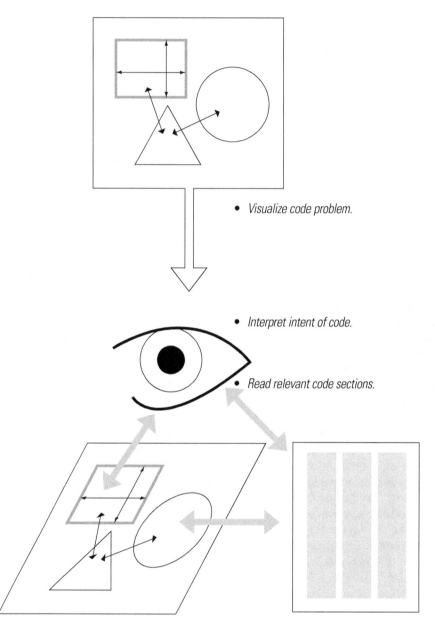

- *Visualize code problem.*

- *Interpret intent of code.*

- *Read relevant code sections.*

- *Re-visualize possible solution that satisfies both design intent and intent of the code.*

Documenting Code Interpretations

Every project should receive a detailed code analysis that is recorded as a permanent part of the permit documents. All code interpretations and citations should have a reference to the code section in question to allow retracing steps in the code analysis. Without a code section citation it is very difficult to have a productive discussion about interpretations. Recording citations focuses code issues for the designer during the design process and facilitates plan reviews by the AHJ.

At a minimum the analysis should contain the following items. We recommend the following format to unify code analysis for all projects. The code section citations used should be specific for the project and sections, not as limited as in our example.

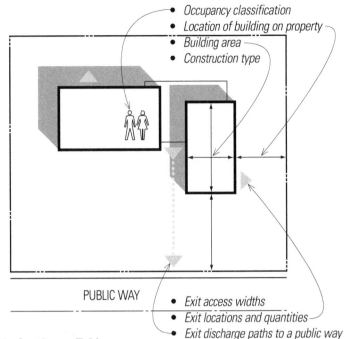

- Occupancy classification
- Location of building on property
- Building area
- Construction type

PUBLIC WAY

- Exit access widths
- Exit locations and quantities
- Exit discharge paths to a public way

Proposed Condition	Allowed per Code	Code Section or Table
Occupancy Classification	Select from Chapter 3	Chapter 3
Fire protection (active)	Select per occupancy	Chapter 9
Building Height (stories/feet)	Allowed per proposed type	Table 503, as adjusted
Building Area	Select per construction type	Table 503, as adjusted
Type of Construction	Determine from design	Chapters 5, 6, 7
Means of Egress	Select per occupancy	Chapter 10

A site and floor plan should be included that describes the location of the building on the property and any height area or construction-type credits or requirements related to location on the site and proximity to streets and other structures. The floor plan should also detail egress requirements, such as exit access widths, exit quantities and locations, and exit discharge paths to the public way.

For the designer, many elements required to determine how the code should apply to a project are a given from the program and the site or zoning constraints:

- *Occupancy classification - the client determines what functions they want;*
- *Location of building on property - determined by the building footprint, zoning, natural features, etc.;*
- *Building height and area - given the scope of the project, the designer will note how large the building needs to be and how many floors will be required.*

- Building height in stories/feet

With these pieces of information it is possible to determine how the code prescribes the minimum for:

- *Construction type - determined by calculation;*
- *Exit locations and quantities;*
- *Exit access widths;*
- *Exit discharge to a public way.*

USING THE CODE

The following procedure is recommended for designers using the International Building Code. Note that most of the major issues are interactive and those iterations of relationships will be required to optimize design solutions. The procedure can be paraphrased as follows.

1. Classify the building according to occupancy, floor area, height and number of stories, fire protection, location on property and type of construction.

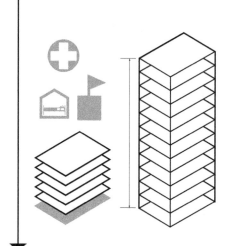

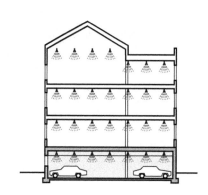

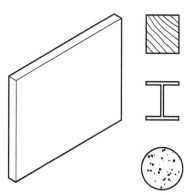

A. Occupancy Classification
Determine the occupancy group that the use of the building or portion thereof most nearly resembles. Compute the floor area and occupant load of the building or portion thereof. See the appropriate sections in Chapters 3 and 5 for buildings with mixed occupancies. See the appropriate sections in Chapter 4 for special design features or occupancy requirements.

B.
Determine if the occupancy is required to be protected by a sprinkler system and identify the threshold(s). Determine if the anticipated height of the building will require fire sprinklers. See the appropriate sections in Chapter 9 for thresholds based on the Occupancy.

C. Type of Construction
Determine the required minimum type of construction based on the occupancy, fire protection, and the designed height and area. This will dictate the materials used and the fire-resistance of the parts of the building as limited in Chapter 6.

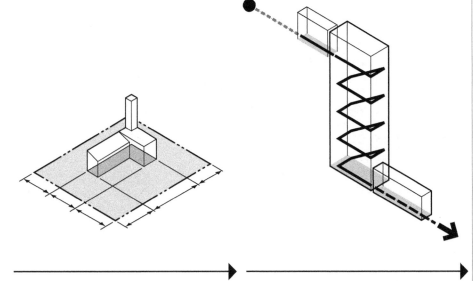

2. Review requirements for fire sprinkler protection.

3. Review the building for conformity with the type of construction requirements.

4. Review the effects on the building based upon its location on the building site.

5. Review the building for conformity with egress requirements.

D. Location on Property
Determine the location of the building on the site and clearances to lot lines and other buildings from the plot plan. Determine the fire-resistance requirements for exterior walls and wall-opening requirements based on fire separation distances to lot lines. The fire resistance requirements for exterior walls and the limitations on their openings are found in Chapter 7.

E. Determine the requirements for means of egress from the building found in Chapter 10.

6. Review the building for other detailed code requirements.

7. Review the building for conformity with structural engineering regulations and requirements for materials of construction.

CODE COMPONENTS

The following section is a review of the critical information required for a project code analysis, based upon the analysis system noted above.

Occupancy Type

Projects must be defined based on occupancy type. A client almost always comes to a designer with a defined need for a facility, or the designer helps a client identify them. The function within the proposed facility determines the occupancy group to which it belongs. Each occupancy group or type has specific requirements related to fire protection, allowable area, height and exiting with minimum construction types growing out of these requirements. The codes are fundamentally occupancy based. Other criteria are derived from the first basic classification by occupancy. Occupancy classifications are defined in Chapter 3.

Building or Floor Area

Once the occupancy group and required fire extinguishing thresholds are known it is important to establish the design area for each floor for each use and for the total building. All types of construction are limited in size based on the size of an occupancy, which may be dependent on the concentration of people. Selection of the allowable area and construction type may require iteration of selections of mixed occupancies and whether separations are provided or not.

Fire Extinguishing Systems

Fire sprinklers, standpipes, fire detection and fire-alarm systems are an integral part of many buildings. Use of such systems, especially automatic fire sprinklers, often results in increases in height and area. A single-story building is permitted a 300% increase and a multiple-story building is permitted a 200% increase in area if protected throughout with an automatic sprinkler system. Similarly, a building may be increased 20 feet and one story in height when an automatic sprinkler system is installed. Fire-protection systems are included in Chapter 9. Note also that sprinkler system requirements are another area where local amendments often include lower thresholds for when sprinklers are required, such as for single-family dwellings or lower thresholds for all buildings. Any such amendments should be verified for each project. It will be critical to a design using the International Building Code to know if a system is required by the code or in the jurisdiction where you are working.

Fire Protection

Fire protection can be divided into two broad categories: passive or active protection. Fire protection can be divided into two broad categories: passive or active protection. Passive protection is that built into the structure, either inherent in the material or added as part of protective membranes. Thus a steel building has more inherent passive protection capability than a wood one because steel is noncombustible whereas wood is not. Active systems are ones where a fire causes a reaction in a system that serves to combat the fire. Sprinklers are a prime example of active systems. A fire causes a sprinkler system to activate and extinguish the fire before exposing the passive systems to a fire. Code analysis and design often includes trade-offs between active and passive systems.

- Fire-resistance standards include:
1. Structure Hour Rating: Requirements for the time it takes for a fire to weaken a structural element to the point of failure. These requirements are minimums based upon providing enough time for fire-fighting and evacuation operations to take place for a specified period without placing emergency responders and occupants in danger.
2. Area or Occupancy Separation Rating. Requirements of how long it will take for a fire to penetrate a wall partition, floor or roof assembly.
3. Flame Spread and Smoke Generation: Requirements of how long it takes for fire to move along the surface of a building material and how much smoke is generated under fire exposure. The density and toxicity of the smoke is also a factor to be considered in these criteria.

- Fire-resistance requirements are found primarily in Chapters 7 and 8.

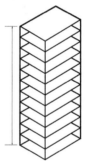

Building Height

Total height in feet and number of stories are typically limited by planning codes, not for technical reasons but as matters of public policy. Other site constraints such as parking requirements, drainage and retention, site amenities and access are other design considerations that will help determine the density that is desirable or appropriate based on the program the client has developed. Building heights are tabulated in Table 503. Study the definition of height and story as noted in the IBC. Also be aware that the definition of height and story is often subject to local amendment. Be certain to check these provisions with the local AHJ to be certain of the exact requirements for your project. This is especially true in older hilly cities like San Francisco, where topography and historical development patterns may generate definitions of height or story different than in other jurisdictions.

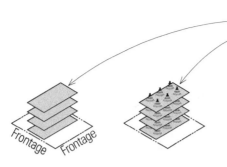

Frontage, maximum increase is 75%
Sprinkler in multi-story, maximum increase is 200%

Construction Type

Construction types are typically categorized by materials based upon their resistance to fire in structural applications and their combustibility. More resistant construction types are allowed to have more occupants, to be of larger area and to have more stories as the resistance increases. As a rough rule of thumb occupancy quantities and construction costs will both decrease with building type from Type I to Type V.

Construction Type

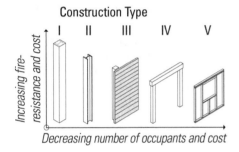

Increasing fire-resistance and cost

I II III IV V

Decreasing number of occupants and cost

Allowable heights and areas are tabulated in Table 503. See section 506 for allowable increases based on location on the property and allowance for installation of automatic fire-sprinkler systems. Types of construction are defined in Chapter 6. Table 601 gives a synopsis of the minimum fire-resistive requirements of each main element of building construction. As you go through a code analysis you will be referring to this table and to area tables to select the optimum balance of construction type, occupancy and area requirements for a specific project. It is typically a budget goal to minimize construction costs by selecting the least costly construction type appropriate for the proposed use of the building.

For the designer to determine the required type of construction, he must know two things: the total building area and the largest design floor area. Because the code limits the total building area to 3x the allowable largest floor area, the first step is to divide the total area of the building by 3.

Area Modifications

§506 allows for area modifications for frontage and for provision of automatic sprinklers

- For example, a 4-story office building that is 12,000 sf (1115 m^2) per floor has a total area of 48,000 sf (4460 m^2). But the largest allowable floor would be the total building area divided by 3.

  $$48,000 \text{ sf} \div 3 = 16,000 \text{ sf } (1486 \text{ m}^2)$$

- Then to determine the Type of Construction, the following formula will provide the answer:

  $$\text{Base Area} = \frac{\text{Largest Allowable Floor Area}}{(1 + SI + FI)}$$

  SI = Sprinkler Increase (300% for one story, 200% for multi-story)
  FI = Frontage Increase (up to 75%)

- In this example the base area would be calculated using the formula as follows:

  $$\text{Base Area} = \frac{16,000 \text{ sf } (1486 \text{ m}^2)}{3.75^*}$$

  $$^*(100\% + 200\% + 75\%) = 3.75$$

  $$\text{Base Area} = 4,267 \text{ sf } (396 \text{ m}^2)$$

- Using the base area the designer can then go into Table 503 and determine which types of construction have an area **equal to or larger** than the base area.

  $$4,267 \text{ sf} \leq \text{Area permitted in Table 503}$$
  $$(396 \text{ m}^2)$$

- See Table 503. In this case for a B occupancy, the building could be of any type based strictly on area, but the height would require it to be of Type V-A or better based on the area and the story increase permitted for sprinklers per §504.

- *Property line*
- *48,000 sf (4460 m^2) program area*

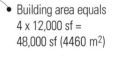

- *For the example assume lot size and site imperatives such as zoning, development controls and parking requirements dictate a multi-story building.*

- Building area equals 4 x 12,000 sf = 48,000 sf (4460 m^2)

- Total Area is 48,000 sf (4460 m^2)

- Largest allowable floor area, 48,000 sf ÷ 3 = 16,000 sf (1486 m^2)

- Largest Allowable Floor Area

- Table 503 (100%)
- Sprinklers (200%)
- Frontage (75%)

- Base area = $\dfrac{16,000 \text{ sf } (1486 \text{ m}^2)}{((1 + (1 + 1) + 0.75))}$

- Base area = 4,267 sf (396 m^2)

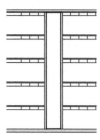

Building Separations and Shafts

Where buildings have mixed occupancies, the IBC allows either mixed use unseparated or separated. In determining the type of construction, if the option for mixed use unseparated is used, the most restrictive of all the unseparated occupancies will control the type of construction. Only a fire wall will allow more area for a particular occupancy on a single floor in certain types of construction where such sizes of use would not otherwise be permitted. If there is some reason that the client is adverse to installation of sprinklers in a facility that otherwise would require it because of the size of the occupancy that requires it, a fire barrier can be designed that will reduce the "fire area" and allow structures that do not require fire suppression. The fire area is an envelope including all interior walls and floors that surround it and the fire resistance characteristics are prescribed in Chapter 7. Sprinklers are not typically required for office buildings unless an occupant floor is over 55' (16 764) above the lowest level of the department's access.

Openings between floors such as for stairs, elevators and mechanical shafts are limited by the code because they can allow the passage of smoke, heat and flames in a fire. Therefore the codes have requirements based on occupancy, building type and building height related to shaft protection. Basic shaft options are contained in Chapter 7. Stair enclosure options are included in Chapter 10.

Exits/Egress

One of the most important functions of building codes is determining egress requirements and provision of safe means of egress for all of the anticipated occupants of a building. There are specific requirements for size, spacing and travel distances for all components of means of egress such as floor plans, doors, corridors and stairs. In simple terms a means of egress consists of three components: an exit access, an exit and an exit discharge. Chapter 10 of the IBC relates to means of egress.

Engineering Requirements

A large portion of the code is devoted to engineering requirements. One of the bases of codes is structural adequacy of buildings for both static loading such as occupants and equipment and dynamic loading such as earthquakes, snow and wind. Requirements for both structural systems and structural materials are contained in the code. Chapters 16–18 deal with forces, inspections and foundations. Chapters 19–23 deal with materials: concrete, lightweight metals, masonry, steel, and wood.

Early Meetings

One advantage of larger projects is that they are often large enough to warrant pre-review and consultation with the building department prior to finalizing design. No matter what size your project is we recommend consulting with the applicable AHJ early in the process wherever it is possible, prior to commencing detailed design, even if a fee is charged. Do not expect the code official to do your work for you. Compliance is the responsibility of the designer. However, codes are subject to interpretation, and it is almost always in your best interest to determine what, if any, interpretations will be needed for any project. This should be done prior to expending a lot of time and energy designing a project that may be deemed not in compliance during plan review.

As noted before, don't be shy about using the table of contents and index to locate sections of the code. DO NOT TRY AND MEMORIZE PARTS OF THE CODE! As sections change and interpretations alter meanings, this is a recipe for trouble in the future. Clients may expect you to be able to rattle off requirements at a moment's notice, but it is not in the best interest of the project to be able to make snap code decisions. Remember where to look up information and check your decisions each time you apply them; do not proceed on memory or analogy from other jobs. With the new code reorganization and the new code language, even seasoned code professionals will be using the index to locate familiar phrases in new locations over the next several years as we all learn the new code. It is worthwhile for designers to remember that for the next few years local code officials will have little more experience with the new IBC than design professionals.

3
Use and Occupancy

The intended use or occupancy of a building is a fundamental consideration for the building code. Typically a client comes to an architect with a defined need for a facility. The desired use of that facility determines the occupancy group to which it is assigned under the code. Occupancy group classifications trigger specific requirements for the allowable area and height of a building, for means of egress, as well as for type of construction. The "I" codes are fundamentally occupancy based. Most other broad sets of code criteria are derived from the basic classification by occupancy.

The codes separate uses into broad groups called Occupancies. Under these groups are subdivisions that further refine the detailed requirements. The building official is directed by the code to classify all buildings or portions of buildings into occupancy groups. The intent regarding classification is best described by the language directing classification of atypical occupancies: "such structure shall be classified in the group which the occupancy most nearly resembles, *according to the fire safety and relative hazard involved.*" This reiterates the intent and purpose of the occupancy classifications that exist in the code. Each of the stated occupancy classifications was determined during code development by using fire-safety and relative hazard performance data to develop criteria.

Determination of the occupancy type follows in almost every case from the program given to the designer by the client. Other code requirements flow from the number of occupants and the hazards to their safety from external and internal factors. As we discussed previously, the code looks at property protection considerations along with life safety concerns. The occupancy's hazards are assessed relative to their impact on adjacent properties as well as on the building occupants. The code also analyzes the hazards posed by adjacent buildings. However, it places the responsibility for protection of the adjacent facilities upon the building under consideration.

Among the considerations for occupancy classification are: how many people will be using a facility; whether there are assembly areas, such as theaters and restaurants; whether people will be awake or asleep in the building; will they be drinking alcohol while using the building, or undergoing medical treatment, which makes them less capable of self-preservation in an emergency? The presence of hazardous materials or processes will also affect the requirements for allowable area, fire separations, and construction type.

Occupancy Groups for Healthcare Facilities

Healthcare facilities will generally be considered as one of two typical occupancy groups. The first is "I" Occupancies, which are institutional uses where people are cared for or live in a supervised environment. The occupants may have physical limitations because of healthcare needs, age or infirmities, or they may have restricted liberty due to penal or correctional restraint. Our focus will be on those occupancy subdivisions which are primarily related to healthcare. These I occupancies include the expected groups such as hospitals, nursing homes and mental hospitals, but they also include such healthcare-related occupancies as residential board and care facilities, assisted living facilities, alcohol and drug treatment facilities, detoxification facilities and convalescent facilities. Smaller care facilities such as board and care homes or small assisted living facilities are classified as "R" Occupancies and while discussed to establish occupancy classification requirements, will not be considered in detail in this book. We will go into the determinations of

occupancy classifications and requirements for each healthcare related "I" Occupancy in detail further on in this chapter.

The second major occupancy classification for healthcare facilities are the offices for healthcare providers, which are typically considered as "B" Occupancies, which contain office, professional or service-type transactions. These B occupancies include such healthcare-related uses as outpatient clinics, physician's offices, dentist's offices and research laboratories.

Many healthcare facilities will have other related uses which are considered as separate occupancies. There are often public food service facilities in hospitals, there are often large assembly rooms used for instruction or public outreach and there are often gift shops or flower stands in a hospital which would be considered as mercantile occupancies. Accordingly, while we will focus on specific requirements for healthcare occupancies, we will address the requirements for mixed occupancy buildings as it is very likely that most healthcare facilities will in actuality have a mix of occupancies.

Note that the healthcare occupancy criteria we will be discussing generally apply only to non-hazardous occupancies. Hazardous occupancies are not addressed in detail by most design professionals and are covered by a separate set of special requirements discussed in §414 and 415. Because they are very specialized and encountered infrequently by most designers, the requirements for hazardous occupancies will thus only be lightly touched on in this book.

Overall Occupancy Classification and Uses of Related Spaces

It is essential to read the detailed requirements for each type of occupancy in a project. There are often cross-references to various other code sections in the detailed occupancy criteria. Another factor that impacts occupancy classification is the mixture of various uses of the building and their relative size in relation to the predominant use of the building. In addition to single or dominant use occupancies, there are three major use relationships that impact occupancy classification. Criteria for mixed uses and for incidental and accessory uses were located in Chapter 3 of the IBC until recently. They have been moved to Chapter 5. See the following sections:

- Incidental use areas (§508.2)
- Mixed occupancies (§508.3)
  - Accessory occupancies (§508.3.1)
  - Nonseparated occupancies (§508.3.2)
  - Separated occupancies (§508.3.3)

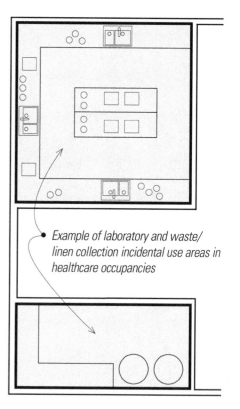

- Example of laboratory and waste/linen collection incidental use areas in healthcare occupancies

Occupancy Groups

There are several ideas common to most occupancy classifications that should be understood. First is the language: "building, or structure, or portion thereof." This distinction allows the use of mixed occupancies in a single building without having to consider the entire building as a single occupancy group. The concept of separated and nonseparated uses allows the designer three options for addressing mixed-use buildings. It also allows rooms within buildings to be considered as distinct occupancies that can then be addressed as accessory, separated, or non-separated uses at the designer's discretion.

The other concept to understand is that the laundry lists of examples in each occupancy group are not the sole definition of what uses are to be classified in which occupancy group. The code recognizes that not all occupancies are included in the lists and gives direction to the building official regarding classification of buildings not included in the examples. The criteria for classification are as noted in the opening of this chapter.

The IBC establishes the following occupancy groups:
- Assembly (A)
- Business (B)
- Educational (E)
- Factory and Industrial (F)
- High Hazard (H)
- Institutional (I)
- Mercantile (M)
- Residential (R)
- Storage (S)
- Utility and Miscellaneous (U)

Healthcare Related Occupancy Classifications

We will address the two primary occupancy groups for healthcare facilities first and in detail. We will then touch in less detail upon the other occupancies that may be encountered in the design and construction of healthcare facilities.

Business Group B

Office buildings, no matter what type of user occupies the offices are typically classified as Group B occupancies. Storage areas for offices, such as back-office file rooms by definition do not constitute a separate occupancy. There are no subdivisions for B occupancies. Outpatient clinics, physician's offices and dental offices are listed in this occupancy group. Even where patients may be rendered incapable of unassisted self-preservation by anesthesia, the use could still be classified as a Group B occupancy. §308.3 states in the definition for I-2 occupancies that if there are more than 5 persons who are not capable of self-preservation then the occupancy is an I-2. This implies that if there are fewer than 5 such persons then the occupancy is something else. One occupancy which fits the description of such occupancy is "B." Also §308.3 requires that care be provided on a 24-hour basis so that a clinic which operates for less than 24 hours could also be classified as a B occupancy. §308.3 complicates the occupancy classification by stating that I-2 facilities, such as hospitals or nursing homes with fewer than five occupants are to be classified as R-3 residential occupancies. We believe that the intent of this section is that facilities which care for fewer than five persons and for more than 24 hours should be considered as R-3 occupancies. Outpatient clinics (the term "outpatient" is undefined) would not be expected to care for more than 5 persons or for more than 24 hours so should be considered a B occupancy. It is not clear whether an outpatient facility which cares for more than 5 persons not capable of unassisted self-preservation would be considered an I-2 or a B occupancy. This is a very good example of why it is critical to meet with the AHJ early in the design process to verify code interpretations.

Testing and research laboratories that do not exceed the quantities of hazardous materials specified in the code are also classified as Group B occupancies. Those that exceed the minimums are classified as Group H occupancies. Educational facilities for junior colleges, universities and continuing education for classes above the 12th grade are considered Group B occupancies, not Group E. Thus classrooms in a teaching hospital would likely be considered as part of the B Occupancy area of a such as healthcare facility. Assembly rooms in healthcare facilities should be examined for conformance with the criteria for Group A occupancies.

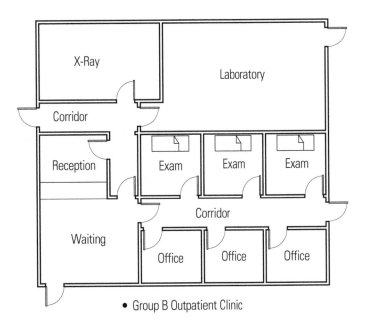

- Group B Outpatient Clinic

Institutional Group I

Institutional occupancies are those where people are cared for, live in a supervised environment, or have special restrictions placed upon them. The occupancy groups in these groups are sub-divided primarily by the abilities of the occupants to take care of themselves in an emergency. The different categories in this occupancy are determined by the number of occupants, their ages, health and personal liberty, and whether they are in the facility all day or part of the day or night. Many of these occupancy groups have a residential character, and as discussed above if they fall outside the designated occupant care or occupant quantity thresholds for Group I occupancies, they will likely be classified as residential occupancies.

Occupancy group distinctions are based on a combination of factors: the number of occupants, time of use of the facility, the ability of the occupants to care for themselves or whether they are restrained in their activities. The different Group I occupancy examples that follow have their distinguishing characteristics specified in the same order to facilitate understanding the differences between them.

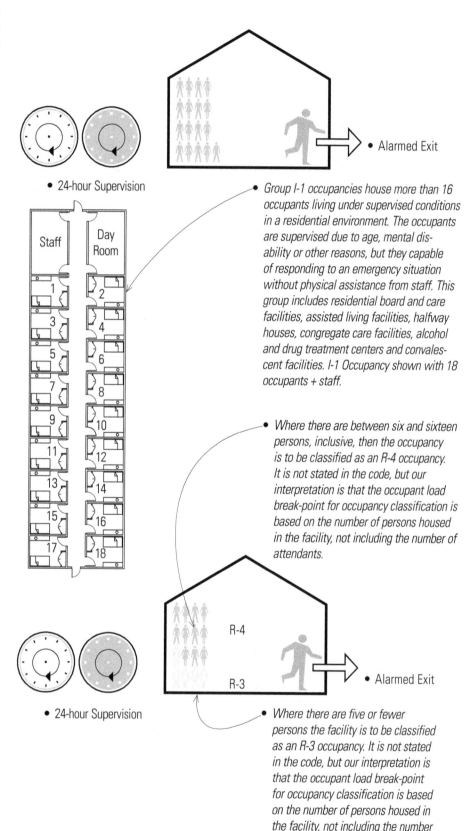

- 24-hour Supervision

Staff

Day Room

- Alarmed Exit

- Group I-1 occupancies house more than 16 occupants living under supervised conditions in a residential environment. The occupants are supervised due to age, mental disability or other reasons, but they capable of responding to an emergency situation without physical assistance from staff. This group includes residential board and care facilities, assisted living facilities, halfway houses, congregate care facilities, alcohol and drug treatment centers and convalescent facilities. I-1 Occupancy shown with 18 occupants + staff.

- Where there are between six and sixteen persons, inclusive, then the occupancy is to be classified as an R-4 occupancy. It is not stated in the code, but our interpretation is that the occupant load break-point for occupancy classification is based on the number of persons housed in the facility, not including the number of attendants.

R-4

R-3

- Alarmed Exit

- 24-hour Supervision

- Where there are five or fewer persons the facility is to be classified as an R-3 occupancy. It is not stated in the code, but our interpretation is that the occupant load break-point for occupancy classification is based on the number of persons housed in the facility, not including the number of attendants.

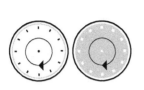

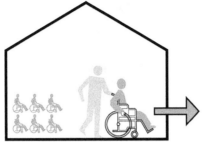

• 24-hour Supervision

• Wide Exit

• Group I-2 occupancies have more than five occupants living under supervised conditions of custodial care. The occupants are not capable of self-preservation and occupy a facility that is used on a 24-hour basis. This group includes buildings used for medical, surgical, psychiatric, nursing or custodial care. This includes facilities such as hospitals, nursing homes (both intermediate care and skilled nursing facilities), mental hospitals and detoxification facilities.

R-3

• 24-hour Supervision

• Wide Exit

• Where there are five or fewer persons using the facility then it is to be classified as an R-3 occupancy. It is not stated in the code, but our interpretation is that the occupant load break-point for occupancy classification is based on the number of persons housed in the facility, not including the number of attendants.

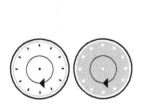

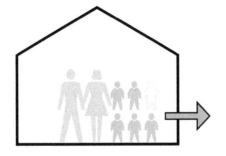

• 24-hour Supervision

• A childcare facility that provides care on a 24-hour basis for more than five children 2-1/2 years of age or less is to be classified as an I-2 occupancy. A child-care facility for fewer than five children of any age is classified as an R-3 occupancy. Childcare facilities used for less than 24 hours per day are to be classified as I-4 occupancies which are further described below.

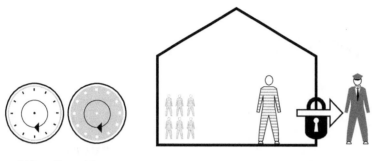

• 24-hour Supervision

• Controlled Exit

• Group I-3 occupancies have more than five people living under supervised conditions and under restraint or security. The primary healthcare facility which would fall into this occupancy would be a prison hospital or a mental hospital. There are five conditions within I-3 occupancies distinguished by the freedom of movement of the secured occupants within and between smoke compartments. They are listed in ascending order of confinement and restraint of movement. See also §408.1 for detailed requirements for Group I-3 occupancies

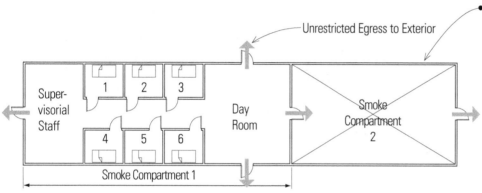

• Condition 1 facilities allow free movement from sleeping areas and other occupied spaces to the exterior via means of egress without restraint. This condition is essentially the same as an R occupancy and may be constructed per the applicable Group R requirements.

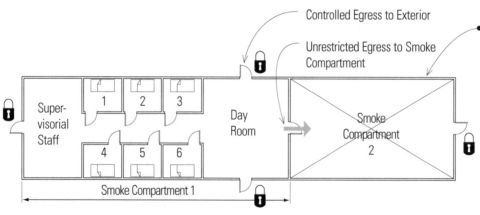

• Condition 2 facilities allow free movement from sleeping areas and other occupied spaces in one smoke compartment to one or more other smoke compartments. Egress to the exterior is impeded by locked exits.

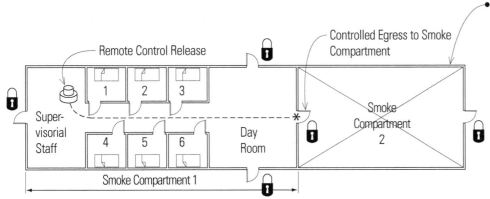

- Condition 3 facilities allows free movements between sleeping areas and related activity spaces, but only within a single smoke compartment. Egress is impeded by remote-control release of means of egress from the smoke compartment to another smoke compartment. It is not stated but it is to be assumed that egress to the exterior is impeded by locked exits.

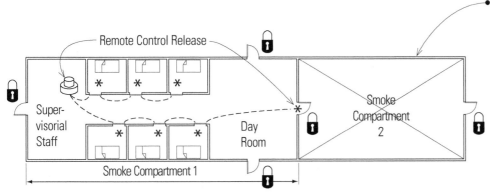

- Condition 4 facilities restrict free movement from an occupied space. Remote-controlled release is provided to permit movement from sleeping units, activity spaces and other occupied areas with a single smoke compartment to other smoke compartments.

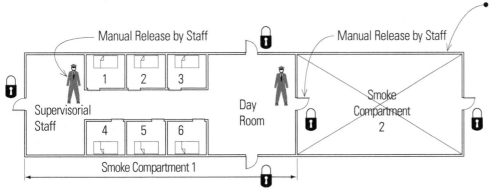

- Condition 5 facilities restrict free movement from an occupied space. Staff-controlled manual release is provided to permit movement from sleeping units, activity spaces and other occupied areas with a single smoke compartment to other smoke compartments. This has a higher degree of personal supervision of restrained movement than does Condition 4.

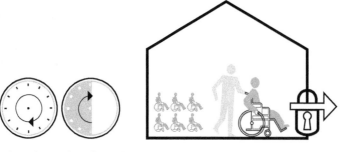

• Less than 24-hour Supervision   • Controlled Exit

• *Group I-4 are daycare facilities used for custodial care for less than 24 hours. The code presumes that the persons are not related by blood, marriage or adoption and that the care does not take place in the home of those cared for. The persons cared for may be children or adults. Facilities that care for five or fewer persons are to be classified as R-3 occupancies. While it is not stated the code seems to assume that persons in these occupancies are not able to respond to emergency situations without assistance from staff, whether due to illness, infirmity, old age or young age.*

• *§308.5.1 addresses adult daycare facilities. The code presumes that persons being cared for need assistance in emergency situations as the exception says that if occupants are capable of responding to an emergency without physical assistance from the staff then the facility is to be classified as an A-3 occupancy.*

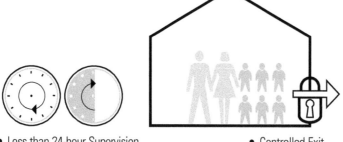

• Less than 24-hour Supervision   • Controlled Exit

• *§308.5.2 covers childcare facilities. The I-4 occupancy covers more than five children, 2-1/2 years of age or less. Where a facility has between 5 and 100 children and is located on the ground floor with exits directly to the exterior then the facility is to be classified as an E occupancy. The code also classifies day care for more than five children older than 2-/2 years of age as an E occupancy. The code thus is assuming that young children, defined as under 2-1/2 years old need physical assistance in emergencies and thus need the added protection which occurs in an I occupancy versus an E occupancy.*

Other Occupancies That May Occur in Healthcare Facilities

The following occupancies are often found in healthcare facilities, especially large facilities that are predominately I occupancies. The presence of mixed occupancies must be acknowledged in the design of the facilities and appropriate occupancy separations and construction classifications provided for code compliance. The list of occupancies that follow is not complete, but edited to address those occupancies which are likely to be found in healthcare facilities.

Assembly Group A

The examples noted in the assembly groups recognize that these uses bring large groups of people together in relatively small spaces. How the spaces are used in relationship to physical features and to human behavior also enter into the distinction between assembly categories, which are meant to serve as cues for assigning buildings or parts of buildings to an occupancy class. The final determination of this classification is made by the building official. Note that the subcategories in the code are examples, not a definitive or exhaustive list of possible uses.

Group A occupancies are defined as having 50 or more occupants, but the use of the space must be examined in relation to the code language stating that these are spaces "for purposes such as civic, social or religious functions, recreation, food or drink consumption...." For instance, classrooms in B or E occupancies may have more than 50 occupants, but are not considered as Group A. Per an exception assembly areas with fewer than 50 occupants are to be classified as Group B Occupancies. Assembly areas with less than 750 sf that are accessory to other uses are also not considered as assembly areas.

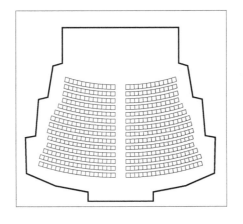

- *Group A-1 is assembly areas, usually with fixed seats, such as might be found in a large auditorium in a teaching hospital or a healthcare conference facility. The presence or absence of a stage is not a distinguishing feature. Most uses classified in this occupancy will have fixed seats. The egress requirements in Group A-1 occupancies recognize that light levels may be low and that people may panic in emergency situations under such circumstances.*

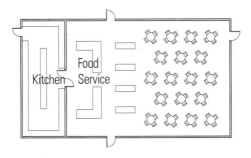

- *Group A-2 is assembly areas where food and drink are consumed, such as in a hospital cafeteria. It presumes that chairs and tables will be loose and may obstruct or make unclear egress pathways for patrons. The requirements also recognize the relatively poor fire history of such occupancies.*

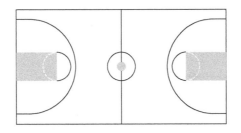

- *Group A-3 occupancies are assembly areas that do not fit into the other Group A categories. It also includes spaces used for worship, recreation or amusement. Examples of these uses in healthcare facilities would be a chapel or a gymnasium for staff or rehabilitation activities. The intent of this classification is that any use that seems to be an assembly occupancy and does not fit the criteria of the other four Group A categories should be classified as an A-3 occupancy.*

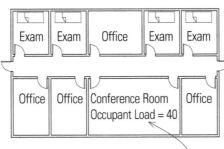

- *Not considered an "A" Occupancy*

- *As noted in §303.1, accessory assembly spaces with less than 50 occupants are to be considered by exclusion as part of the overall occupancy. For example, having a conference room or a lunchroom with fewer than 50 occupants serving a larger use does not trigger classifying that space as an Assembly Group A occupancy.*

Educational Group E

Group E occupancies are used by six or more people for classes up to the 12th grade. Uses for the day care of five or more children over 2-1/2 years of age make up another set of Group E occupancies. Those uses with fewer than five children are classified as Group R-3.

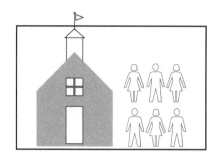

High-Hazard Group H

Hazardous occupancies could easily be the subject of another book and will only be touched on in an introductory fashion in this text. The uses classified under this occupancy group are very specialized and require careful code and design analysis. Understanding the products, processes, hazard levels of materials used in the occupancy and their quantities is essential. Variations in material quantities and hazards interact to set the design criteria for hazardous occupancies. The classification of uses in this category will almost undoubtedly require consultation with the building official at an early stage of design.

There are two sets of criteria for hazardous occupancies. The first set is related to the hazard of the materials in use and the quantities of those materials in use. High-Hazard Groups H-1 through H-4 fall in this category. The second set relates to the nature of the use as well as the quantity and nature of hazardous materials in use. This is High-Hazard Group H-5.

Areas that contain limited quantities of hazardous materials may occur in other occupancy groups when the amounts are less than the designated limits for exempt quantities. For example, small amounts of flammable cleaning fluids or paints might be stored in a room in a business occupancy. A mercantile occupancy can sell specified quantities of materials that may be considered hazardous without being designated a Group H occupancy as long as the amount of material is below the limit for exempt quantities. This exemption pertains only to occupancy classification related to quantities; it does not waive compliance with any other code provisions. Note also that the International Fire Code sets forth many additional construction and use requirements for Group H.

Control Areas

The other basic concept in the code provisions for High-Hazard Group H is that of control areas. This applies both to subdivisions of buildings classified as hazardous occupancies and to areas where hazardous materials occur within other occupancies. The definition of this concept bears stating verbatim from §307.2:

- Control areas are "spaces within a building that are enclosed and bounded by exterior walls, fire walls, fire barriers and roofs, or a combination thereof, where quantities of hazardous materials not exceeding the maximum allowable quantities per control area are stored, dispensed, used or handled."

- The control area concept allows multiple parts of a building to contain an array of hazardous materials when the areas are properly separated and the quantity of materials within each area meet the specified maximums for each type of material. These criteria reinforce the concept that hazard levels are mitigated by passive and/or active fire-protection measures. This concept is based on two primary considerations. The first is the nature of the hazard of the material in question. The second is that the level of hazard is primarily related to the quantity of materials within a given area.

- Control areas must be separated from one another by 1-hour fire-barrier walls and floors having a minimum fire-resistance rating of 2 hours. An exception to §414.2.4 allows floors to be 1-hour fire-resistance rated in buildings of Type IIA, IIIA and VA construction when the building is fully sprinklered and three stories or less in height. For the fourth and succeeding floors above grade, fire-barrier walls must have a 2-hour fire-resistance rating.

- Note that both the percentage of maximum allowable quantity of hazardous materials and the number of control areas decrease above and below the first floor of a building.

- Higher than nine floors above grade, one control area with 5% of allowable quantity is permitted per floor.

- The seventh through ninth floors above grade may have two control areas per floor with 5% of allowable quantity per control area.

- The fourth through sixth floors above grade may have two control areas per floor with 12.5% of allowable quantity per control area.

- Third floor above grade may have two control areas with 50% of allowable quantity per control area.

- The second floor above grade and first floor below grade may have three control areas with 75% of allowable quantity per control area.

The special detailed requirements for the separation of control areas are contained in Chapter 4 and outlined in Table 414.2.2. Chapter 4 must therefore be read in concert with Chapter 3 to determine all applicable code requirements.

Mercantile Group M

Mercantile uses ancillary to B or I occupancies could include a hospital gift shop, a florist or a coffee bar in the medical center lobby. The occupancy group includes incidental storage of up to 10% of the total floor area in accessory areas, per §508.3.1 and Footnote b. to Table 508.3.3. The spaces are still to be considered as different occupancies. Larger storage areas would be classified as Group S. Most retail facilities, no matter what merchandise they sell, fall into this occupancy. There are limits to the quantities of hazardous materials that may be stored in mercantile occupancies without being classified as a Group H occupancy.

Residential Group R

Residential occupancies include typical housing units, distinguished mainly by the total number of occupants. This group also includes smaller-scale institutional occupancies that fall below certain thresholds for the number of occupants.

S	M	T	W	T	F	S
1	2	3	4	5	6	7
8	9	10	11	12	13	14
15	16	17	18	19	20	21
22	23	24	25	26	27	28
29	30					

S	M	T	W	T	F	S
		1	2	3	4	5
6	7	8	9	10	11	12
13	14	15	16	17	18	19
20	21	22	23	24	25	26
27	28	29	30			

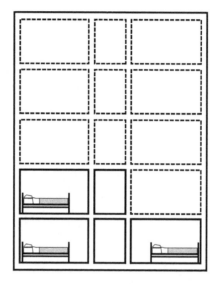

- Note that although definitions of terms related to residential occupancies are contained in the residential occupancy section, they apply not only to this section but also throughout the code.

- Group R-2 occupants are permanent, sleeping in buildings containing more than two dwelling units for more than 30 days. These include apartments, dormitories and long-term residential boarding houses. Congregate living facilities with 16 or fewer occupants (note the cut-off is 16 or fewer) are permitted to comply with requirements for R-3 occupancies.

- Permanent residency

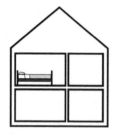

- Group R-3 occupants are permanent, not transient, and the group is defined as not meeting the criteria for R-1, R-2 or Group I occupancy groups. These are primarily single-family residences and duplexes. Day-care facilities for five or fewer people using the facility for less than 24 hours also fall into this occupancy group. In many jurisdictions these occupancies are regulated under the International Residential Code when it is adopted by the local jurisdiction.

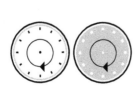
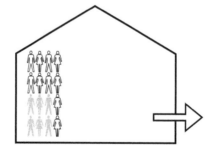

- Group R-4 occupancies are used for residential care or assisted-living uses with more than five but not more than 16 occupants. Facilities for more than 16 persons are classified as I-1 occupancies.

Storage Group S

Storage for materials with quantities or characteristics not considered hazardous enough to be considered a Group H occupancy is classified as Group S. The two subdivisions are similar to the distinctions made for F occupancies: Moderate-Hazard S-1 occupancies and Low-Hazard S-2 occupancies. The lists of examples for each category are quite lengthy and detailed. When occupancies contain mixed groups of various products it can be quite difficult to determine which occupancy group to use. Careful consideration of projected uses for the facility and potential changes in use over time must be considered. It is useful to confer with the building official early in the design process to get concurrence on the proposed classification. As in the distinction between F-1 and F-2 occupancies, the distinction between S-1 and S-2 is that S-2 is used for the storage of noncombustible materials.

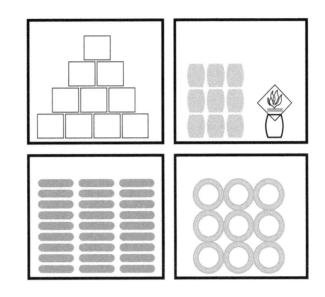

Utility and Miscellaneous Group U

The group is for incidental buildings of an accessory nature. These structures are typically unoccupied except for short times during a 24-hour period and are typically separate from and subservient to other uses. This occupancy group is used sparingly. It is not meant to be a catch-all for occupancy types that are not readily categorized. Note also that this group contains items that are not buildings, such as fences over 6′ (1829) in height and retaining walls.

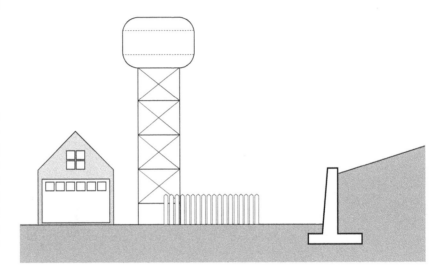

4

Special Uses and Occupancies

In addition to the requirements for typical occupancies, the Code addresses detailed requirements for specific building configurations in Chapter 4. The occupancy of these buildings usually fits into one or more of the typical occupancy groups, but characteristics of these special buildings require additional code provisions. Examples of such building configurations or uses that may occur in healthcare occupancies are high-rise buildings, atriums, parking garages, I-2 occupancies, and I-1, R-1, R-2 and R-3 occupancies. Note that healthcare facilities may have underground facilities, but they are not likely to be so deep as to qualify as underground buildings and accordingly such underground facilities will not be addressed here.

High-rise buildings for healthcare may be Group B, Group I, or Group R occupancies. They typically have very large occupant loads and are defined as having occupied floors located above fire department ladder access. Atriums have large interior volumes open to pedestrian pathways and to occupied spaces. An entry lobby in a large multi-use hospital or medical office building could conceivably be considered as an atrium, so we will address the general requirements for atria in this chapter as well. It is very likely that healthcare facilities will contain parking facilities, so these will be addressed in this chapter. There are also additional requirements for I and R occupancies that appear in Chapter 4 of the IBC and are supplementary to the occupancy requirements in Chapter 3. The building uses addressed in Chapter 4 have added code consideration above and beyond more typical uses in the same occupancy group.

The development of new types of buildings often happens in advance of code provisions specific to them. Code officials must respond to requests by owners to build such structures by addressing them on a case-by-case basis. As these new types of buildings or new uses become more prevalent the code responds by collecting information about how different jurisdictions have addressed these new buildings and the code-development process generates new code provisions to address them. These provisions are meant to apply over and above the other provisions applicable to their occupancy group classification. After the designer has classified the building by occupancy the building type must be examined to see if it meets the definitions for these specialized use groups and thus must also meet the added criteria for them. The process of analysis of use and occupancy should commence with an analysis of how the proposed building fits into the uses and occupancies described in Chapter 3 of the Code and then should be analyzed against the criteria in Chapter 4 to see which, if any, are applicable.

It is worth remembering that the detailed provisions of Chapter 4 of the Code relate to and coordinate with the more basic requirements spelled out in Chapter 3 for use and occupancy requirements. The designer as code user should make a progression from the general to the specific in analyzing a building. Begin with the general categorization of uses and occupancies. Then proceed to review the detailed requirements of Chapter 4 for provisions applicable to the building in question. While one may begin the analysis of a specialized use by looking up the detailed requirements, the Code is organized to proceed from determining the occupancy first and then applying detailed criteria. Just as one should not read up from footnotes to find table sections that may be misapplied, the user should not work backwards in these analyses as this may lead to erroneous code interpretations.

We will go through the specialized building types that may occur in healthcare occupancies to describe their distinguishing criteria and touch upon the major added code provisions applicable to them. We will discuss those special uses most likely to be encountered by designers. Note that several very specialized uses have been omitted, as they are not seen frequently. Most are related to Group H Occupancies or to process-related special uses. Most designers do not frequently encounter these or other very specialized structures, such as amusement buildings. It is possible that certain medical processes or procedures may necessitate use of underground facilities, but as noted these are not likely to be so deep or so extensive as to require consideration of the facility as an underground building so we will not examine those requirements. The designer should however, be aware that requirements for underground buildings are contained in Chapter 4 of the IBC and are in addition to other occupancy or construction type requirements found elsewhere in the code.

Many of the sections in Chapter 4 contain definitions of the uses or occupancies to which the special provisions apply. These definitions are listed in Chapter 2 of the Code and cross-referenced for location and explanation in the section indicated. Note that although the definitions may seem familiar and similar to common construction terminology, they have very specific meanings in the Code. Examine the building design conditions and definitions carefully to determine the applicability of the definition to the building design. This analysis of the definition may also point out necessary modifications to the design to make it code-compliant or reveal the need to reclassify it.

- *First identify the general use and occupancy group to which the building belongs, and review the requirements contained in Chapter 3.*

- *Group A Assembly*
- *Group B Business*
- *Group E Educational*
- *Group F Factory*
- *Group H Hazardous*
- *Group I Institutional*
- *Group M Mercantile*
- *Group R Residential*
- *Group S Storage*
- *Group U Utility*

- *Then refer to Chapter 4 and review the requirements for special detailed requirements based on use and occupancy. Those more likely to be found in healthcare-related projects are noted in **bold type***

- *Covered Mall Buildings*
- ***High-Rise Buildings***
- *Atriums*
- ***Motor-Vehicle Related Occupancies (Parking Garages)***
- *Underground Buildings*
- ***Institutional Groups I–2 and I–3***
- *Motion-Picture Projection Rooms*
- *Stages and Platforms*
- *Special Amusement Buildings*
- *Aircraft-Related Occupancies*
- *Combustible Storage*
- *Hazardous Materials*
- *Application of Flammable Finishes*
- *Drying Rooms*
- *Organic Coatings*
- ***Group I-1, R-1, R-2, R-3***
- *Hydrogen Cutoff Rooms*

As building technology allowed advances in high-rise construction, buildings often out-stripped code provisions needed to address the new conditions impacting fire and life safety. High-rise buildings, made possible by innovative structural technology and elevators for transporting occupants, exceed the capabilities of fire-fighting procedures used for shorter buildings. They have occupied floors above the reach of even the longest ladders carried by fire department vehicles.

Note that high-rise egress systems are based on the occupant loads and egress requirements spelled out elsewhere in the code (see Chapter 10). Stairways are one of the primary means of egress, with elevators providing an accessible means of egress per §1007. Note that per §1007 areas of refuge cannot be eliminated by providing automatic sprinkler systems.

- *The definition of a high-rise building in §403 is based on the height that typical fire-department extension ladders and hose streams can effectively fight a fire. Thus a building with an occupied floor more than 75' (22 860) above the lowest level of fire-department access is defined as a high-rise. Fire fighting in a high-rise assumes that the firefighters must enter the building and go up inside the building to fight a fire.*

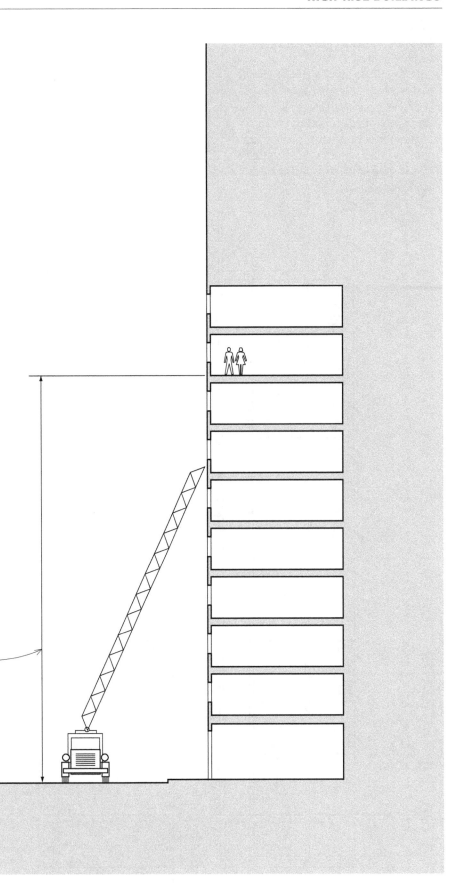

HIGH-RISE BUILDINGS

- *Emergency phone every fifth floor in each required stairway where the doors to the stairway are locked.*
- *§403.7 requires a two-way communications system for fire-department use per §907.2.12.3, operating between the fire command center and elevators, lobbies, standby power rooms, areas of refuge, and inside enclosed stairways. A communication device is to be installed at each floor level within enclosed stairways.*

Code requirements for high-rises are a combination of passive and active measures.

- *The buildings must be constructed of non-combustible materials.*
- *Shafts and vertical penetrations must be enclosed to prevent the spread of smoke and fire.*
- *See page 88 for an illustration of the requirements of §707.14 regarding elevator lobbies.*
- *§403.2 requires automatic fire sprinklers be installed throughout high-rise buildings.*
- *In Seismic Design Categories C, D, E or F, a secondary water supply capable of supplying the required hydraulic sprinkler demand for up to 30 minutes is required per §903.3.5.2.*
- *§403.3 allows a reduction in fire-resistance rating if the automatic sprinkler system has control valves equipped with supervisory initiating devices and water-flow initiating devices for each floor.*
- *§403.5 requires smoke detectors connected to an automatic fire-alarm system per §907.2.12.1.*
- *§403.6 requires an emergency voice/alarm communication system to be activated with the operation of any automatic fire detector, manual fire-alarm box or sprinkler device.*
- *§403.8 requires a fire-fighting command center per §911 in a location approved by the fire department.*
- *§403.10 and 11 require a standby and emergency power systems to be located in a separate room enclosed with 2-hour fire-resistance-rated fire barriers and fueled by a natural gas pipeline or a 2-hour supply of on-site fuel. These power systems are required for the operation of systems necessary for evacuation or for fire fighting, such as elevators, smoke control systems, emergency lights and fire-fighting systems and operations. Confer with the AHJ to determine the Fire Code distinctions and requirements for standby and emergency power systems.*
- *§403.12 requires stairway doors that are locked from the stairway side to be capable of being unlocked simultaneously from the fire command center.*

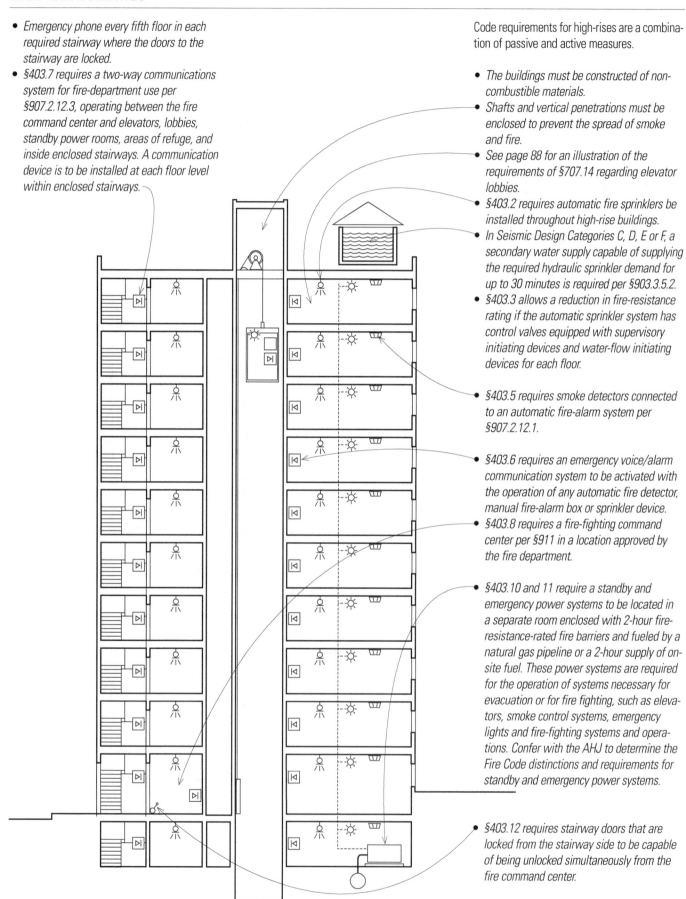

Atriums were an innovative building type that required a specific building-code response as their use became more prevalent. The code requirements in §404 for atriums combine aspects of malls and high-rise fire and life safety provisions.

As for covered malls and high-rise buildings, the Code requires that a mixture of active and passive fire-protection measures be provided in atriums. Atrium buildings, however, rely more heavily on active systems for fire and life safety.

- The Code requires that means of egress using the atrium must have shorter lengths than in other building types.
- Buildings containing atriums must be fully sprinklered throughout the building. There are limited exceptions to §404.3 for certain areas, but the basic design assumption should be that such buildings are fully sprinklered.
- Openings into the atrium may be glazed if special fire-sprinkler protection is provided.

The greatest concern about atrium fire and life safety involves the control of smoke. Atriums are required to have smoke management systems conforming to §909. These interconnect elaborate systems of detectors, fans and controls to contain or move smoke to allow safe egress for occupants of atriums.

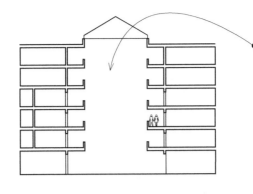

- *By definition, an atrium is an opening connecting two or more floor levels that is closed at the top. In other words, it is considered to be an interconnected series of floor openings inside of a building that create a physical connection and a common atmosphere between floor levels of the building. The code definition assumes that an atrium has an enclosed top.*

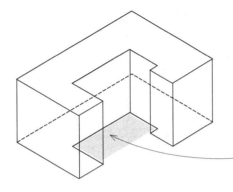

The definition of an atrium is partly determined by what it is not.

- *A building wrapped around an open space creates a court, not an atrium.*

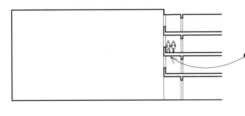

- *Balconies in assembly areas or mezzanines do not create an atrium.*

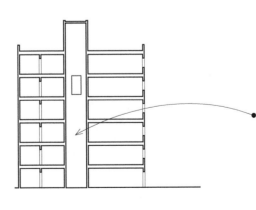

- *A series of openings through floors that are enclosed by a shaft such as an enclosed stair, elevator or ductwork does not create an atrium.*

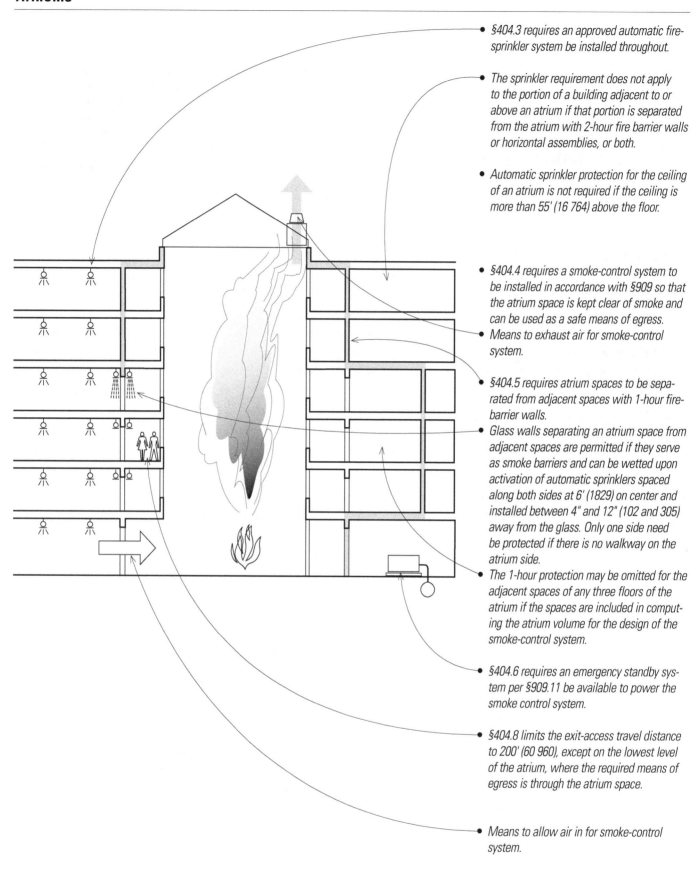

§404.3 requires an approved automatic fire-sprinkler system be installed throughout.

The sprinkler requirement does not apply to the portion of a building adjacent to or above an atrium if that portion is separated from the atrium with 2-hour fire barrier walls or horizontal assemblies, or both.

Automatic sprinkler protection for the ceiling of an atrium is not required if the ceiling is more than 55' (16 764) above the floor.

§404.4 requires a smoke-control system to be installed in accordance with §909 so that the atrium space is kept clear of smoke and can be used as a safe means of egress.

Means to exhaust air for smoke-control system.

§404.5 requires atrium spaces to be separated from adjacent spaces with 1-hour fire-barrier walls.

Glass walls separating an atrium space from adjacent spaces are permitted if they serve as smoke barriers and can be wetted upon activation of automatic sprinklers spaced along both sides at 6' (1829) on center and installed between 4" and 12" (102 and 305) away from the glass. Only one side need be protected if there is no walkway on the atrium side.

The 1-hour protection may be omitted for the adjacent spaces of any three floors of the atrium if the spaces are included in computing the atrium volume for the design of the smoke-control system.

§404.6 requires an emergency standby system per §909.11 be available to power the smoke control system.

§404.8 limits the exit-access travel distance to 200' (60 960), except on the lowest level of the atrium, where the required means of egress is through the atrium space.

Means to allow air in for smoke-control system.

Building codes have traditionally had a great deal of difficulty classifying and addressing parking structures and garages. There is a long-standing perception among code writers that there are inherent dangers for vehicles carrying flammable fuels and emitting noxious gases. Yet buildings housing these types of use have had a generally good safety record. Because of special concerns regarding these occupancies expressed in predecessor codes, §406 has grouped these uses into a separate category that can be considered a distinct occupancy group.

Motor-vehicle-related uses break down into five general categories:

1. Private garages or carports
2. Open parking garages
3. Enclosed parking garages
4. Motor-fuel dispensing stations
5. Repair garages

Private Garages or Carports

Private garages or carports are for private use and meet the common definition of what we all think of as parking for housing. These are typically considered Group U occupancies. Separation requirements are contained in Chapter 4, in §406.1.4. These requirements are cross referenced in Table 503, which regulates heights and areas of buildings. Note that one- and two-family dwellings and their accessory structures such as garages are intended to be regulated by the International Residential Code.

- *An enclosed garage abutting a house is to have a minimum of one layer of ½" gypsum board on the garage side. Where rooms extend over the garage a minimum of ⁵/₈" type "x" gypsum board is to be used*
- *Doors between the garage and house are to be solid wood or solid core doors at least 1³/₈" thick or complying with Section 715.4.3.*

Parking Garages

§406.2 classifies parking garages into open parking garages and enclosed parking garages.

Open Parking Garages

Open parking garages are multiple-vehicle facilities used for parking or the storage of vehicles where no repairs take place. These are typically classified as Group S-2 occupancies. To meet this definition, the amount and distribution of openings are specified in §406.3.3 of the Code. Because these openings are distributed in a manner that provides cross-ventilation for the parking tiers, no mechanical ventilation is required.

- *§406.2.2 requires a minimum clear height of 7' (2134) at each floor level for vehicular and pedestrian travel in all parking garages.*
- *§406.2.3 requires guards in accordance with §1012.*
- *§406.2.4 requires vehicle barriers in accordance with §1607.7.*

- *§406.3.3.1 requires uniformly distributed openings on two or more sides for the natural ventilation of open parking garages.*
- *The total area of openings on a tier must be at least 20% of the total perimeter wall area of that tier.*
- *The aggregate length of openings on a tier must be at least 40% of the total perimeter of the tier. Interior walls are to be at least 20% open, with uniformly distributed openings.*
- *Although openings on a third side are not required, openings on opposing sides are preferred for cross-ventilation.*

- *Table 406.3.5 determines the allowable floor areas and heights of open parking garages according to type of construction. Note that this table is cross referenced in Table 503, which regulates the heights and areas of buildings.*
- *§406.3.6 permits area and height increases for open parking garages having a greater percentage of openings along a greater perimeter of the building and under specific conditions of use.*

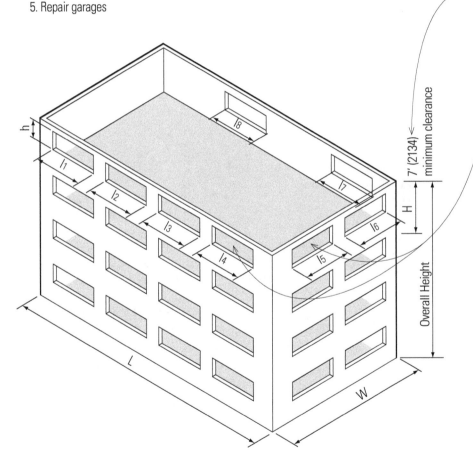

$$h \times [l_1 + l_2 + l_3 + l_4 + l_5 + l_6 + l_7 + l_8] \geq 20\% \text{ of } [H \times (2W + 2L)]$$

$$[l_1 + l_2 + l_3 + l_4 + l_5 + l_6 + l_7 + l_8] \geq 40\% \text{ of } [2W + 2L]$$

Enclosed Parking Garages

• Enclosed parking garages are similar to open parking garages except that the amount of wall enclosure relative to the building area does not allow them to be considered as open garages. Because they do not meet the criteria for open parking garages and are considered enclosed, mechanical ventilation is required to compensate for the lack of cross-ventilation.

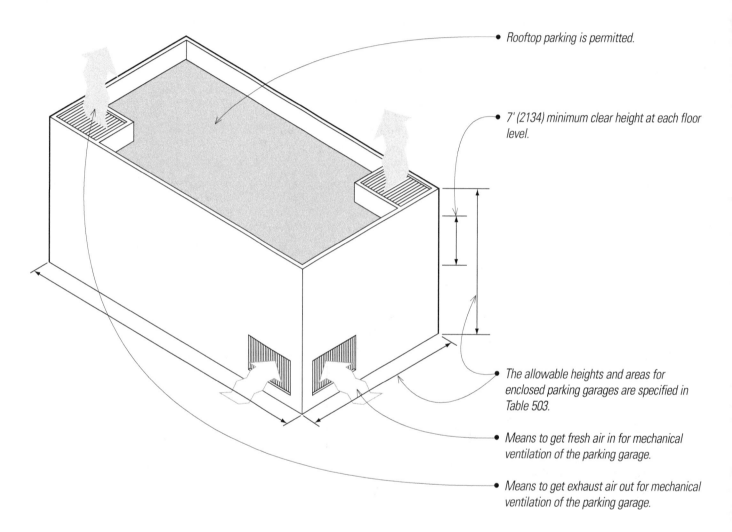

• Rooftop parking is permitted.

• 7' (2134) minimum clear height at each floor level.

• The allowable heights and areas for enclosed parking garages are specified in Table 503.

• Means to get fresh air in for mechanical ventilation of the parking garage.

• Means to get exhaust air out for mechanical ventilation of the parking garage.

Institutional Groups I-2 and I-3

The code adds provisions related to Group I-2 occupancies in Sections 407 and Group I-3 occupancies in Section 408 that must be read in concert with those in Chapter 3. Chapter 3 is intended to address occupancy classification of the building and Chapter 4 is intended to address detailed requirements for certain uses and occupancies. Code analysis of these occupancy groups requires looking at both sets of provisions together.

The life safety provisions of I-2 Occupancy requirements are based upon the difficulty of evacuating the occupants of these facilities. There is a hierarchy of protection, beginning with the patient room, connected to protected corridors, then to adjacent smoke compartments and finally to evacuation of a floor or of the entire building. The requirements also recognize the low fuel load in these facilities and the constant presence of staff. The requirements are also based on a group of protection features such as fire and smoke detection, fire containment barriers, horizontal evacuation and fire extinguishing systems.

- Corridors in I-2 occupancies are to be continuous to exits
  - per §407.3 corridor walls are to be constructed as smoke partitions per §710.
  - The doors into the corridor, when not located in a rated partition do not need to be fire rated or to have a closer. However, these doors must have positive latching and when closed must be an effective barrier to limit the transfer of smoke. This provision allows for doors to hospital rooms to remain open without closers, but be smoke barriers when closed and latched. Doors considered to be fire doors, such as those into an exit enclosure, are to be fire rated and are to conform with §715.4.
  - Locking devices may be used on the corridor side of the door to restrict access to patient rooms, but the patient side must provide ready egress except in mental health facilities where the patients are restrained from egress unless under supervision by the staff.

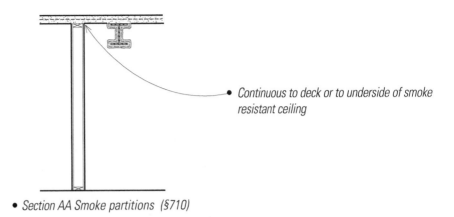

- Section AA Smoke partitions (§710)

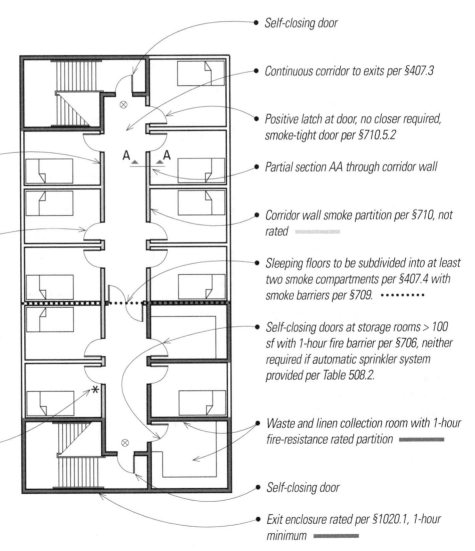

- Continuous to deck or to underside of smoke resistant ceiling

- Self-closing door

- Continuous corridor to exits per §407.3

- Positive latch at door, no closer required, smoke-tight door per §710.5.2

- Partial section AA through corridor wall

- Corridor wall smoke partition per §710, not rated

- Sleeping floors to be subdivided into at least two smoke compartments per §407.4 with smoke barriers per §709.

- Self-closing doors at storage rooms > 100 sf with 1-hour fire barrier per §706, neither required if automatic sprinkler system provided per Table 508.2.

- Waste and linen collection room with 1-hour fire-resistance rated partition

- Self-closing door

- Exit enclosure rated per §1020.1, 1-hour minimum

- For functional reasons it is desirable to have some areas open to corridors. §407.2.1 through §407.2.4 set out criteria that allow certain areas to be open to corridors.
- Spaces such as waiting areas constructed as required for corridors may be open to the corridor and of unlimited area if they meet **all** of the following criteria:
  - The spaces are not used for patient sleeping units, treatment rooms or hazardous or incidental use areas as defined by §508.2. In other words , the waiting area must be just that, an area with no other function than awaiting an appointment or treatment contained within it
  - The open space must have an automatic fire detection system per §907. This is designed to detect a fire before it compromises the viability of the corridor as a means of egress.
  - The corridors which open in the unlimited space are part of the same smoke compartment and provided with automatic fire detection per §907 or if in a different smoke compartment then the added area must have quick-response sprinklers in accordance with §903.3.2.
  - The space must be arranged to not obstruct access to the required exits. This is to be read in concert with the requirement in §407.2 that the corridors are to be continuous to the exits. The corridor width required for egress should be maintained through these contiguous areas. This applies to both plan layouts during design and to furniture arrangements after occupancy.

- Per §407.2.2 nurses stations, used for such things as clerical, charting or communications may open to a corridor if they are constructed as required for corridors. This is based upon functional requirements for observation of patient needs from the nursing station. This allowance is not intended to apply to nursing offices, supply storage areas or drug distribution areas with a higher fuel load which do not require patient observation.

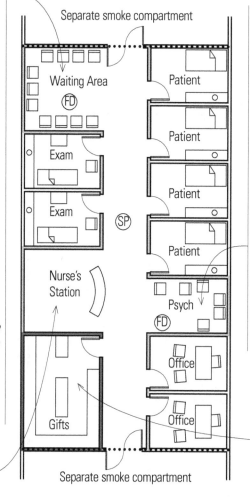

Separate smoke compartment

Waiting Area

FD

Exam

Exam

SP

Nurse's Station

Psych

FD

Office

Office

Gifts

Separate smoke compartment

Patient

Patient

Patient

Patient

Corridor

Fire-resistance rated partition per §508.2

Smoke barrier per §709

- Per §407.2.3 mental health treatment areas where mental patients not capable of unassisted self-preservation are housed or treated in groups, and not used for incidental uses as defined in §508.2 may be open to the corridor when **all** of the following criteria are met:
  - Each area does not exceed 1,500 sf (140 m²)
  - The area permits supervision by the facility staff
  - The area does not obstruct any access to required exits. This is also to be read in concert with the requirement in §407.2 that the corridors are to be continuous to the exits.
  - The area is equipped with an automatic fire detection system per §907.2
  - No more than one such space occurs in any one smoke compartment
  - Both the walls and ceilings of the space are constructed as required for corridors

- Per §407.2.4 gift shops less than 500 square feet (46.5 m²) in area may open to the corridor if the gift shop and any related storage areas are fully sprinklered and the storage areas are protected per §508.2. That section requires a 1-hour separation or provision of an automatic fire-extinguishing system for storage rooms. This should be provided at the gift shop.

- Per §407.4 all stories used by patients for sleeping or treatment are to have smoke barriers dividing the floors into at least two smoke compartments. Other stories with 50 or more occupants are also to be similarly subdivided. Smoke compartments are to be no larger than 22,500 sf (2092 m²) and the travel distance from any point in a smoke compartment to a smoke barrier door is not to exceed 200 feet (60.96 m). Smoke barriers are to comply with §709.
  - There is to be an independent means of egress from each smoke compartment created by smoke barriers without having to return through the smoke compartment from which the means of egress originated. This requirement is for one means of egress and does not appear to be tied to occupant load, but to configuration of the smoke compartments.

- Per §407.6 corridors in nursing homes, both intermediate care and skilled nursing, detoxification facilities and spaces permitted to open to corridors per §407.2 (such as waiting rooms and nurse's stations as described above) must have automatic fire detection systems. This is a cross reference from requirements in §407.2 Hospitals (note that Hospital is not given a definition in the IBC) are to have smoke detection per §407.2. Section §407.2 refers to §907 for smoke detection requirements. The specific requirements for I-2 occupancies are contained in §907.2.6.2. Corridor smoke detection is not required per the exceptions to §407.6 when patient rooms have smoke detectors or automatic door closing devices activated by smoke detectors in the patient sleeping units.

- There is to be a refuge area of 30 net sf (2.8 m²) per patient on each side of each smoke barrier. This refuge area may be part of a corridor, patient room, lounge area or similar low-hazard area. Where patients are not confined to a bed or litter the refuge area need only be 6 net sf (0.56 m²) per occupant on each side of the smoke barrier for the total number of occupants in adjoining smoke compartments.

- Per §407.4 smoke compartments with patient sleeping units are to be sprinklered per §903.3.1.1 and the sprinklers are to be quick-response or residential sprinklers per §903.3.2.

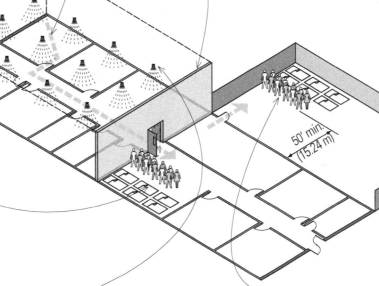

• Smoke barrier

50' min. (15.24 m)

Group I-2 Smoke Barrier at Smoke Compartment

- Per §407.7 secured yards with fencing and locks may be provided if there is sufficient space provided in similar fashion to the refuge requirement for smoke compartments. There is to be a refuge area of 30 net sf (2.8 m²) per patient for bed or litter patients and 6 net sf (0.56 m²) per occupant for ambulatory patients located in the exterior fenced area between the building and the fence. These safe dispersal areas are to be located at least 50 feet (15.24 m) from the building they serve.

Group I-3 Occupancies

- *Per §408.2 I-3 occupancies, which are detention and correction facilities may be part of mixed use occupancies. The healthcare facilities which would fall under the I-3 occupancy group would be prison hospitals, mental hospitals or correctional medical facilities. The criteria in §408 apply in concert with other applicable portions of the code. For instance, egress provisions from this section are to be read together with the egress requirements of Chapter 10 in the IBC. For brevity in our discussion we will use the term detention to apply to any I-3 occupancy, even though it may be in actuality a correctional facility. Because this occupancy group is a very specialized subset of buildings, crossing over between healthcare and detention facilities we will touch on the most relevant provision of these special requirements, not upon every one.*

- *Occupants are to have access to means of egress at all times, but the means of egress that traverse other uses are to be designed to conform to requirements for detention facilities. An exception allows the use of a horizontal exit into a space not conforming with detention egress provisions as long as the use into which the exit occurs is not a high-hazard use.*

- *The means of egress requirements of §408.3 modify those in Chapter 10, allowing narrower doors and sliding doors.*

- *Exits may discharge into a fenced or walled courtyard. The exit discharge does not need to lead to a public way. The yards or courts are to be sized to accommodate all occupants (which is presumed to include staff and inmates). The area is to be calculated as beginning 50 feet ((15.24 m²) from the building and provide a net area of 15 square feet (1.4 m²) per person.*

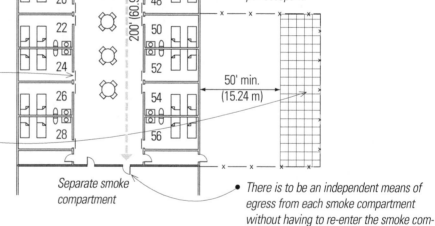

- *Egress doors are permitted to be locked with either key release or remote release depending upon the Occupancy Condition level of confinement. The level of confinement enters into the inter-relationship between security and ease of egress.*

- *Every story occupied by residents for sleeping is to have a smoke barriers per §709 to divide every story into at least two smoke compartments unless the floor has direct access to a public way, a secured yard or a separate building with a 2 hour fire resistance assembly or a 50 foot separation distance.*

- *The maximum number of occupants of each smoke compartment is to be 200. The maximum travel distance from an exit access door to a door in the smoke barrier is 150 feet; the maximum travel distance from any point in a room to a door in the smoke barrier is 200 feet.*

- *There are to be refuge areas on each side of the smoke barrier for the occupants of the adjoining smoke compartment. The space shall provide at least 6 square feet (0.56 m²) per occupant.*

- *There is to be an independent means of egress from each smoke compartment without having to re-enter the smoke compartment from which the means of egress originates.*

Other Specialized Uses

It is possible that designers will encounter H occupancies in healthcare facilities, especially in research institutions where labs may be part of the building or campus under consideration. Chapter 4 contains several sections related to H occupancies, which must be read in concert with the requirements in Chapter 3. We will not touch on these detailed requirements for hazardous occupancies as that is the proper subject for another volume. However, as noted in Chapter 3, it is worth remembering that there

are two basic sets of criteria for determining the requirements for hazardous occupancies. The first set of criteria is related to the hazard of the materials in use and the second set is related to the nature of the use and the quantities of those materials in use. The designer must work with the client to ascertain what materials will be used, where they will be used and in what quantities the materials will be stored to help the client determine the proper occupancy classification for potential H occupancies.

The remaining groups of uses in Chapter 4 relate to specific uses of buildings or parts of buildings that are infrequently encountered in most healthcare occupancies. These include motion-picture projection rooms, stages and platforms, special amusement buildings, aircraft-related facilities and high-piled combustible storage. These are specialized uses not often encountered in the normal course of healthcare work and will not be addressed in this book.

5
Building Heights and Areas

As noted in Chapter 3, building designers almost invariably start a project with a given occupancy. After a building's occupancy classification is determined, the code analysis task becomes one of determining what heights and areas are allowable for the occupancy classification, given various types of construction. These interactive criteria are set forth in Table 503 of the IBC. Economics and utility generally dictate that buildings be built using the least costly and complicated type of construction that will meet the criteria set forth in the Code.

There are two types of design choices that typically impact the use of Table 503 for determining allowable heights and areas. The first is when the design of a given occupancy must provide enough area to contain the known uses. The second choice comes into play in speculative buildings built for economic gain. The goal in these facilities, such as office buildings or retail uses, is to maximize the allowable height and area using the most economic construction type. Most healthcare occupancies will be of the first type, but it is conceivable that a medical office building could be built speculatively so we will address both alternatives in our discussion.

When addressing the question of maximizing the economically viable building size, the designer must use an iterative process to maximize the space for a given building. One must make assumptions regarding construction type and analyze the relative cost and return for various construction types before the owner can make a decision. This iterative process may also be used to maximize the economic efficiency of a building where the program size requirements are the primary consideration. In either case, the goal is to achieve the maximum area with the minimum investment in construction materials while still meeting or exceeding the code mandated requirements to protect public health, safety and welfare.

The organization of Chapter 5 in the code is based upon a set of basic criteria that are then modified by mitigating factors to allow increases or tradeoffs between heights, areas, construction types, fire protection and life safety systems. Upon first reading Table 503 may seem very restrictive, but there are modifications contained throughout the rest of the chapter that give the designer greater flexibility. Note the footnotes to the table which refer to various exceptions.

The definitions in Chapter 5 of the Code are cross-referenced from the list of definitions in Chapter 2. As with other similar definitions, they apply throughout the code, not just within Chapter 5. The definitions have very specific code-related criteria that may be different from the colloquial meanings of the terms. They should be studied carefully for applicability when determining allowable heights and areas.

Height limitations for buildings contained in planning and zoning regulations do not necessarily use the same definitions or criteria for determining heights. The definitions in the code are developed to facilitate uniform application of regulations. Read the documents that pertain to the regulations in question. Do not ever apply building-code criteria to planning issues, or vice versa.

Building Area is usually considered to include the outside face of the exterior walls of a building. This is based on the language "area included within surrounding exterior walls." The word *included* implies that the wall is part of the area to be considered, as opposed to using the word *enclosed*, which would imply the area begins at the inside face of the exterior wall. Another real-estate term often used for this area is gross building area. The safest way to calculate building area is to measure from outside face of wall to outside face of wall. This generates the most building area, and is the most conservative way to determine a building's area when nearing the upper limit of allowable area.

The definitions of Basement, Grade Plane and Building Height are correlated and are based upon the relationship of building parts below or above the grade plane as it is defined. Note that building height, as measured in feet, takes roof profiles into account. Building heights are measured to the average height of the highest roof, thus allowing for pitched roofs, varying parapet heights and rooftop equipment enclosures. The height is calculated using different criteria than those used to determine the height of a building in stories. "Story" is intended to mean occupiable or usable space located inside the building contained by a floor below and a plane (next story or roof) above. This definition has been moved to Chapter 2 in the newest edition of the code.

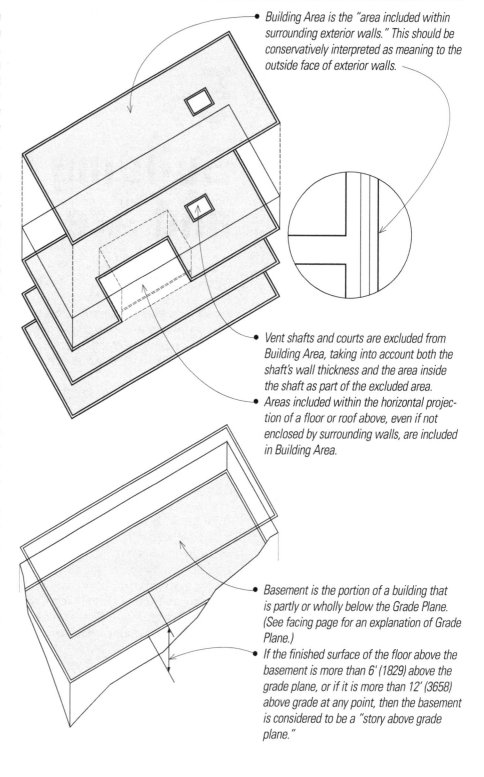

Building Area is the "area included within surrounding exterior walls." This should be conservatively interpreted as meaning to the outside face of exterior walls.

Vent shafts and courts are excluded from Building Area, taking into account both the shaft's wall thickness and the area inside the shaft as part of the excluded area.

Areas included within the horizontal projection of a floor or roof above, even if not enclosed by surrounding walls, are included in Building Area.

Basement is the portion of a building that is partly or wholly below the Grade Plane. (See facing page for an explanation of Grade Plane.)

If the finished surface of the floor above the basement is more than 6' (1829) above the grade plane, or if it is more than 12' (3658) above grade at any point, then the basement is considered to be a "story above grade plane."

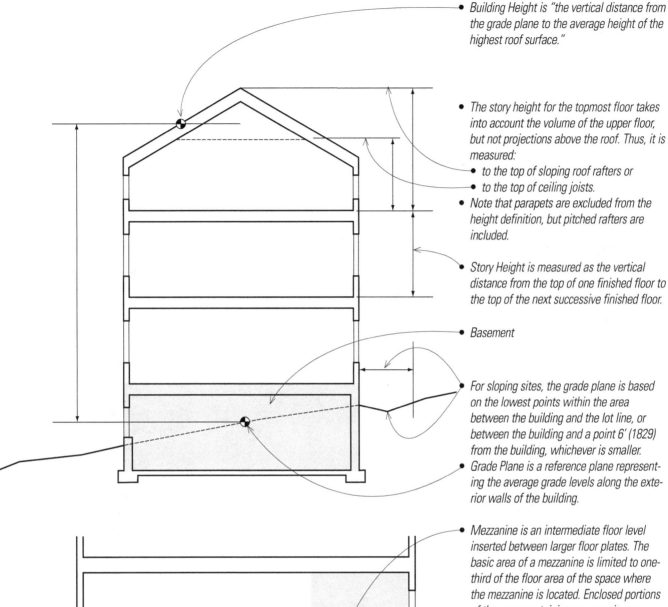

Building Height is "the vertical distance from the grade plane to the average height of the highest roof surface."

The story height for the topmost floor takes into account the volume of the upper floor, but not projections above the roof. Thus, it is measured:

- to the top of sloping roof rafters or
- to the top of ceiling joists.

Note that parapets are excluded from the height definition, but pitched rafters are included.

Story Height is measured as the vertical distance from the top of one finished floor to the top of the next successive finished floor.

Basement

For sloping sites, the grade plane is based on the lowest points within the area between the building and the lot line, or between the building and a point 6' (1829) from the building, whichever is smaller.

Grade Plane is a reference plane representing the average grade levels along the exterior walls of the building.

Mezzanine is an intermediate floor level inserted between larger floor plates. The basic area of a mezzanine is limited to one-third of the floor area of the space where the mezzanine is located. Enclosed portions of the room containing a mezzanine are not to be included in determining the floor area of the room in which the mezzanine is located. Also, the area of the mezzanine itself is not to be included in the floor area of the room. Also, exceptions allow larger mezzanines in Type I and II construction. Mezzanines may be up to two-thirds of the room floor area for special industrial occupancies per §503.1.1. In Type I and II buildings equipped with automatic sprinklers and an emergency voice/alarm communication system, mezzanines may be up to one-half the room floor area. Floor areas greater than the allowable area will be considered as creating another story. Note that the definition language allows multiple levels of mezzanines.

HEIGHT AND AREA LIMITATIONS

We will first discuss the standards in Table 503 as general limitations. We will then examine the allowable modifications contained in Chapter 5 of the Code.

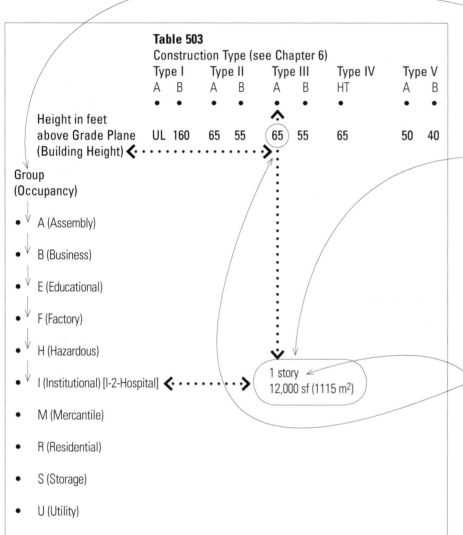

Table 503
Construction Type (see Chapter 6)

	Type I		Type II		Type III		Type IV	Type V	
	A	B	A	B	A	B	HT	A	B
Height in feet above Grade Plane (Building Height)	UL	160	65	55	(65)	55	65	50	40

Group (Occupancy)

- A (Assembly)
- B (Business)
- E (Educational)
- F (Factory)
- H (Hazardous)
- I (Institutional) [I-2-Hospital]
- M (Mercantile)
- R (Residential)
- S (Storage)
- U (Utility)

1 story
12,000 sf (1115 m²)

- *Allowable height and building area, as defined in §502, are determined by the intersection of occupancy group and construction type.*
- *For healthcare facilities an occupancy will be determined before heights and areas, the table will typically be entered by reading down the list of occupancy groups to find the occupancy that fits the building design.*

- *Then reading across leads to the allowable heights and building areas based on types of construction.*
- *Note that the distinction between A and B categories of construction types is the level of fire resistance as described in Table 601. A is of higher fire resistance, thus Type A buildings of any construction type have higher allowable heights and areas than Type B buildings. Using the principle of classifying occupancies by degree of hazard and building types by fire-resistance, the higher the level of fire and life safety, the larger and taller a building can be.*
- *Heights are expressed in two ways, both of which refer to defined terms. The first is height in feet above the grade plane and is generally independent of occupancy, but tied to fire-resistance; the second is height in stories and is tied to occupancy. Both sets of criteria apply to each analysis. This is to avoid having high floor-to-floor heights between stories that could generate a building exceeding the height limit in feet above grade plane if heights were not tabulated also. This could be an issue in hospitals with tall floor-to-floor heights for interstitial mechanical equipment.*
- *The illustration taken from Table 503 illustrates the relationship of occupancy and construction type to allowable heights and building areas. The examples are chosen from building types typically encountered by designers with the construction types chosen to highlight the differences as one proceeds from Type I fire-protected construction to Type V unrated construction. Entries in the table represent heights in feet/number of stories/allowable basic floor area of the ground floor.*

TABLE 503

Excerpt from IBC Table 503
(Showing Allowable Number of Floors and Proportionate Floor Areas for occupancies likely to occur in healthcare facilities)

Construction Type (See Table 601)	Type I A Fire-Rated	Type II A Fire-Rated	Type III A Partially Rated	Type IV Heavy Timber	Type V B Nonrated
Occupancy					
I-1 (Board and Care)	UL/UL/UL	65/4/19,000 sf 21.3 m/4/(1765 m²)	65/4/16,500 sf 21.3 m/4/(1533 m²)	65/4/18,000 sf 21.3 m/4/(1672 m²)	40/2/4,500 sf 13.1 m/2/(418 m²)
I-2 (Hospital)	UL/UL/UL	65/2/15,000 sf 21.3 m/2/(1394 m²)	65/1/12,000 sf 21.3 m/1/(1115 m²)	65/1/12,000 sf 21.3 m/1/(1115 m²)	NP/NP/NP
B (Medical Office)	UL/UL/UL	65/5/37,500 sf. 21.3 m/5/(3484 m²)	65/5/28,500 sf 21.3 m/5/(2648 m²)	65/5/36,000 sf 21.3 m/5/(3344 m²)	40/2/9,000 sf 13.1 m/2/(836 m²)
A-3 (Assembly)	UL/UL/UL	65/3/15,500 sf 21.3 m/3/(1440 m²)	65/3/14,000 sf 21.3 m/3/(1301 m²)	65/3/15,000 sf 21.3 m/3/(1394 m²)	40/1/6,000 sf 13.1 m/1/(557 m²)

§503.1.4 states that Type I buildings, permitted to be of unlimited height and area per Table 503, do not require the mitigations for unlimited-area building imposed upon other construction types by other sections of this chapter. Note, however, that certain Group H occupancies have height and area restrictions even for Type I buildings.

Having set the allowable limits for basic heights and areas in Table 503, the rest of the chapter addresses modifications and exceptions to the basic criteria. It is essential that the designer read the entire chapter after making the initial determination of heights and areas. Only by reading the chapter can all of the factors affecting heights and areas be determined.

Note that one set of occupancies that occurs relatively often in practice is not directly addressed in Table 503, but has a cross-reference to a separate set of height and area criteria. This building type/occupancy group is the stand-alone open parking garage. The height and area tables for this use are contained in Table 406.3.5. This use is also restricted to the types of construction listed in that table.

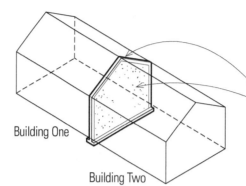

Building One

Building Two

- §503.1 allows one or more fire walls (defined and regulated by the provisions of Chapter 7) to divide a single structure into a number of "separate" buildings.
- Fire wall per §702.1 and §705
- Three-hour fire-resistance rated wall at I occupancy per Table 705.4

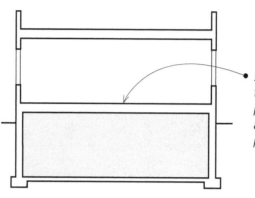

- §506.1.1 allows single level basements that are not considered a story above grade plane to be excluded from overall building areas if they do not exceed the areas permitted for a one-story building.

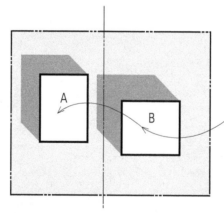

A

B

- §503.1.2 states that multiple buildings on a single site may be considered as separate buildings (see also §704.3 regarding assumption of an imaginary line between them), or as a single building for determining building areas.

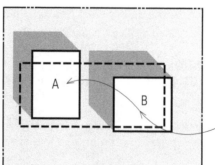

A

B

- For a group of buildings to be considered as a single building, the group must meet the aggregate area limitations based on the most restrictive occupancy. Means of egress must also be carefully examined when multiple buildings are treated as one.

- E.g., Areas of A + B must not exceed the single building area of the more restrictive occupancy (shown dashed).

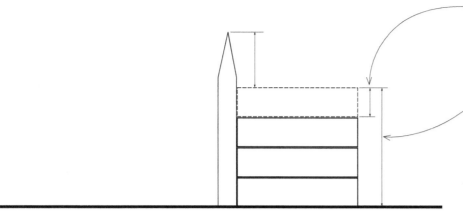

- *§504.2 permits buildings that are provided with sprinklers to increase 20' (6096) or one story in height. Residential buildings are limited to increase 20' (6096) and one story in height. Residential buildings protected by an NFPA "13R" sprinkler system per §903.3.1.2 are limited to no more than 4 stories and an overall height of 60' (18 288). These increases do not apply to the more restrictive Group H occupancies or to Group 1-2 occupancies of less than Type IIA construction.*

- *Per §504.3 Towers and steeples may project up to 20' (6096) above the allowable building height if of combustible construction and may rise to any height if of noncombustible construction. Note that these criteria are based on the presumption that such spaces are not used for habitation or storage. They must be essentially decorative or for operational purposes only.*

MEZZANINES

§505 considers mezzanines that meet the definition in §502 to be part of the floor below them. If they meet the criteria limiting their area to one-third of the floor below, then they are not considered part of the overall building area, or as an additional story. However, the area must be counted toward the overall fire area as defined in §702.

Room Area

1/3 Room Area

- *Mezzanines are limited in area to one-third the area of the floor area of the room in which they are located, but exceptions to §505 allow larger mezzanines in Type I and Type II buildings. Such a mezzanine is not only excluded from the overall area but also from the total floor-area calculation for the room containing the mezzanine. Thus a 1,000 sf (93 m²) floor space can have a 333 sf (31 m²) mezzanine per the basic area allowances.*

- *Type I and II buildings housing special industrial occupancies get a bonus allowing the mezzanine to be up to two-thirds of the room floor area.*

- *Type I and II buildings with sprinklers and voice/alarm systems may have mezzanines of up to one-half the area of the room floor area.*

- *Mezzanines must be of habitable height, having a minimum of 7' (2134) clear headroom at the mezzanine level as well as in the floor area under the mezzanine.*

- *Enclosed portions of a room are not included when determining the size of the room in which a mezzanine is located.*

- *Mezzanines are conceived of by the Code as open areas set above other spaces in a room. The code makes an absolute sounding statement that all mezzanines shall be open and unobstructed to the room in which they are located, except for a railing-height wall at the edge. The statement is then followed by numerous exceptions. The basic idea is that if the mezzanine is small in area or occupant load, or is furnished with a clearly defined separate exit path, it may be enclosed.*

- *Mezzanines are required to have two separate means of egress. This is typical except in certain conditions where there is a low occupant load, or if common paths of egress travel are very short. These limitations must be examined together in light of egress requirements in Chapter 10 of the Code.*

AREA MODIFICATIONS

§506 specifies when the areas limited by Table 503 may be increased. The two basic factors that permit increased building areas are frontage on a public way as defined in §1002.1, or provision of a sprinkler system.

§506.1 allows the basic areas limited in Table 503 to be increased due to building frontage and/or automatic sprinkler system protection. Thus, the total allowable area per floor is equal to the basic tabular area from Table 503 plus any increase due to frontage on a public way or open space, plus any increase for the installation of an automatic sprinkler system.

- *§506.4 permits the total allowable building area for buildings of three or more stories above grade plane to be three times the maximum area permitted per floor as determined per §506.1. Two-story buildings are limited in total area to two times the maximum area permitted per floor. No single story is to exceed the allowable floor area as determined per §506.1.*

- Every building must adjoin or have access to a public way in order to receive area increases. When a building has more than 25 percent of its perimeter opening onto a public way or open space at least 20' (6096) wide, the building's area may be increased based on the added width of the public way(s) and/or open space(s) and the extent of the building perimeter surrounded by them.

- Equation 5-2:

$$I_f = [F/P-0.25] \, W/30$$
$$= [250/250-0.25] \, 40/30*$$
$$= [1.0-0.25]1*$$
$$= [0.75] = 75\%$$

where

I_f = Area increase due to frontage
F = building perimeter fronting on a public way or open space of 20' (6096) minimum width
P = total perimeter of the entire building
W = width of public way or open space (feet) per 506.2.1

*W/30 cannot exceed 1.0 per §506.2.1 except it may be up to 2 when the building meets all criteria of §507 except for compliance with the 60' (18 288) public way or yard requirement

$$I_f = [F/P-0.25] \, W/30$$
$$= [125/250-0.25] \, 50/30*$$
$$= [0.5-0.25]1*$$
$$= [0.25] = 25\%$$

- Equation 5-1 combines the increase for frontage with the increase for sprinklers.
$$A_a = \{A_t + [A_t \times I_f] + [A_t \times I_s]\}$$

where

A_a = allowable area per story
A_t = allowable floor area §Table 503
I_f = allowable increase factor for frontage calculated per §506.2
I_s = allowable increase factor for sprinklers per §506.3

- Perimeter meeting criteria for frontage increase as per Equation 5-2. ------

Application of Equations 5-1 and 5-2 to an example building: Group B, Type III-A construction, three stories, fully sprinklered, location on property as depicted below, fully sprinklered. Note that this is an iterative process and will typically be used out of sequential order from the way it is presented in the code.

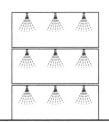

- A_t: Allowable Area per Table 503: 28,500 sf (2648 m²)
- H: Maximum Building Height per Table 503: 5 stories or 65' (19 812)

- Frontage Increase per Section 506.2 (Equation 5-2)

- I_f = [F/P – 0.25] W/30
  = [(50' + 50')/(50' x 4 sides) – 0.25] 50/30*
  = [100/200 – 0.25]1*
  = [0.5 – 0.25]1*
  = [0.25] = 25%
  = 28,500 sf (2648 m²) x 0.25
  = 7,125 sf (662 m²)

- Perimeter meeting criteria for frontage increase as per Equation 5-2.

    * W/30 cannot exceed 1.0 per §506.2.1 (assumes does not comply with all of §507)

- Sprinkler Increase (Section 506.3)
  Three story building = 200% (2x)

  Thus
  A_t = 28,500 sf (2648 m²) x 2
      = 57,000 sf (5292 m²)

- Area Modification per Section 506.1 (Equation 5-1)
  This equation ties the frontage and sprinkler increases together.

  A_a = {A_t + [A_t x I_f] + [A_t x I_s]}

  where
  A_a = allowable area per story
  A_t = allowable floor area §Table 503
  I_f = allowable increase factor for frontage
  I_s = allowable increase factor for sprinklers
  A_a = 28,500 + [28,500*.25] + [28,500*2]
      = 28,500 + 7,125 + 57,000
  A_a = 92,625 sf

- Area Determination (Per Section 506.4)
Per §506.4 for a three story building the area A_a determined per §506.1 may be increased by 3 times. For the example the maximum area of the building is thus:

92,625 sf x 3 = 277,875 sf (25,815 m²)

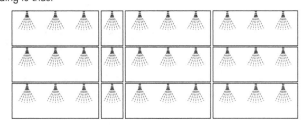

§508 Mixed Use and Occupancy

This section was moved from its prior location in Chapter 3 in the new 2006 edition of the code.

- Incidental uses   (§508.2)
- Accessory occupancies (§508.3.1)
- Mixed occupancies   (§508.3)

Incidental Use Areas

Uses or occupancies that are incidental to the main occupancy are not considered to have enough impact to warrant their classification as a mixed occupancy. The structure or portion thereof must be classified into one of the occupancy groups outlined in §508.2.

When the occupancy of a building does contain one or more incidental use areas as defined in Table 508.2 the areas are considered part of the main occupancy but require fire-resistance rated separations from the rest of the occupancy, or fire extinguishing systems, according to the requirements listed in the table. Note that incidental use area separation requirements are not applicable to dwelling units.

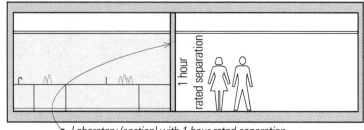

- *Laboratory (section) with 1 hour rated separation*

or

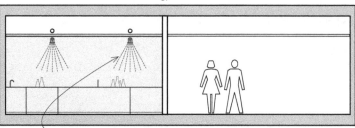

- *Laboratory (section) with automatic sprinkler system*

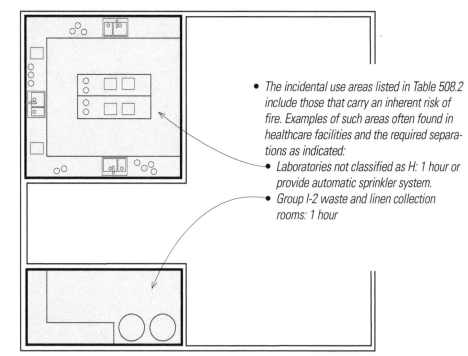

- *The incidental use areas listed in Table 508.2 include those that carry an inherent risk of fire. Examples of such areas often found in healthcare facilities and the required separations as indicated:*
  - *Laboratories not classified as H: 1 hour or provide automatic sprinkler system.*
  - *Group I-2 waste and linen collection rooms: 1 hour*

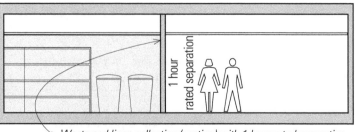

- *Waste and linen collection (section) with 1 hour rated separation*

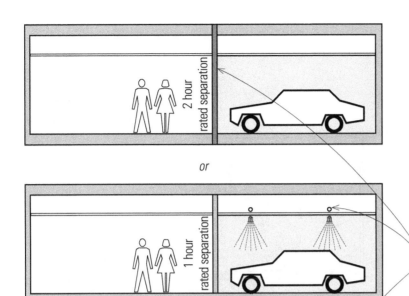

or

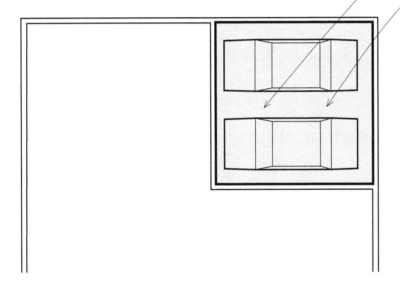

- Additional examples of incidental use areas per Table 508.2 often found in healthcare facilities and the required separations as indicated:
  - Parking garages per §406.2: 2 hours, or 1 hour and provide automatic fire-extinguishing system.
  - Note that per §508.2.3 where automatic sprinkler systems are provided to meet the requirements of Table 508.2 the system need only be provided in the incidental use area.

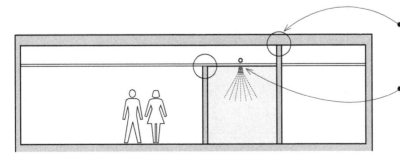

- Fire barriers per §706 or horizontal assemblies per §711 should extend from the floor to the underside of the fire-resistance-rated floor/ceiling or roof/ceiling assembly.
- In some cases, an automatic fire-extinguishing system may be substituted for the fire-rated separation, in which case the separation must still be able to resist the passage of smoke. Where an automatic fire-extinguishing system is provided to meet the requirements of Table 508.2 it need only be provided in the incidental use room or area.

Accessory Occupancies

When the occupancy contains a distinctly different accessory use that takes up less than 10% of the area of any story in which they are located or less than the height or area allowed for various occupancies by Table 503, the accessory use need not be separated from the primary occupancy. Exceptions exist for hazardous uses or when required by §508.2 as an incidental use area.

Per exceptions to §508.3.1 accessory assembly occupancies are not considered separate occupancies if under 750 sf, or as accessory to a Group E (education) occupancy. Also, religious education classrooms and religious auditoriums with under 100 occupants are not considered separate occupancies.

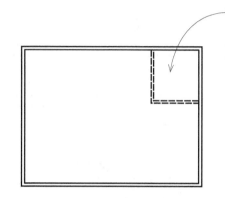

• Accessory occupancy less than 10% of the areas of the primary occupancy need not be separated from the primary occupancy.

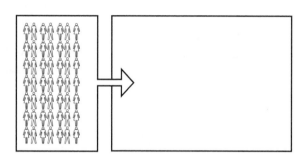

Mixed Occupancies

When a building has a mix of occupancies that are each distinct or extensive enough as separate uses, each use is considered a separate and distinct occupancy. The mix of occupancies is addressed in one of three ways:

1. Accessory Occupancies
   As described above.
2. Separated Occupancies
   Separated occupancies require that an occupancy separation with a fire-resistance rating as defined by Table 508.3.3 be provided between the occupancies.
3. Nonseparated Occupancies
   For nonseparated occupancies, the entire building is regulated according to the most restrictive of the height, area and fire-protection requirements for each of the multiple occupancies under consideration.

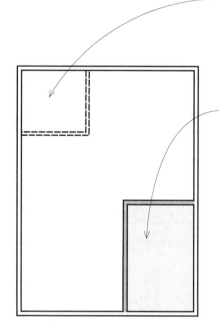

• When using nonseparated occupancy criteria per §508.3.2 uses need not be separated from each other or from the remainder of the building, but the allowable height, area and construction type for the entire building is governed by the most restrictive criteria for each of the unseparated uses.
• When using separated occupancy criteria per §508.3.3 different uses must be separated from each other with a fire barrier, but the allowable height area and construction types for the building as a whole are determined by the requirements for each separate area.

§509 is devoted entirely to exceptions to the provisions of Table 503 and the other sections in chapter 5. These exceptions apply only when all in the conditions in the subsections are met. These conditions are based on specific combinations of occupancy groups and construction types. These special provisions were code responses developed over time to meet construction conditions found in the jurisdictions of the model codes that preceded the IBC. These special provisions will usually give the designer greater flexibility in meeting the requirements of the building program than will Table 503 alone. These conditions should be annotated in the table for reference by the designer to be certain that they are not overlooked when commencing design and code analysis.

Group S-2 Parking Garage with Groups A, B, M, R or S Above

The special provisions of §509.2 apply to a set of conditions where an enclosed or open parking garage is built with other uses above. This is a very common building type. It is a mixed-use facility where the uses occur above the parking provided for the uses. The special provisions applicable to this building type have been developed over the years in response to the proliferation of apartment buildings using wood-frame construction over a concrete garage structure. As more of these mixed-construction structures came into being, they required provisions to make code interpretations more uniform.

Group S-2 Enclosed Parking Garage with S-2 Open Parking Garage Above

§509.3 section uses the same principle as that for mixed-use buildings in IBC §509.2. The criteria are different but basically allow an open garage to be built over an enclosed garage while treating the transition between them as a new ground plane when fire-resistance requirements are met.

Specific requirements for special provisions of §509.2 include:

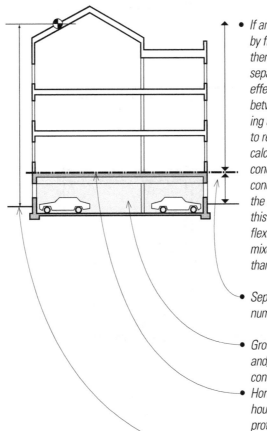

- Stairs and shafts enclosures going through the horizontal assembly have a rating of at least 2 hours
- The use above the horizontal assembly contains A Occupancies each with less than 300 occupants, or Groups B, M or R or S.

- The basement or first story above grade plane is of Type I-A construction and has a 3 hour horizontal separation from the building above

- The building below is an S-2 occupancy for storage of private motor vehicles

- Refer also to §509.4 for parking provisions beneath Group R occupancies.

- If an enclosed parking garage is separated by fire-resistant barriers of defined ratings, then the building above is considered a separate building for heights and areas. In effect, by creating a horizontal separation between the podium below and the building above, the code allows the designer to redefine where the ground plane is for calculation of heights and areas. All of the conditions listed in §509.2 must be met in concert for the provisions to apply. Even with the combination of requirements, applying this provision will usually give much greater flexibility for the design of this type of mixed-use and mixed-construction building than applying Table 503 as written.

- Separate buildings for allowable area, number of stories, and types of construction

- Group S-2 occupancy in a basement and/or first story above grade of Type IA construction
- Horizontal assembly having a minimum 3-hour fire-resistance rating, with openings protected by enclosures having a 2-hour fire-resistance rating
- Single building for height limitation in feet

Open Parking Garage
Beneath Groups A, I, B, M and R

§509.7 allows constructing the occupancies as indicated above an open parking garage and treating them as separate buildings for allowable height and area purposes. The height and area of the open garage are regulated by §406.3, and the heights and areas of the groups above the garage are regulated by Table 503. There is a restriction that heights in feet and stories for the part of the building above the garage be measured from the grade plane for the entire group of stacked uses. Also this section requires fire resistance to be provided for the most restrictive of the uses. It also requires egress from the uses above the garage be separated by fire-resistance-rated assemblies of at least two hours.

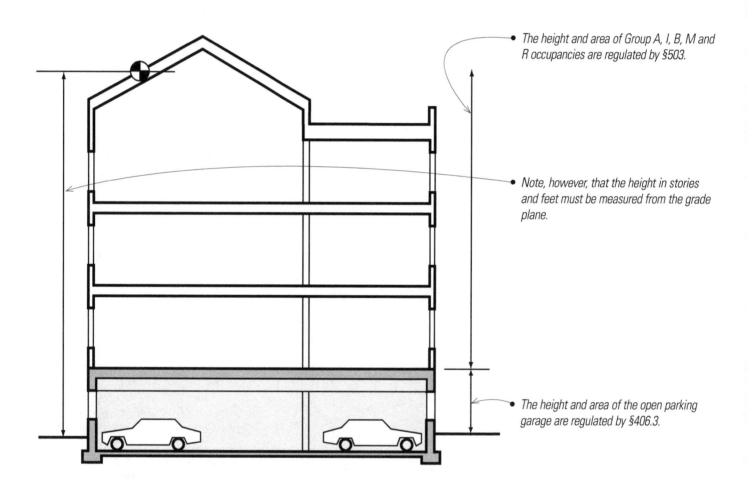

• The height and area of Group A, I, B, M and R occupancies are regulated by §503.

• Note, however, that the height in stories and feet must be measured from the grade plane.

• The height and area of the open parking garage are regulated by §406.3.

6
Types of Construction

The classification of buildings by types of construction has been a part of most model codes from their inception. The Uniform Building Code contained these classifications in the first 1927 edition. The IBC also recognizes the relationships of occupancy and construction type contained in the predecessor codes, but the provisions of the IBC are organized in a much different manner and with less direct definitions than the older model codes. Many code sections became footnotes, and some provisions now appear in definitions in other chapters. Fire-resistance ratings that had more extensive written descriptions in the legacy codes, that aided understanding in relation to known building materials have, in the IBC, briefer descriptions in §602, with many distinctions made in relation to test criteria alone. However, those familiar with construction will still understand the basic distinctions between construction type classifications.

The simplification of the definitions and reduced description of the respective responsibilities of designers and building officials may make determinations easier. On the other hand, there may be more confusion in the application of these provisions as users of the code apply the abstract provisions to real projects. The development of precedents for interpretations comes from the use of the Code to generate a body of applied knowledge to set the tone for future interpretations.

Definitions of building construction in the older model codes went into greater detail about construction materials for such elements as structural framework and stair treads. The new criteria define the subdivision of materials between "combustible" and "noncombustible" by their test performance under given conditions. This is a more precise definition, but more obscure for the casual code user who understands intuitively that steel or concrete are noncombustible materials versus how they perform in testing per ASTM E 119, *Standard Test Methods for Fire Tests of Building Construction and Materials*. This change takes some adaptation by those code users steeped in the old model codes to adjust to the IBC way of doing things.

TYPES OF CONSTRUCTION

The IBC classifies all buildings into five broad categories based on the fire-resistance capabilities of the predominant materials used for their construction. The five types of construction are given Roman-numeral designations, and progress downward in fire-resistance from Type I, the most fire-resistive construction, to Type V, the least fire-resistive.

The five types of construction classes are subdivided into two broad categories, A and B, based upon the inherent fire-resistance or combustibility of the materials; these are further subdivided according to the fire-resistance gained by the application of protection to major elements of the construction systems. This fire-resistance rating is predicated on the protection of building elements from exposure to fires both from within the building and from adjacent structures.

Table 503 correlates the five types of construction with the allowable heights and areas for buildings, based on their occupancy. The designer thus works with three sets of dependent variables when making the initial code analysis. The desired occupancy and the desired building height and area will determine the type of construction allowable under the Code. Typically the occupancy is pre-determined and the area will be dictated by the anticipated use or by site constraints. Thus the construction type is typically the last variable to be explored after the other variables are fixed.

Table 601 defines the required fire-resistance of major building elements for each type of construction. See pages 68–72 for a more detailed discussion. Note that Chapter 7 governs the actual materials and assemblies used in fire-resistance-rated construction. There is a direct relationship between occupancy type, the character of the occupants and the quantity of occupants to the construction classification. Higher occupant quantities, more hazardous occupancies and occupants with special needs, such as hospital patients, the aged and the infirm all require more fire-resistance or additional levels of active fire-suppression systems.

Table 503

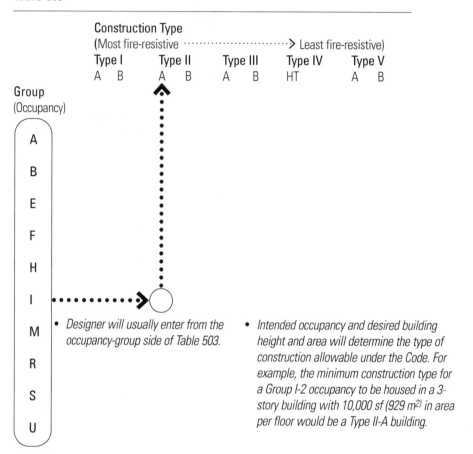

- Designer will usually enter from the occupancy-group side of Table 503.

- Intended occupancy and desired building height and area will determine the type of construction allowable under the Code. For example, the minimum construction type for a Group I-2 occupancy to be housed in a 3-story building with 10,000 sf (929 m²) in area per floor would be a Type II-A building.

The broadest distinction between the various types of construction can be summarized in the table below.

Materials	Protected Elements	Less Protected Elements	Unprotected Elements
Noncombustible	Type I-A, II-A	Type I-B	Type II-B
Combustible	—	—	—
Mixed Systems	Type III-A	Type III-B	—
Heavy Timber	—	Type IV	—
Any Materials	Type V-A	—	Type V-B

• Note that levels of fire-resistance decrease from left to right and top to bottom of this table.

Noncombustible Materials

The principal elements of construction Types I and II are made of noncombustible materials. The Uniform Building Code defines "noncombustible" as "Material of which no part will ignite and burn when subjected to fire" (1997 UBC §215). The IBC definition of noncombustibility is contained in §703.4, and states that materials required to be noncombustible must meet the test criteria prescribed in the American Society for Testing and Materials (ASTM) Standard E 136: "Standard Test Method for Behavior of Materials in a Vertical Tube Furnace at 750°C."

Combustible Materials

The elements of Types III, IV and V construction allow the use of combustible materials in varying degrees. Additional levels of fire protection can increase the fire-resistance rating of these three types of construction. Note that construction types with mixed elements of noncombustible and combustible construction are considered combustible and are of Types III, IV or V.

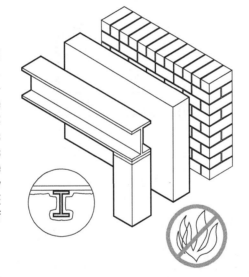

• *Noncombustible materials include masonry, concrete and steel.*

• *Note that application of additional fire protection materials to the noncombustible elements of Types I and II construction yields higher Type I-A and II-A ratings, above the basic Type I-B and II-B classification for unprotected noncombustible construction.*

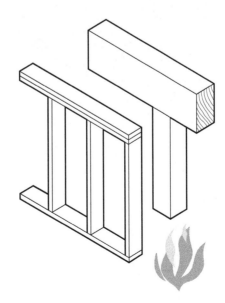

• *There is no definition of "combustible construction" contained in the IBC. Technical dictionaries of construction or mining terms define it as: "capable of undergoing combustion or of burning. Used esp. for materials that catch fire and burn when subjected to fire." By inference these materials would be those that do not comply with the requirements for noncombustibility contained in ASTM E 136, such as wood and plastics.*

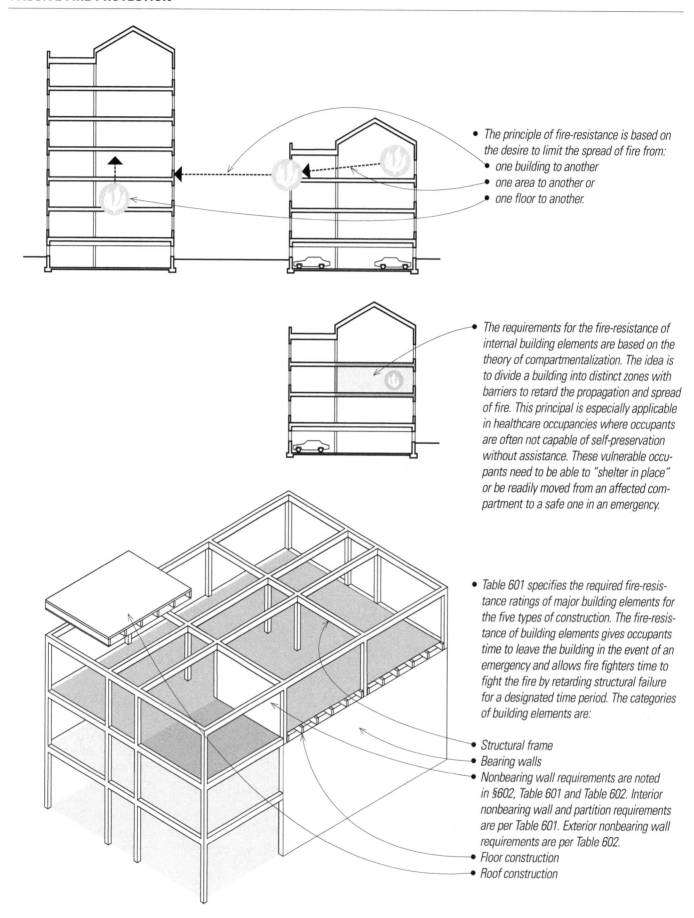

- The principle of fire-resistance is based on the desire to limit the spread of fire from:
  - one building to another
  - one area to another or
  - one floor to another.

- The requirements for the fire-resistance of internal building elements are based on the theory of compartmentalization. The idea is to divide a building into distinct zones with barriers to retard the propagation and spread of fire. This principal is especially applicable in healthcare occupancies where occupants are often not capable of self-preservation without assistance. These vulnerable occupants need to be able to "shelter in place" or be readily moved from an affected compartment to a safe one in an emergency.

- Table 601 specifies the required fire-resistance ratings of major building elements for the five types of construction. The fire-resistance of building elements gives occupants time to leave the building in the event of an emergency and allows fire fighters time to fight the fire by retarding structural failure for a designated time period. The categories of building elements are:

- Structural frame
- Bearing walls
- Nonbearing wall requirements are noted in §602, Table 601 and Table 602. Interior nonbearing wall and partition requirements are per Table 601. Exterior nonbearing wall requirements are per Table 602.
- Floor construction
- Roof construction

The classification by types of construction determines the level of passive fire-resistance that is inherent in the building's structure and envelope. This is distinct from active fire-suppression systems such as sprinklers. As we will discuss in Chapters 7 and 9, there can be tradeoffs under the code between passive and active systems. The idea is to look at the building as a whole and provide a balance of fire protection to achieve a predetermined level of structural protection and occupant safety.

It is important for the designer to remember that most building owners will opt for the least fire-resistance rating that meets the requirements of the code, as there is a direct relationship between providing fire-resistance and cost of construction. The determination of construction type usually entails an analysis of both the desired occupancy and the requirements for fire-separation distance between buildings or parts of buildings. This helps determine the necessary minimum fire-resistance of the building elements. Conversely, using higher fire-rated building systems for a portion of a building for reasons such as utility or aesthetics does not require the entire building to be of a higher fire rating than is required by the occupancy or location on the property. The standards applied to classification are minimum standards, but the minima and maxima apply to the whole building, not isolated components.

The building official will examine the construction classification assigned by the designer and make the final determination of classification. The building is to be looked at as a whole system and considered in the aggregate. §602.1.1 notes that although portions of the building may exceed the requirements for the type and for the building occupancy, the whole building need not meet requirements higher than those necessary for the intended occupancy.

- *Passive fire-resistance results from the use of construction materials and assemblies that can be expected to withstand exposure to fire without collapsing or exceeding a certain temperature on the side facing away from a fire. Examining Table 503 discloses that the application of fire-resistive materials to construction materials to move within a construction classification from Category "B" to "A" can allow increases in the allowable heights and areas. This recognizes that combustible materials may be made more fire-resistive by the application of fire-retardant coverings to accomplish the goal of retarding the spread of fire and to extend the durability of the building structure in a fire. The application of passive fire-resistive materials may increase the durability of even noncombustible materials in a fire. Even steel or concrete, if unprotected, can lose strength under fire exposure.*

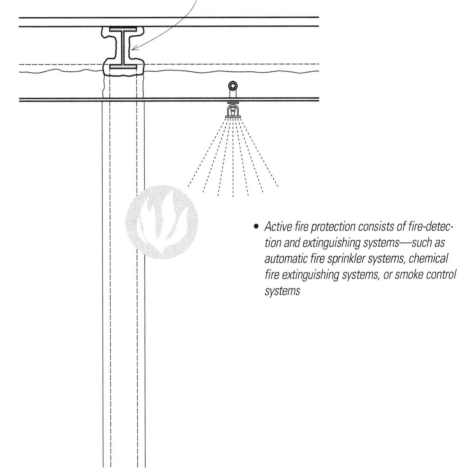

- *Active fire protection consists of fire-detection and extinguishing systems—such as automatic fire sprinkler systems, chemical fire extinguishing systems, or smoke control systems*

TABLE 601

Table 601 specifies the fire-resistance rating requirements for the major building elements, based on a specified type of construction or construction classification. A building may only be classified as a single type of construction unless a fire wall divides it into separate structures. Even if some building elements satisfy the fire-resistance rating requirements for a higher type of construction, the building as a whole need only conform to the lowest type of construction that meets the minimum requirements of the Code based on occupancy. This also means that a building can only be classified as high on the scale of noncombustibility as its least fire-resistive elements. Below is an abbreviated version of Table 601.

Required Fire-Resistance Ratings in Hours

Construction Type	Type I		Type II		Type III		Type IV	Type V	
	A	B	A	B	A	B	HT	A	B
Building Element*									
Structural Frame	3	2	1	0	1	0	HT	1	0
Bearing Walls									
Exterior	3	2	1	0	2	2	2	1	0
Interior	3	2	1	0	1	0	1/HT	1	0
Nonbearing Walls	Requirements for nonbearing exterior walls and interior partitions are noted in §602 and Table 602.								
Floor Construction	2	2	1	0	1	0	HT	1	0
Roof Construction	1½	1	1	0	1	0	HT	1	0

Under the previous model codes, elements such as stairways were also defined in terms of construction materials under the sections addressing types of construction. The new test-based criteria called out in §1009.5 states that stairways be "built of materials consistent with the types permitted for the type of construction of the building" This new definition may lead to confusion when applied in conjunction with §603, which allows the use of combustible materials in Type I and Type II construction.

The footnotes to Table 601 contain essential information for the designer to consider when making design decisions and when reviewing Table 601 in conjunction with Table 503, which correlates allowable height and building areas with types of construction and occupancies. The designer should read the footnotes carefully as they contain notable exceptions and trade-offs. For example, sprinkler trade-offs for combustible construction, once contained in separate code provisions in the Uniform Building Code, are now in the footnotes to Table 601.

TABLE 601

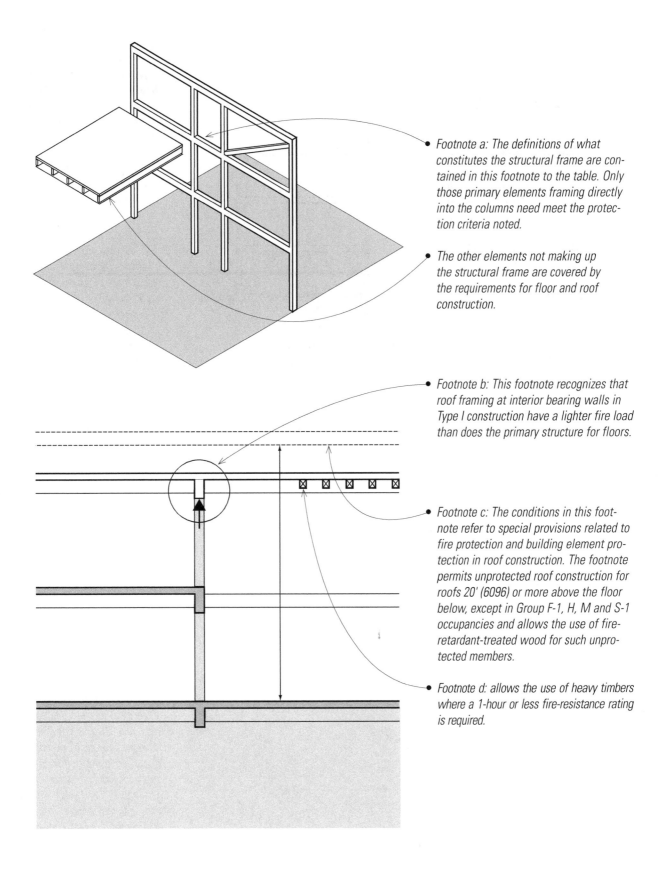

Footnote a: The definitions of what constitutes the structural frame are contained in this footnote to the table. Only those primary elements framing directly into the columns need meet the protection criteria noted.

The other elements not making up the structural frame are covered by the requirements for floor and roof construction.

Footnote b: This footnote recognizes that roof framing at interior bearing walls in Type I construction have a lighter fire load than does the primary structure for floors.

Footnote c: The conditions in this footnote refer to special provisions related to fire protection and building element protection in roof construction. The footnote permits unprotected roof construction for roofs 20' (6096) or more above the floor below, except in Group F-1, H, M and S-1 occupancies and allows the use of fire-retardant-treated wood for such unprotected members.

Footnote d: allows the use of heavy timbers where a 1-hour or less fire-resistance rating is required.

TABLE 601

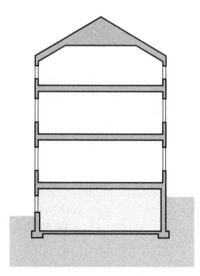

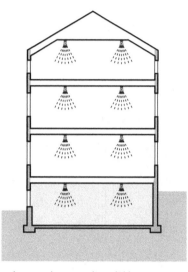

- 1-hour fire-resistance-rated construction

- Approved automatic sprinkler system
- Substitution not allowed for exterior walls

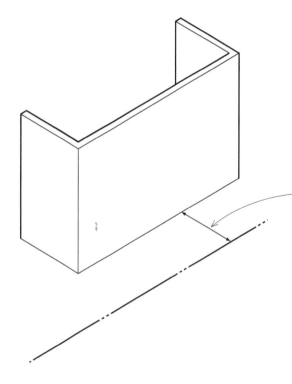

- Footnote e: This footnote calls out trade-off provisions for the use of automatic fire sprinkler systems to allow diminution of protection of the building elements in buildings of Type II, II and V construction. This is based on the interaction between passive fire-resistance that is built into the structure of a building and active fire suppression that is designed to put out a fire before it can grow large enough to threaten the structural integrity of the building elements needing protection. These trade-offs can be used only when the sprinkler system is not otherwise required by the code and is not being used to augment other considerations such as increasing allowable building area or height. The design team, when deciding to make the trade-offs allowed, should carefully consider the nature of passive and active systems. For example, consideration of which approach is best for a building in a seismically active area may influence these decisions. Active systems may depend on a source of off-site water that may not be functioning after an earthquake. On the other hand, applied fire-resistive materials may be damaged by building movement and not adequately protect the structure in a post-earthquake fire that may be extinguished using an on-site water supply with an emergency generator-driven pump.

- Footnote f: This footnote refers back to a specific provision for interior nonbearing partitions for Type IV buildings as spelled out in the detailed parts of §602.4.6.

- Footnote g: This footnote advises the designer to compare the requirements of both Tables 601 and 602 to determine the fire-resistive requirements for exterior bearing walls. It is possible that the requirements of Table 602 will require a higher fire-resistance-rating for an exterior wall than 601; in the event of overlapping requirements, the most restrictive is to govern per §102.1. For example, a Group I-2 occupancy in a Type II-B building located within 5' (1524) of a property line would require a 1-hour wall rating per Table 602 and need no rating per Table 601. In this case the 1-hour requirement would govern, as the code requires the most restrictive provision to govern per §102.1.

TABLE 602

Table 602 specifies the fire-resistance rating requirements of exterior walls based on fire-separation distance as well as type of construction and occupancy groups. Use of the table requires that the occupancy group be known in order to determine the fire-resistance requirements for exterior walls. Note also that the exterior bearing wall requirements contained in Table 601 need to be compared to the requirements in Table 602. Where such comparisons result in a potential conflict, per §102.1 the most restrictive requirement will govern.

- The term "fire separation distance" is defined in §702. This is new terminology for some users of prior model codes. This definition replaces terminology related to "location on property" with a more general idea that encompasses not just the spatial relationship of buildings to their site boundaries but to adjacent buildings on the same site as well. There is an interrelationship between building occupancy, construction type and location on the site. These are dependent variables with each impacting the other as a building design develops. Fire separation distance is to be measured perpendicular to the face of the building as illustrated.

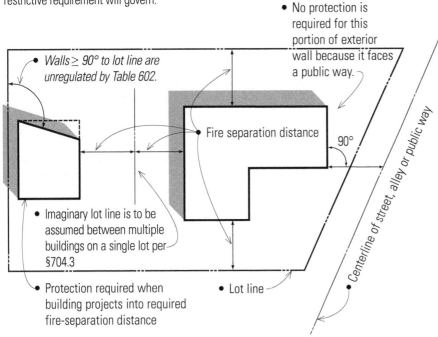

- Walls ≥ 90° to lot line are unregulated by Table 602.
- No protection is required for this portion of exterior wall because it faces a public way.
- Fire separation distance
- Imaginary lot line is to be assumed between multiple buildings on a single lot per §704.3
- Protection required when building projects into required fire-separation distance
- Lot line

- The fire-resistance requirements decrease with increasing distance between buildings. They also decrease in relation to decreasing construction-type requirements. Table 602 also builds-in assumptions about the hazards of occupancies relative to fire-separation distances. The guiding principle is that increased distances offset the hazards presented by various occupancies. Distance also mitigates the reduction in resistance to external fires presented by less resistive construction types.

Table 602
Required Fire-Resistance Rating of Exterior Walls based on Fire-Separation Distance

Fire-Separation Distance (feet)	Type of Construction	Group H	Groups F-1, M, S-1	Groups A, B, E, F-2, I, R, S-2, U
		• Decreasing hazard →		
<5	All	3	2	1
	I-A			
		Increasing fire-resistance requirements ←		
	V-B			
>30	All	0	0	0

• Increasing distance
• Decreasing construction type

- The amount of opening area in exterior walls is governed by the area limitations set forth in Table 704.8. The opening protection levels are correlated between the wall protection requirements of Table 602 and the opening protection requirements of Table 715.4. The designer must first ascertain the required fire-resistance rating of exterior walls to determine the required level of opening protection in the exterior walls. The designer can then determine the allowable amount of protected and unprotected openings in the exterior wall by using these two sections together.

TYPES I AND II CONSTRUCTION

The building elements of Type I and Type II construction are of noncombustible materials. As noted on page 49, the definition of noncombustible is contained in §703.4 and requires meeting the criteria of ASTM E 136.

- Type I-A construction, providing the highest level of fire-resistance-rated construction, requires passive protection for all elements of the structure.
- Type I-B construction is similar to Type I-A construction, but permits a 1-hour reduction in fire-resistance rating for the structural frame, bearing walls and floor construction, and a ½-hour reduction for roof construction.
- Type II-A construction requires 1-hour protection for most elements, but allows for the use of either active or passive protection of all elements of the structure to achieve equivalent protection.
- Type II-B construction allows unprotected noncombustible building elements. This type of construction was described as "Type II-non-rated" in some previous model codes.

Combustible Materials in Types I and II Construction

The key to the use of combustible materials in a noncombustible construction type is understanding that these uses must be ancillary to the primary structure of the building. The premise for allowing the use of these combustible materials is that they will be of limited quantity and used under defined conditions where they will not contribute in any large measure to compromising the desired level of fire-resistance in the structure. As one can see, the rules for Construction Types I and II, while calling for noncombustible materials, allow for a number of exceptions when all the requirements are taken together.

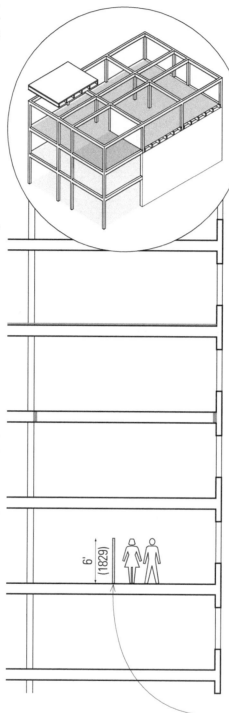

§603 contains a list of notes specifying which combustible materials can be used in buildings of Types I and II construction.

1. Note 1 reiterates allowances for the use of fire-retardant-treated wood for non-bearing walls and partitions and for the roof construction noted in the Table 601 footnotes.
2. Materials with a flame spread of less than 25 have ratings corresponding to Class A interior finishes per ASTM E 84. These indices may be higher, as noted in the exception, where insulation is encapsulated between layers of noncombustible materials without an air space.
3. Foam plastics are allowable if compliant with the provisions of Chapter 26 of the Code. Note especially the provisions of §2603.5 regarding the use of foam plastics on the exterior of Type I, II, III and IV buildings.
4. Most roof coverings have a classification of A, B or C, so this note should almost always be useable.
5. The Code recognizes that combustible decorative and utilitarian interior finishes, such as wood floors, will be applied over the noncombustible structural elements.
6. Wood trim at or near grade level is acceptable where not beyond ready fire-fighting access up to a level of 15' (4572) above grade.
7. This note requires fire-stopping in wood floors in Type I and II construction. This is similar to Note 5.
8. The key item in this section is that the area in question be occupied and controlled by a single tenant. These provisions do not apply for multi-tenant spaces. Also, these partitions must not define exit access passages that could be construed as corridors for the use of more than 30 occupants. This note also recognizes that single-tenant floors can have partitions of wood or similar light construction if they are lower than 6' (1829) and allow standing occupants to generally survey the occupied space in the event of an emergency.

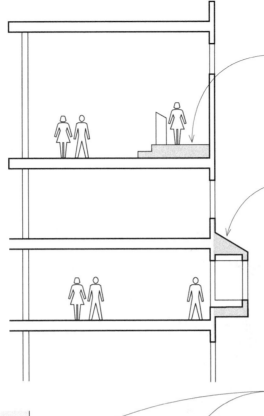

9. Stages and platforms are defined in §410 as raised areas used for worship, lectures, music, plays or other entertainment.

10. See specifically the provisions of §1406 regarding the relationship of fire separation to the fire-resistive properties of combustible veneers. Note also that plastic veneers are to comply with Chapter 26 of the Code. Projections from exterior walls are specified in §704 and these too are related to the type of construction. §704 in turn refers also to §1406 regarding the use of combustible materials on the exterior of buildings.

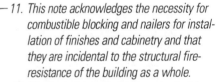

11. This note acknowledges the necessity for combustible blocking and nailers for installation of finishes and cabinetry and that they are incidental to the structural fire-resistance of the building as a whole.

12. Plastic glazing materials are permissible in conformance with the requirements of Chapter 26 of the Code.

13. This is similar to the allowance for use of combustible materials per Note 11.

14. Decorative plastic veneer is allowable if installed in accordance with §2605.2

15. This is similar to the allowance for use of combustible materials per Note 11.

16. This is a cross-reference to footnote d of Table 601 and is redundant.

17. These aggregates are allowed to be combustible by §703.2.2 with the proviso that the assembly meets the Code's fire-resistance test criteria.

18. Fire resistance materials may not be strictly be considered noncombustible and must be applied under special inspection per §1704.10 and 1704.11.

19–22. These exceptions are similar to Note 18 and allow for combustible materials to be used if they meet specific criteria spelled out in other sections of the IBC or in other I codes.

TYPE III CONSTRUCTION

Type III buildings are a mix of noncombustible and combustible elements, having noncombustible exterior walls and combustible interior construction. These building types arose in the U.S. at the end of the 19th century out of a desire to end the kind of conflagrations that struck congested business districts such as in Chicago. The buildings were designed to prevent a fire from spreading from building to building by requiring noncombustible building exterior walls.

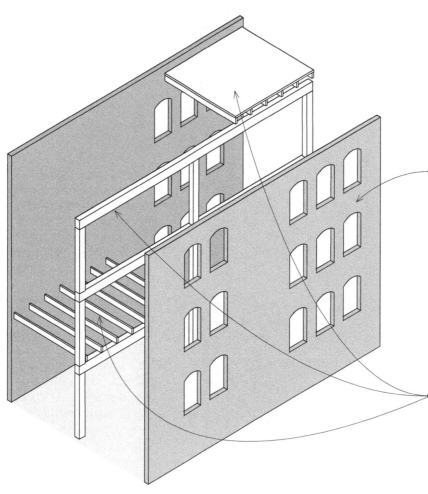

The construction materials on the exterior of a Type III building are required to be of non-combustible materials. Fire-retardant treated wood is allowable in exterior walls where the required fire-resistance rating is 2 hours or less. Table 601 requires 2-hour walls for Type III-A buildings; thus fire-retardant-treated wood is acceptable except where this is superseded by the requirements of Table 602 based on fire-separation distance. This would only occur in Group H occupancies with a fire-separation distance of less than 5' (524).

Type III buildings are considered combustible since the Code allows their interior building elements to be of combustible materials and also to be of unprotected construction if allowed by the building height and area allowances based upon occupancy.

Type IV buildings came about to address fire-safety conditions for traditional methods of constructing manufacturing and storage buildings, as Type III buildings did for office and residential occupancies. The type of construction used in Type IV buildings is known as "mill construction," or "Heavy Timber" construction. These buildings utilize heavy timber structural members and heavy wood floor decking inside exterior walls of noncombustible construction. Many of these buildings also have movable heavy metal shutters to close off exterior openings to prevent a fire outside the building from propagating into the building through unprotected openings.

The criteria for "heavy timber" (HT) construction are based on the past performance of historical construction, not on the scientific rationales contained in ASTM E 136 and E 119. These buildings have a good empirical performance record in fires. The insurance industry promoted construction of these types of buildings during the late 1800s and into the 1900s to limit their fire losses.

Type IV buildings generally burn slowly under fire conditions. The heavy timber members begin to flame and char at about 400°F (204.4°C). As the charring continues, it retards further deterioration of the wood members by insulating the interior of the wood members from the fire.

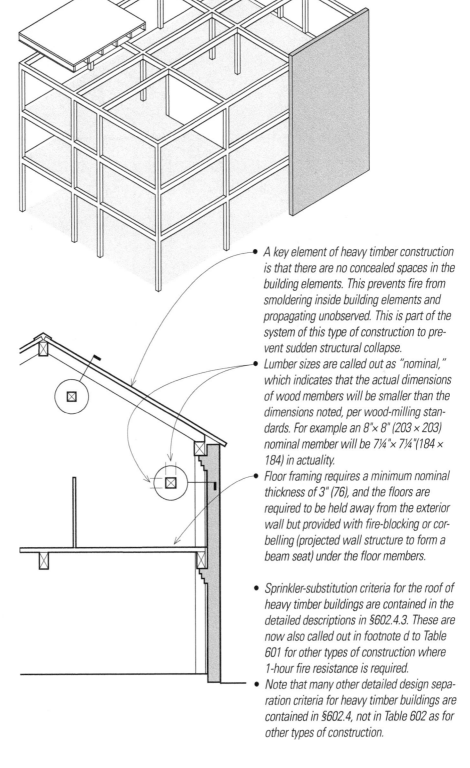

- *This type of construction utilizes heavy-sawn timbers in older buildings but will likely use glued-laminated structural wood members in modern construction.*

- *A key element of heavy timber construction is that there are no concealed spaces in the building elements. This prevents fire from smoldering inside building elements and propagating unobserved. This is part of the system of this type of construction to prevent sudden structural collapse.*
- *Lumber sizes are called out as "nominal," which indicates that the actual dimensions of wood members will be smaller than the dimensions noted, per wood-milling standards. For example an 8"× 8" (203 × 203) nominal member will be 7¼"× 7¼"(184 × 184) in actuality.*
- *Floor framing requires a minimum nominal thickness of 3" (76), and the floors are required to be held away from the exterior wall but provided with fire-blocking or corbelling (projected wall structure to form a beam seat) under the floor members.*

- *Sprinkler-substitution criteria for the roof of heavy timber buildings are contained in the detailed descriptions in §602.4.3. These are now also called out in footnote d to Table 601 for other types of construction where 1-hour fire resistance is required.*
- *Note that many other detailed design separation criteria for heavy timber buildings are contained in §602.4, not in Table 602 as for other types of construction.*

TYPE V CONSTRUCTION

Type V construction is the least restrictive construction type. It allows the use of any materials permitted by the Code. A typical example of Type V construction is the conventional light-wood-framed single-family residence.

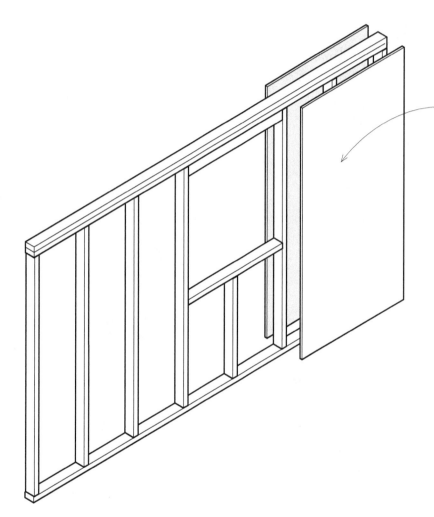

• *Because any element of Type V construction may be combustible, the fire-resistance of building elements is typically provided by the application of fire-resistant materials to the building parts.*

• *Type V-A construction is protected construction and all major building elements must therefore have a 1-hour fire-resistance rating. The only exception to this is for non-bearing interior walls and partitions contained in Table 601. Footnote e allows use of an automatic sprinkler system to substitute active protection for 1-hour passive protection.*

• *Type V-B construction is unprotected and requires no fire-resistance ratings except where Table 602 requires exterior-wall protection.*

7
Fire-Resistive Construction

Fire-resistance is the major factor in determining classification of construction types. Structural materials are broadly classified as combustible or noncombustible. Noncombustible materials provide greater resistance to fire by their nature. But even noncombustible structural materials can be weakened by exposure to fire. Additional materials that have capabilities to resist fire of a designated intensity for a length of time as determined by fire tests can be applied to structural materials to achieve required fire-resistance. As we saw in the discussion of types of construction, there is a direct relationship between fire-resistance requirements by construction type to occupancy type and to the allowable number of occupants.

The Code recognizes two basic methods for providing fire-resistive protection to ensure life safety in buildings. These can be classified as either passive or active protection. The differences between these approaches lie in the way they respond to the effects of fire on a building structure. Passive protection is built into the building structure and provides a barrier between the structure and the fire. Active protection such as fire sprinklers responds to fire by activation of systems to contain or suppress fire and smoke to allow the structure to remain intact for a longer period of time than without protection while allowing the occupants to escape. The code allows trade-offs between the provision of active versus passive fire protection. For example, 1-hour structural protection requirements may be offset by provision of fire sprinklers under certain circumstances in certain occupancies.

In this chapter, we consider code requirements for passive fire-resistance; Chapter 9 discusses active measures of fire-resistance. For designers and owners the trade-offs between passive and active fire-resistance are part of the design and economic analyses that go into deciding which systems are most suitable for a given project. The consideration of passive versus active systems is part of the iterative process of comparing occupancy and site requirements to allowable heights and areas for various construction types. Again, as noted in previous chapters, design goals typically involve using the most economical construction type that meets the needs of the occupancy.

The Code recognizes the efficacy of trade-offs between types of construction and types of fire protection. It also recognizes that there are limits to the value of the trade-offs between active and passive fire-resistance as they relate to types of construction and uses. Where active systems are required by the Code in relationship to given criteria, such as to increase heights and areas, then the provision of active systems in lieu of passive protection is generally not allowed. The idea of a trade-off implies a voluntary selection by the designer and owner of how to provide the required degree of fire-resistance. When code provisions otherwise require active systems, they are typically precluded by the code from being available to offset passive requirements.

FIRE-RESISTIVE CONSTRUCTION

Fire-Resistance Ratings

§702 defines fire-resistance rating as the period of time "that a building element, component or assembly maintains the ability to confine a fire, continues to perform a given structural function, or both," as determined by tests or methods prescribed in §703. The time-rating in hours indicates how long a building material, element or assembly can maintain its structural integrity and/or heat-transfer resistance in a fire, and corresponds to the construction type designations in Chapter 6 of the Code.

§703.2 prescribes that fire-resistance ratings be assigned on the basis of a fundamental fire test promulgated by ASTM International (ASTM) [formerly known as the American Society for Testing and Materials]. ASTM Test E 119 exposes materials and assemblies to actual fire tests. The material or assembly being tested is installed in a furnace in a condition similar to the anticipated exposure—i.e., vertical for walls, horizontal for floors or ceilings—and then exposed to a fire of a known intensity. The fire exposure is governed by a standard time-temperature curve whereby the fire grows in intensity over a given period of time, reaching a predetermined temperature at a given rate and maintaining that temperature thereafter. The sample is then exposed to the fire until failure occurs or until the maximum desired duration of protection is exceeded. This determines the fire-resistance rating in hours for the material or assembly in question.

- Fire-resistive construction, whether passive or active, has two primary purposes:

- The first is the protection of the building structure. Where passive protection is provided such protection is typically applied directly to structural members.

- The second is the separation of spaces to prevent the spread of fire or smoke within a building and the spread of fire between buildings. The protection of spaces addresses fire or smoke impacts on larger-scale building systems, such as floors, walls and ceilings as well as openings in these systems.

- §702 defines a fire area as the aggregate floor area enclosed and bounded by
  - fire walls
  - exterior walls or
  - fire-resistance-rated horizontal assemblies of a building.

- Fire-resistance ratings are based on the performance of various materials and construction assemblies under fire-test conditions as defined by the American Society for Testing and Materials (ASTM). Tests typically are based upon assumptions about the side of the assembly where the fire is likely to occur.

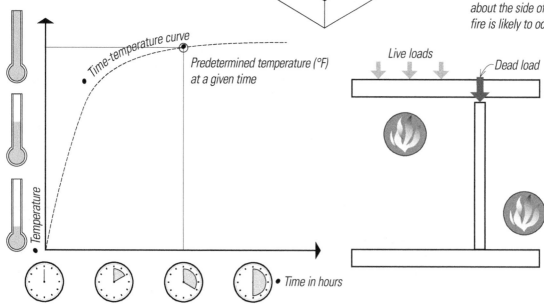

Predetermined temperature (°F) at a given time

Time-temperature curve

Temperature

Time in hours

Live loads

Dead load

The exception to §703.2 allows exterior bearing wall rating requirements to equal those for nonbearing walls when all factors such as fire separations and occupancy are considered. This exception recognizes that fire-resistance in exterior walls is concerned with stopping the spread of fire beyond the structure as well as protecting the structure. Since the governing criteria for this condition is preventing the spread of fire outside the building, the code recognizes that there is no point in protecting the structure to a higher level than is required for the walls enclosing the space.

§703.2.3 assumes that tested assemblies are not restrained under the definitions contained in ASTM E 119. "Restrained" refers to the ability of structural members to expand or contract under fire conditions. Assemblies considered as restrained typically have a higher hourly rating with less application of fire protection and are thus attractive to use in design. However, the Code requires that such assemblies be identified on the plans. The difficulty of designing and proving that assemblies are truly restrained very often outweighs any advantages gained in reducing the quantities of fire-protection materials used. We recommend that designers follow the lead of the code section and assume *all* assemblies to be *unrestrained* when determining fire-resistance requirements.

Alternative Methods

While fire ratings are fundamentally based on the ASTM E 119 test to determine hourly ratings, §703.3 allows designers to use several alternate methods to demonstrate compliance with fire-resistive criteria. One method allows the use of ratings determined by such recognized agencies as Underwriters Laboratory or Factory Mutual. The Code itself contains a "cookbook" of prescriptive assemblies in Table 720, which gives the designer a list of protection measures that can be applied to structural members, to floor and roof construction, and to walls to achieve the necessary ratings. §721 allows the designer to calculate the fire-resistance of assemblies by combining various materials. This gives much greater flexibility to meet actual design conditions than does the very specific set of assemblies listed in Table 720.

§703.3 also allows engineering analysis based on ASTM E 119 to be used to determine projected fire-resistance. This typically requires use of a consultant familiar with extrapolations from data acquired from similar ASTM E 119 fire tests to predict the performance of systems without the time and expense of performing a full-scale fire test.

§703.3 also acknowledges the testing measures prescribed in §104.11, which allows the building official to approve alternate ways of meeting the code when new technologies or unusual situations are encountered. The reality of using this clause is that the building official will require testing or a consultant's verification of the efficacy of a proposed fire assembly rating in order to grant approval to alternate fire-resistance systems. The designer will need to offer convincing evidence in some form to allow the building official to determine if the proposed system is code-compliant.

§703.4 defines noncombustibility in terms of test criteria. The characteristics that determine noncombustibility must not be affected by exposure to age, moisture or atmospheric conditions. The Code also recognizes that certain combinations of combustible and noncombustible materials may be considered as noncombustible if they meet test criteria.

Chapter 7 analyzes various construction components and conditions in light of their fire-resistance capabilities. Once again, as in other chapters, the code sections take the form of statements and exceptions.

The Code is organized to move from the exterior of the building to areas inside the building and then to the structure. The first set of assemblies can best be thought of as planes, both vertical and horizontal, arranged around the structural system. These planes may be bearing walls and part of the structural system, or they may be curtain walls or interior partitions independent of the structure.

Various interrelated conditions impact the fire-resistance requirements of the systems considered. Openings (and their protection), location on the property, relationships of exterior walls facing each other (as in courts), separations of interior spaces by fire walls, vertical circulation, vertical openings, protection of egress paths, smoke barriers, penetrations by utility systems, the abutment of floor systems with curtain-wall systems and fire-resistive protection of structural systems—all must be considered. We will explore each of these sets of requirements in the same order they are presented in the Code.

The contents of §704 apply more broadly than its title "EXTERIOR WALLS" would suggest. The relationship of exterior walls to the lot line as well as openings within and projections from the walls are covered in this section. The wall criteria also interact with the type of construction to dictate the fire-resistance of the elements of the wall. This section should be read in conjunction with §1406, which governs the use of combustible materials on the exterior face of exterior walls.

Projections

§704.2 governs the extent of allowable projections according to their relationship to the lot line. The combustibility of the projections is governed by the wall construction type (which as we have seen is related to heights, areas and occupancy types). Where openings are prohibited, or opening protection is required by location on property, any combustible projection must be of 1-hour construction, even in unrated buildings. Heavy-timber projections, or those in compliance with §1406.3 are also permitted.

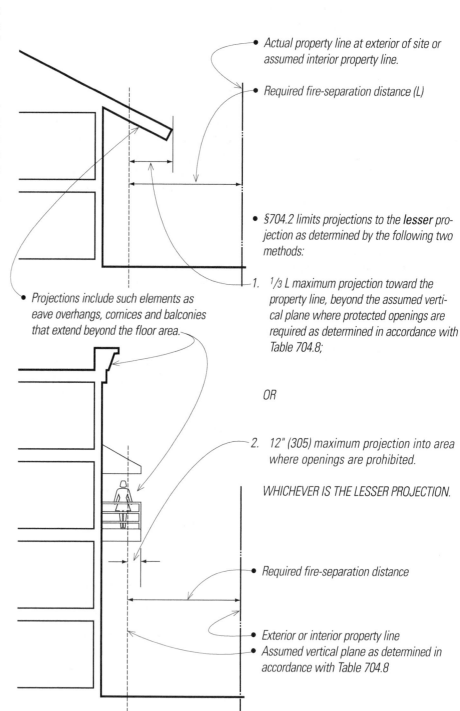

- Actual property line at exterior of site or assumed interior property line.

- Required fire-separation distance (L)

- §704.2 limits projections to the *lesser* projection as determined by the following two methods:

1. $^1/_3$ L maximum projection toward the property line, beyond the assumed vertical plane where protected openings are required as determined in accordance with Table 704.8;

   OR

2. 12" (305) maximum projection into area where openings are prohibited.

   WHICHEVER IS THE LESSER PROJECTION.

- Projections include such elements as eave overhangs, cornices and balconies that extend beyond the floor area.

- Required fire-separation distance

- Exterior or interior property line
- Assumed vertical plane as determined in accordance with Table 704.8

Multiple Buildings

§704.3 assumes that, when determining the protection requirements for multiple buildings on the same property, an assumed line exists between the buildings or elements. The Code does not specify that the assumed line be located midway between the elements so the designer is free to locate the lot line at any point between the elements in question as long as the wall protection requirements are met based on the distance to the assumed lot line. The intent of §704.3 is to prevent the spread of fire by radiant heating or convection. The impact of these conditions may be diminished by distance or by wall treatments, such as having openings in one wall face a solid wall on the opposite side of the assumed property line.

Fire-Resistance Ratings

§704.5 requires that the fire-resistance ratings for exterior walls be as prescribed by Tables 601 and 602. Based on the intent of the Code to prevent the spread of fire from one property to another, when an exterior wall is located more than 5' (1524) from the lot line the fire exposure is assumed to be from the inside. When an exterior wall is located 5' (1524) or less from the lot line, the exposure must be assumed to come from either inside or outside the building. This recognizes that another building may be built on the lot line on the adjacent lot.

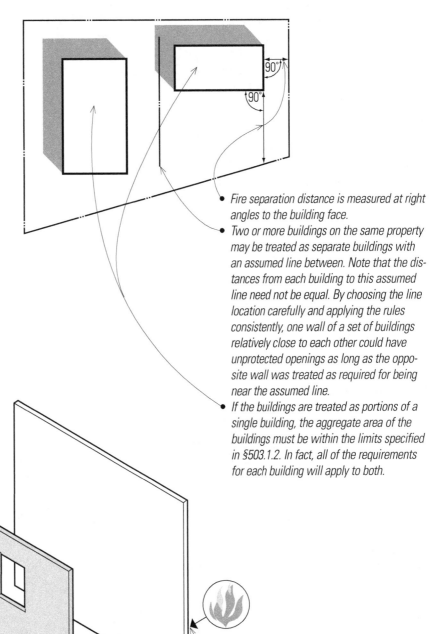

- Fire separation distance is measured at right angles to the building face.
- Two or more buildings on the same property may be treated as separate buildings with an assumed line between. Note that the distances from each building to this assumed line need not be equal. By choosing the line location carefully and applying the rules consistently, one wall of a set of buildings relatively close to each other could have unprotected openings as long as the opposite wall was treated as required for being near the assumed line.
- If the buildings are treated as portions of a single building, the aggregate area of the buildings must be within the limits specified in §503.1.2. In fact, all of the requirements for each building will apply to both.

- §704.5 specifies that exterior walls having a fire-separation distance greater than 5' (1524) be rated for exposure to fires from the inside.

- Exterior walls having a fire-separation distance of 5' (1524) or less must be rated for exposure to fire from both sides.

Openings

While §704.7 contains detailed calculations (Equation 7-1) for determining the fire-resistance rating of protected openings, we will focus on the simpler calculation contained in §704.8 and Table 704.8. These relate location on the property to the percentage of wall openings and whether the openings are protected or not. For windows in exterior walls to be protected they must comply with the opening protection requirements of §715, which we will discuss later.

Note that §704.12 requires opening protections per §715.5. The exception to this section also allows opening protection to be provided by an approved water-curtain exterior sprinkler system when the building also has a conventional interior automatic fire-sprinkler system. This is an active measure that provides a substitution for passive protection. Additional passive or active fire-protection measures can also allow increases in the area of openings. For example, the provision of a fire-sprinkler system (adding an active fire-protection system to passive measures) per §704.8.1 allows more unprotected openings in buildings.

Vertical Separation of Openings

§704.9 regulates the vertical relationship of openings to each other more precisely than in previous model codes. These requirements do not apply to buildings less than three stories in height, or when fire sprinklers are provided. Again, provision of an active system allows more freedom in determining whether passive systems must also be incorporated.

Vertical Exposure

For multiple buildings on the same property, §704.10 requires the protection of openings in any wall that extends above an adjacent roof. This protection can be provided in various ways; by distance, by opening protectives, or by protection of the roof framing and its supporting structure. The principle is one of reduction of the likelihood of fire spreading from one location to another. The buildings must be separated by a minimum distance, some method of fire protection applied to the openings, or protection of the construction facing the openings.

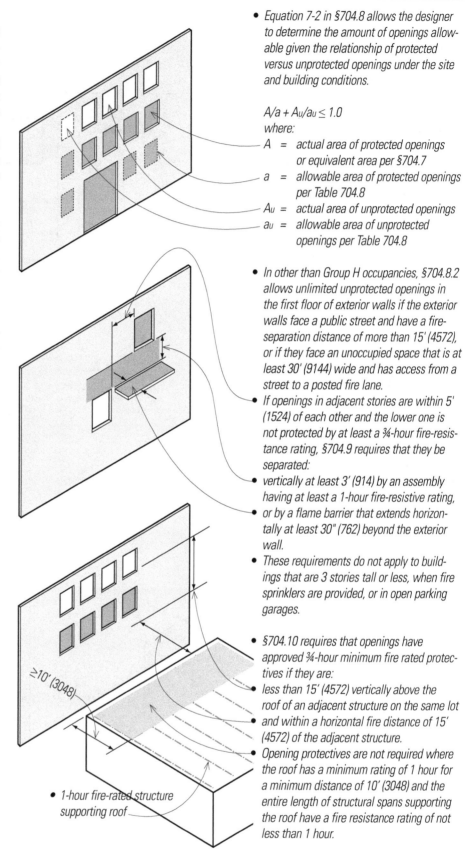

- *Equation 7-2 in §704.8 allows the designer to determine the amount of openings allowable given the relationship of protected versus unprotected openings under the site and building conditions.*

  $A/a + A_u/a_u \leq 1.0$
  where:
  - A = *actual area of protected openings or equivalent area per §704.7*
  - a = *allowable area of protected openings per Table 704.8*
  - A_u = *actual area of unprotected openings*
  - a_u = *allowable area of unprotected openings per Table 704.8*

- *In other than Group H occupancies, §704.8.2 allows unlimited unprotected openings in the first floor of exterior walls if the exterior walls face a public street and have a fire-separation distance of more than 15' (4572), or if they face an unoccupied space that is at least 30' (9144) wide and has access from a street to a posted fire lane.*

- *If openings in adjacent stories are within 5' (1524) of each other and the lower one is not protected by at least a ¾-hour fire-resistance rating, §704.9 requires that they be separated:*

- *vertically at least 3' (914) by an assembly having at least a 1-hour fire-resistive rating,*

- *or by a flame barrier that extends horizontally at least 30" (762) beyond the exterior wall.*

- *These requirements do not apply to buildings that are 3 stories tall or less, when fire sprinklers are provided, or in open parking garages.*

- *§704.10 requires that openings have approved ¾-hour minimum fire rated protectives if they are:*

- *less than 15' (4572) vertically above the roof of an adjacent structure on the same lot and within a horizontal fire distance of 15' (4572) of the adjacent structure.*

- *Opening protectives are not required where the roof has a minimum rating of 1 hour for a minimum distance of 10' (3048) and the entire length of structural spans supporting the roof have a fire resistance rating of not less than 1 hour.*

≥10' (3048)

- *1-hour fire-rated structure supporting roof*

Parapets

§704.11 makes a general statement that parapets shall be provided at exterior walls of buildings. The purpose of parapets is to impede the spread of fire from one building to another by providing a barrier to fire and radiant heat transfer if fire breaks through the roof membrane.

The exceptions that follow reduce or eliminate the need for parapets if any of the conditions are met. The list of exceptions becomes the code criteria, not the more general opening statement that parapets will always occur. Note, however, that there may be extensive construction work involved to avoid a parapet. The converse is also true that the provisions noted in the exception can be avoided by providing a parapet. The exceptions, where parapets are not required, can be summarized as follows. No parapets are needed where:

1. The wall satisfies the fire-separation distance criteria in accordance with Table 602.
2. The building area is less than 1,000 sf (93 m²) on any floor.
3. The roof construction is entirely noncombustible or of at least 2-hour fire-resistive construction.
4. The roof framing is protected against fire exposure from the inside, as illustrated to the right.
5. In residential occupancies, a fire barrier is provided by sheathing the underside of the roof framing, or the roof sheathing is of noncombustible materials for 4' (1219) back from the roof/wall intersection and the entire building has a Class C or better roof covering.
6. Where §704.8 allows the wall to have 25% or more of its openings unprotected due to the building's location from a property line.

Fire protection of openings in exterior walls other than windows and doors must be addressed as part of the design as well. Expansion and seismic joints, wall and floor intersections, ducts, louvers and similar air-transfer openings must comply with the detailed requirements of sections discussed later in this chapter.

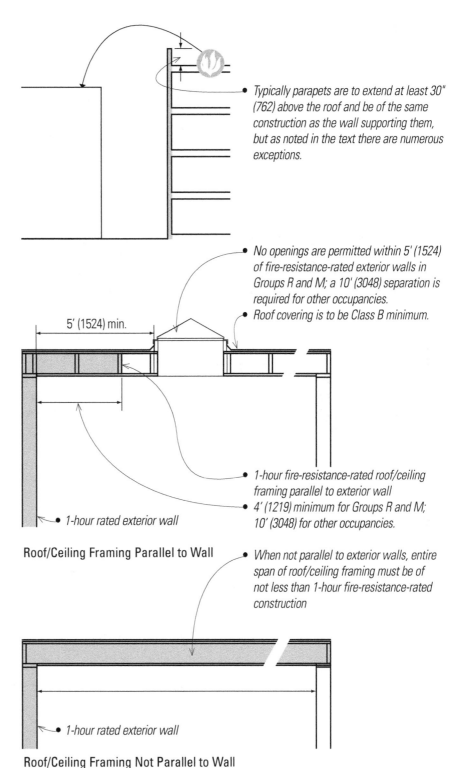

• *Typically parapets are to extend at least 30" (762) above the roof and be of the same construction as the wall supporting them, but as noted in the text there are numerous exceptions.*

• *No openings are permitted within 5' (1524) of fire-resistance-rated exterior walls in Groups R and M; a 10' (3048) separation is required for other occupancies.*
• *Roof covering is to be Class B minimum.*

5' (1524) min.

• *1-hour fire-resistance-rated roof/ceiling framing parallel to exterior wall*
• *4' (1219) minimum for Groups R and M; 10' (3048) for other occupancies.*

• 1-hour rated exterior wall

Roof/Ceiling Framing Parallel to Wall

• *When not parallel to exterior walls, entire span of roof/ceiling framing must be of not less than 1-hour fire-resistance-rated construction*

• 1-hour rated exterior wall

Roof/Ceiling Framing Not Parallel to Wall

Proportional Examples of Fire-resistance-Rated Construction at Interior Walls
(Not to Scale, for illustration only)

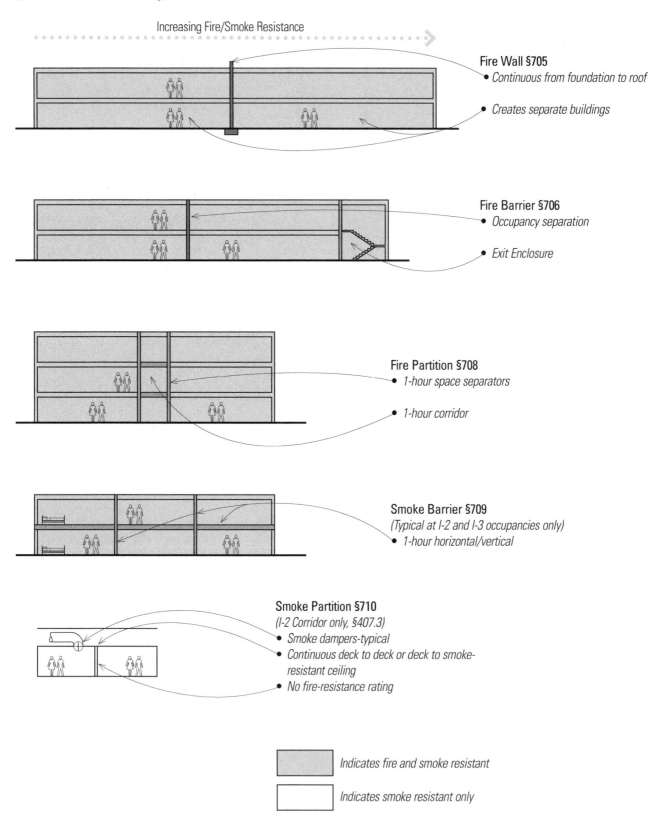

Increasing Fire/Smoke Resistance

Fire Wall §705
- *Continuous from foundation to roof*
- *Creates separate buildings*

Fire Barrier §706
- *Occupancy separation*
- *Exit Enclosure*

Fire Partition §708
- *1-hour space separators*
- *1-hour corridor*

Smoke Barrier §709
(Typical at I-2 and I-3 occupancies only)
- *1-hour horizontal/vertical*

Smoke Partition §710
(I-2 Corridor only, §407.3)
- *Smoke dampers-typical*
- *Continuous deck to deck or deck to smoke-resistant ceiling*
- *No fire-resistance rating*

Indicates fire and smoke resistant

Indicates smoke resistant only

§702 defines a fire wall as a fire-resistance-rated wall whose purpose is to restrict the spread of fire. To perform this function, a fire wall must extend continuously from the building foundation to and through the roof, and have sufficient structural stability to withstand collapse if construction on either side of it collapses.

Structural Stability

§705.2 requires that fire walls have a structural configuration that allows for structural collapse on either side while the wall stays in place for the time required by the fire-resistance rating.

Materials

§705.3 requires fire walls to be constructed of noncombustible materials, except in Type V construction.

Fire-Resistance Ratings

§705.4 bases the required fire-resistance ratings of fire walls on occupancy. For healthcare occupancies, fire walls are typically required to have 3-hour fire-resistance ratings. Table 705.4 allows 2-hour ratings in Type II or V buildings for certain occupancies. Where different occupancies or groups are separated by a fire wall, the more stringent requirements for separation rating will apply.

Horizontal Continuity

§705.5 specifies that fire walls are to be continuous horizontally from exterior wall to exterior wall, and extend at least 18" (457) beyond the exterior surface of the exterior walls.

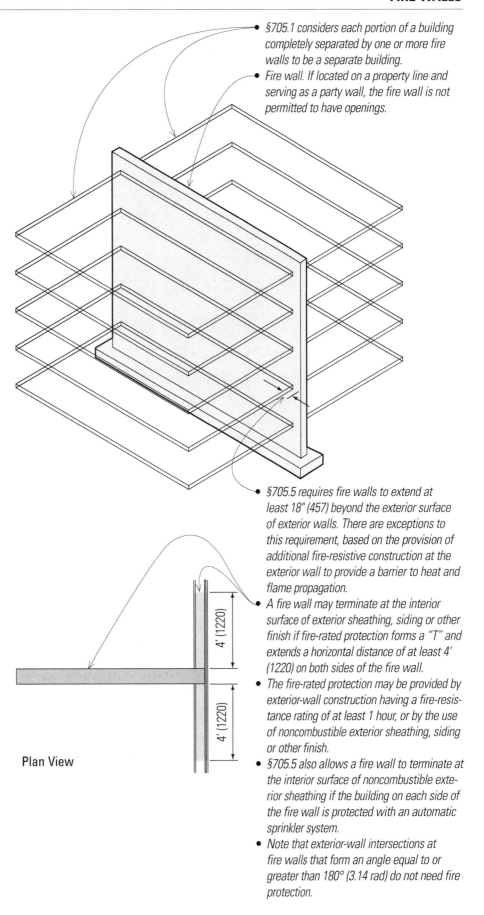

§705.1 considers each portion of a building completely separated by one or more fire walls to be a separate building.

Fire wall. If located on a property line and serving as a party wall, the fire wall is not permitted to have openings.

§705.5 requires fire walls to extend at least 18" (457) beyond the exterior surface of exterior walls. There are exceptions to this requirement, based on the provision of additional fire-resistive construction at the exterior wall to provide a barrier to heat and flame propagation.

A fire wall may terminate at the interior surface of exterior sheathing, siding or other finish if fire-rated protection forms a "T" and extends a horizontal distance of at least 4' (1220) on both sides of the fire wall.

The fire-rated protection may be provided by exterior-wall construction having a fire-resistance rating of at least 1 hour, or by the use of noncombustible exterior sheathing, siding or other finish.

§705.5 also allows a fire wall to terminate at the interior surface of noncombustible exterior sheathing if the building on each side of the fire wall is protected with an automatic sprinkler system.

Note that exterior-wall intersections at fire walls that form an angle equal to or greater than 180° (3.14 rad) do not need fire protection.

Plan View

4' (1220)

4' (1220)

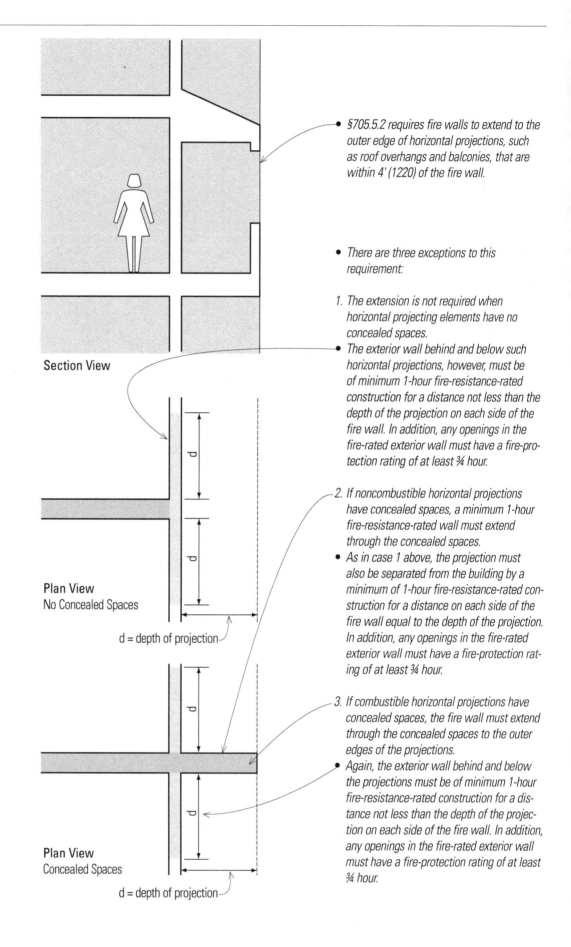

Section View

Plan View
No Concealed Spaces

d = depth of projection

Plan View
Concealed Spaces

d = depth of projection

- §705.5.2 requires fire walls to extend to the outer edge of horizontal projections, such as roof overhangs and balconies, that are within 4' (1220) of the fire wall.

- There are three exceptions to this requirement:

1. The extension is not required when horizontal projecting elements have no concealed spaces.
- The exterior wall behind and below such horizontal projections, however, must be of minimum 1-hour fire-resistance-rated construction for a distance not less than the depth of the projection on each side of the fire wall. In addition, any openings in the fire-rated exterior wall must have a fire-protection rating of at least ¾ hour.

2. If noncombustible horizontal projections have concealed spaces, a minimum 1-hour fire-resistance-rated wall must extend through the concealed spaces.
- As in case 1 above, the projection must also be separated from the building by a minimum of 1-hour fire-resistance-rated construction for a distance on each side of the fire wall equal to the depth of the projection. In addition, any openings in the fire-rated exterior wall must have a fire-protection rating of at least ¾ hour.

3. If combustible horizontal projections have concealed spaces, the fire wall must extend through the concealed spaces to the outer edges of the projections.
- Again, the exterior wall behind and below the projections must be of minimum 1-hour fire-resistance-rated construction for a distance not less than the depth of the projection on each side of the fire wall. In addition, any openings in the fire-rated exterior wall must have a fire-protection rating of at least ¾ hour.

Vertical Continuity

§705.6 requires that fire walls be continuous vertically from the building foundation to a point at least 30" (762) above adjacent roofs.

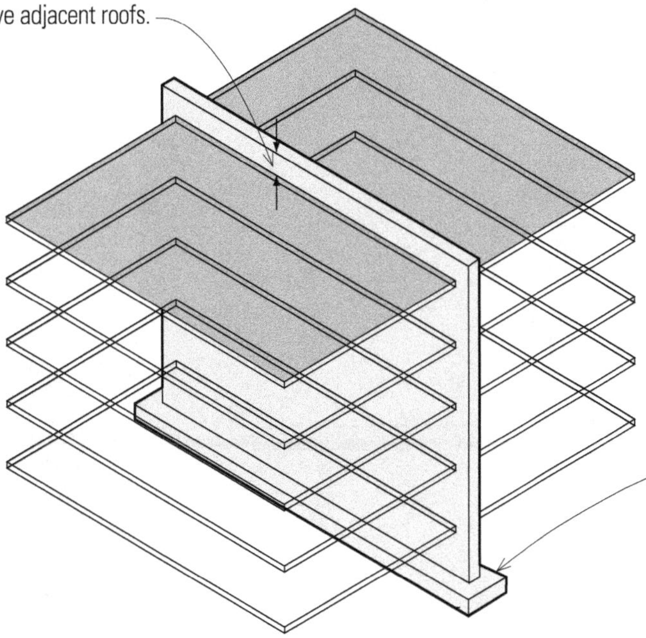

The same principles for turning the fire barrier perpendicular to the fire wall at exterior walls apply to where fire walls meet roof construction. Roof construction is to be noncombustible, of fire-retardant-treated wood or have fire-resistive materials applied on the inside to protect the roof framing. For design purposes, it is best to presume that this fire protection of roof construction extends a minimum of 4' (1220) on each side of the fire wall when framing is parallel to the wall and the full length of the span when the framing is perpendicular to the fire wall.

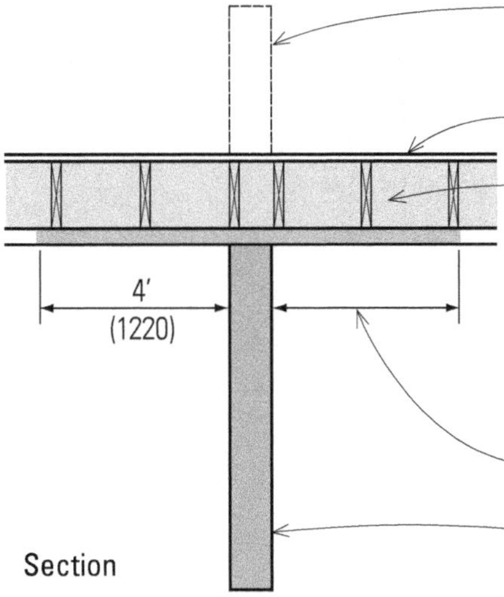

- No horizontal offsets in the fire wall are permitted.

- Note that a fire wall for a building located above a parking garage in accordance with §509.2 extends from the horizontal separation between the parking garage and the building above.

- Exceptions to 30" (762) extension of fire wall above roof:

- Not less than Class B roof covering

- Roof framing parallel to a fire wall must have not less than a 1-hour fire-resistance, rating within 4' (1220) of fire wall.
- Roof framing must be supported by bearing walls or columns and beams also having a fire-resistance rating of 1 hour.

- No openings are permitted within 4' (1220) of fire wall.
- 2-hour fire wall

4'
(1220)

Section

- Not less than Class B roof covering
- Noncombustible roof sheathing, deck or slab

- The entire span of roof framing perpendicular to a fire wall must have a 1-hour fire resistance rating.

- No openings within 4' (1220) of fire wall

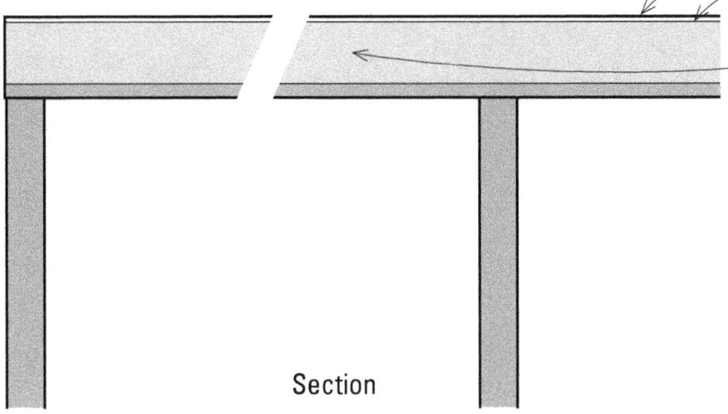

Section

Penetrations and Joints

Openings, penetrations and joints are required to have fire protection per sections occurring later in Chapter 7. Per §705.11, ducts are not allowed to penetrate fire walls on property lines. Exceptions allow ducts at fire walls not located at property lines where fire assemblies per §712 and 716 protect the duct penetrations and the aggregate area of openings does not exceed that permitted under §705.8.

Stepped Roofs

When a fire wall serves as the exterior wall of a building and separates buildings having different roof levels, the criteria of §705.6.1 apply to the lower roof.

Stepped Roofs

- *The exterior wall is of not less than 1-hour fire-resistance-rated construction for a height of 15' (4572) above the lower roof.*
- *Within this wall, ¾-hour fire protective-assemblies are required in all openings.*

- *The fire wall may terminate at the underside of the roof sheathing, deck or slab of the lower roof if the lower roof assembly has not less than a 1-hour fire-resistance rating within 10' (3048) of the fire wall.*

- *The fire-resistance rating of the fire wall must extend 30" (762) above the lower roof.*

Combustible Framing

- *Where combustible framing rests on concrete or masonry fire walls, §705.7 requires the framing members to be separated by at least 4" (102) between the ends of the members. This may require enlarging the bearing surface where this occurs to provide adequate bearing for the framing along with the required separation.*
- *All hollow spaces are to be filled with non-combustible fire-blocking material for not less than 4" (102) above, below and between the structural members.*

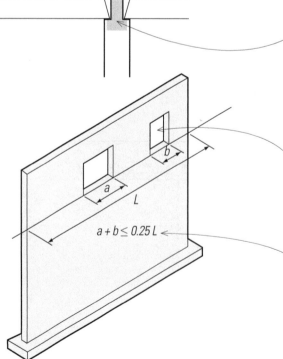

$$a + b \leq 0.25\,L$$

Openings

- *Openings in fire walls are allowed but are restricted by §705.8 to 120 sf (11 m^2) for each opening. This can be increased if the buildings on both sides of the fire wall are equipped with an automatic sprinkler system.*
- *The aggregate width of openings may not exceed 25% of the length of the fire wall.*
- *Openings are not permitted in party walls constructed per §705.1.1.*

Fire barriers are similar to fire walls but with simpler criteria. They are used to separate vertical exit enclosures from other egress components, to separate different occupancies, or to divide a single occupancy into different fire areas. Required fire-resistance ratings for fire barriers are determined by their use. Barriers used in means of egress protection are rated per the applicable sections in Chapter 10. Occupancy separation ratings are per Table 508.3.3. Fire resistance ratings between fire areas are to be per Table 706.3.9.

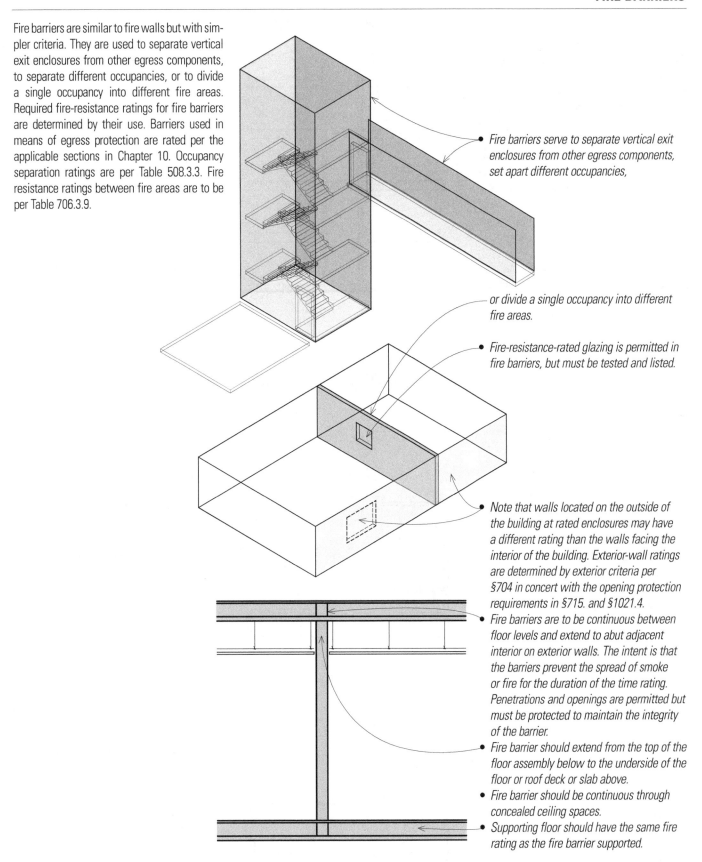

- Fire barriers serve to separate vertical exit enclosures from other egress components, set apart different occupancies,

or divide a single occupancy into different fire areas.

- Fire-resistance-rated glazing is permitted in fire barriers, but must be tested and listed.

- Note that walls located on the outside of the building at rated enclosures may have a different rating than the walls facing the interior of the building. Exterior-wall ratings are determined by exterior criteria per §704 in concert with the opening protection requirements in §715. and §1021.4.

- Fire barriers are to be continuous between floor levels and extend to abut adjacent interior on exterior walls. The intent is that the barriers prevent the spread of smoke or fire for the duration of the time rating. Penetrations and openings are permitted but must be protected to maintain the integrity of the barrier.

- Fire barrier should extend from the top of the floor assembly below to the underside of the floor or roof deck or slab above.

- Fire barrier should be continuous through concealed ceiling spaces.

- Supporting floor should have the same fire rating as the fire barrier supported.

SHAFT AND VERTICAL EXIT ENCLOSURES

Shaft enclosures differ from fire barriers in that they typically enclose shafts extending through several floors. Shafts are often constructed of the same materials as fire barriers, but the required ratings for shafts and vertical exit enclosures are usually determined by the number of floors they interconnect. The general statement in §707.2 is that all openings through floor/ceiling or roof/ceiling assemblies will be protected. Shafts are to be of 2-hour construction if extending four stories or more and 1-hour otherwise. Shaft ratings are to equal those of the floor assembly, but need not be greater than 2 hours.

§707.2 states the basic requirement for shaft protection, followed by a long list of exceptions. These exceptions must be read carefully, especially when designing open stairways, escalators and open elevators. The exceptions are based on the extent of the opening that may need to be enclosed, its use and the occupancy or use of the space.

Exceptions to §707.2:

1. Openings in residential units extending less than four stories need not be enclosed.

3. Pipe, vent or conduit penetrations protected at the floor or ceiling according to §712.4 need not be protected.

4. Ducts protected per §712.4 for through penetrations need not be protected.

6. Chimneys need not be in a shaft, but concealed spaces must be fire-blocked.

8. Ramps for automobiles in parking garages may be open when the criteria of §406.3 and 406.4, for open or enclosed parking garages, are met.

9. Openings connecting mezzanines and the floor below need not be enclosed.

10. Expansion or other floor and wall joints need not be enclosed, but they must be protected per §713 as for similar penetrations.

11. Shaft enclosures are not required for stairs or ramps that comply with exceptions 8 or 9 in §1020.1

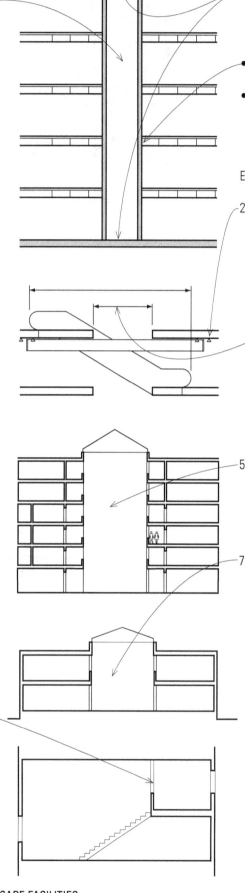

- Shaft enclosures should be continuous from the top of the floor assembly below to the underside of the floor or roof deck or slab above, including concealed spaces such as those above suspended ceilings.
- Vertical spaces within shaft enclosures should be fire-blocked at each floor level.
- See the next page for enclosing shafts at the top and bottom when they do not extend to the top or bottom of the building or structure.

Exceptions to §707.2:

2. Escalators and stairs not used as means of egress need not be enclosed in fully sprinklered buildings when a draft curtain and closely spaced sprinklers are provided around the floor opening. This exception applies for four or fewer stories except for Group B and M occupancies, where there is no limit in either case. The opening dimension cannot exceed twice the size of the escalators or stairs.

Where power-operated shutters with a rating of 1½ hours are provided at each level with an opening there is no limit to the number of floors that can be penetrated.

5. Floor openings in buildings considered as atriums per §404 need not have shaft protection, but many other specific criteria for fire protection and/or smoke control for these occupancy types must be met.

7. Openings that do not connect more than two stories and that are not part of a means of egress may be open in other than I-2 or I-3 occupancies, even in buildings not equipped with a fire-sprinkler system. The way the exception is worded it seems clear that in order for the exception to be used the building in question must meet all of the criteria in the exception. This exception is further limited for use in a number of healthcare occupancies. Exception 7.4 does not allow the exception to be used in I or R occupancies where the opening in question is open to a corridor in those occupancies. The requirements for smoke compartments in I occupancies may also preclude the use of this exception if the condition does not comply with Exception 7.7 which requires the opening to be in the same smoke compartment.

The Code requires that shafts not have multiple purposes. Penetrations in shaft walls are limited to those related to the purpose of the shaft. For example, ducts serving occupied spaces should not run through vertical exit enclosures, but ducts supplying air to the enclosure may be provided as long as penetration protection requirements are met to maintain the fire rating of the enclosure.

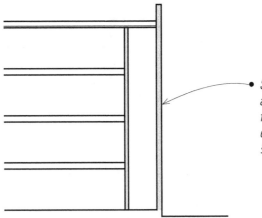

• Shaft continuity and rating requirements at exterior walls are similar to those for fire barriers. Exterior walls serving as shaft enclosure must satisfy the requirements of §704.

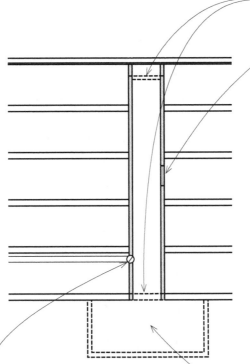

• Shafts must be fully contained. Thus, a shaft that does not fully extend to the floor below, or to the floor or roof above, must be enclosed at the top or bottom.

• Openings into shafts are to be rated in accordance with the requirements of §715. Specific requirements for opening protectives fire-resistance ratings are contained in Table 715.4.

• Shafts may be enclosed at the top or bottom with construction having the same fire-resistance rating as the last floor penetrated but not less than the required rating of the shaft.

• These enclosures must match the fire rating of the shaft wall. The designer must pay careful attention in these circumstances as assemblies respond to test fires of the same intensity much differently based upon their orientation in relation to heat flows. It may not be possible to achieve equivalent ratings when a wall assembly tested vertically is installed horizontally. The code requires equal fire ratings as verified by testing in the actual horizontal or vertical orientation.

• Note that fire dampers are required where ducts enter or exit shafts under these conditions. The integrity of the shaft is to be maintained continuously around all surfaces of the shaft.

• Shafts may also terminate in a room related to the purpose of the shaft where the room is enclosed by construction having the same fire-resistance rating of the shaft enclosure.

SHAFT AND VERTICAL EXIT ENCLOSURES

§707.13 addresses refuse and laundry chutes. The Code recognizes that these shafts may interconnect many floors of a building. Requirements are included to provide a shaft, but also included are requirements to separate the chutes from the rest of the building with 1-hour-rated access rooms on each floor and 1-hour-rated termination rooms at the bottom of the shaft. Active fire protection is also required for these shafts per §903.2.10.2.

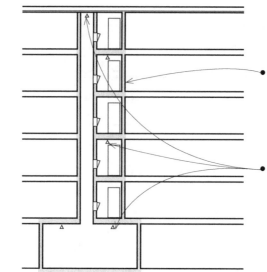

• Refuse and laundry chutes must be separated from the rest of the building by 1-hour-rated access rooms at each floor level.

• §903.2.10.2. requires an automatic sprinkler system at the top of a chute, in the terminal rooms and on alternate floors when the chute passes through three or more floors.

§707.14 requires that elevator shafts be constructed as other shafts. This section adds requirements for elevator lobbies where an elevator shaft enclosure connects more than 3 stories. The lobbies are to be constructed as for other fire barriers with opening protectives and a means of egress.

The exceptions to the elevator lobby requirements are:

1. When the building lobby level is sprinklered, the street-level lobby may be open without elevator lobby enclosures.
2. Where shafts are otherwise not required per the exceptions to §707.2. As indicated these exceptions are modified to exclude their use in certain conditions in healthcare occupancies.
3. Doors that enclose elevator shaft to seal off the shaft from smoke and fire may be substituted for lobbies when they comply with §3002.6.
4. In fully sprinklered buildings which are not I-3 occupancies and buildings with occupied floors less than 75' above the lowest level of fire-department access (i.e. not a high rise), lobbies may be omitted.
5. Smoke partitions may be used in lieu of fire partitions where the building is fully sprinklered (there are proprietary products which are listed to meet these criteria).
6. Lobbies are not required where the elevator hoistway is pressurized per §707.14.2.

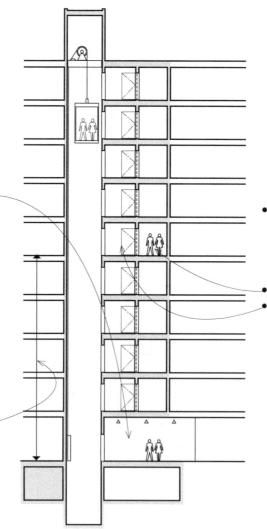

• Elevator lobbies are required per §707.14.1 where an elevator shaft enclosure connects more than three stories.

• Corridor
• Per §707.14.1 the lobby separates elevator shaft enclosure doors from each floor by fire barriers and opening protectives equal to the fire resistance rating of the corridor. Lobbies are to be provided with at least one means of egress complying with Chapter 10.

Fire partitions are the next level of fire-resistive wall construction below fire walls and fire barriers. They typically have 1-hour fire-resistance ratings and are primarily used for separations between dwelling units and guest rooms in I-1, R-1 and R-2 occupancies; per §402.7.2 for separation of tenants in a covered mall; and per §1017.1 for corridors. Some corridors need not be fire-rated under certain specific circumstances as described in Table 1017.1 There is also an exception that allows fire partitions in dwelling and sleeping units of Type IIB, IIIB or VB construction where there is an automatic sprinkler systems to be of minimum ½-hour fire-resistance-rated construction.

Fire partitions have the same relationship to exterior walls as fire walls and fire barriers. They also have similar requirements for penetrations, openings and ductwork as for the other groups of partition types.

The exceptions with the most relevance for designers are contained in §708.4 related to the extent and continuity of partitions. The partitions are to extend to the structure above and below with exceptions as listed on the right.

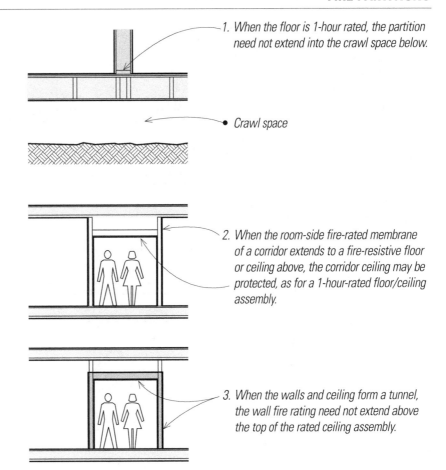

1. When the floor is 1-hour rated, the partition need not extend into the crawl space below.

• Crawl space

2. When the room-side fire-rated membrane of a corridor extends to a fire-resistive floor or ceiling above, the corridor ceiling may be protected, as for a 1-hour-rated floor/ceiling assembly.

3. When the walls and ceiling form a tunnel, the wall fire rating need not extend above the top of the rated ceiling assembly.

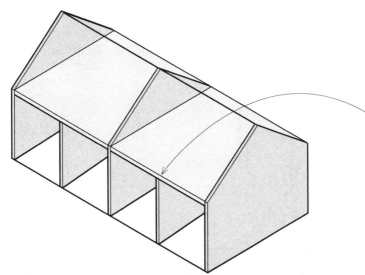

5. The fire partition forming the separation between dwelling units in Group R-2 occupancies need not extend into the attic if the attic is draft-stopped at every 3,000 sf (279 m²) or every two units, whichever is smaller.
6. If combustible floor/ceiling or roof/ceiling spaces are sprinklered, fire-blocking or draft-stopping will not be required.

Smoke Barriers and Smoke Partitions

§709 and 710 treats smoke barriers as fire barriers, but with emphasis upon restricting the migration of smoke. Smoke barriers are required at I-2 occupancies by §407.4 to subdivide stories where patient sleeping rooms occur. Smoke barriers are required to have a 1-hour rating with 20-minute-rated opening protectives. There is an exception which allows smoke barriers in I-3 occupancies to be constructed of minimum 0.10 inch-thick (2.5) steel. Smoke partitions are required to restrict smoke movement but are not required to have a fire-resistance rating.

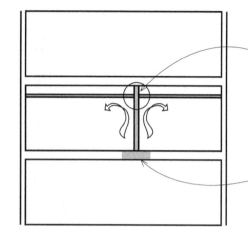

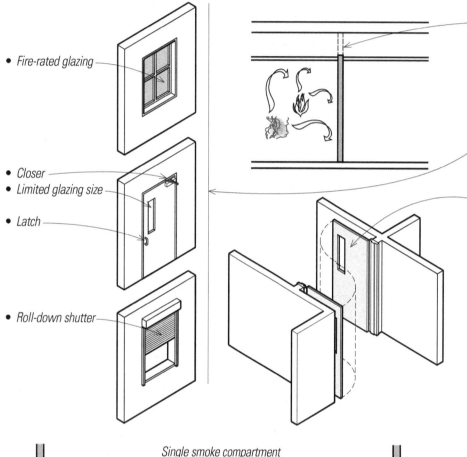

- Fire-rated glazing

- Closer
- Limited glazing size

- Latch

- Roll-down shutter

Single smoke compartment

- Steel duct

- Smoke barriers inside smoke compartment

Smoke Barriers

- Smoke barriers are to be continuous from outside wall to outside wall and from the top of the foundation or floor below to the underside of the floor above or the roof sheathing. There is to be continuity through concealed spaces such as those found above suspended ceilings.

- In other than IIB, IIB or VB construction the support construction for the smoke barrier is to be protected to the same level as the walls or floors.

- The Exception to §709.4 allows the smoke barrier to be omitted in interstitial spaces when the spaces are protected by ceilings that provide resistance to the passage of fire and smoke equivalent to the smoke barrier.

- Openings in smoke barriers are to be protected per §715.

- Per §709.5 where doors are installed in I-2 occupancies across corridors, which is a common practice in hospitals, a pair of opposite-swinging doors without a mullion and with vision panels is to be installed in the corridor. The doors are to have fire-protection-rated glazing materials and in fire-protection-rated window frames. The size of the glazing shall not exceed the tested area. The doors are to be close fitting and may not have undercuts, louvers or grilles. The doors are to have head and jamb stops, astragals or rabbets at meeting edges and be self-closing on activation of a smoke detector. This exception recognizes the occurrence of a typical field condition and sets criteria for such openings in smoke barriers.

- Ducts and air transfer openings that penetrate a smoke barrier are to comply with §716. Per §716.5.5 penetrations at smoke barriers will typically have a smoke damper except when the openings in the ducts are limited to a single smoke compartment and the ducts are made of steel.

- Where ducts penetrate the boundary of a smoke compartment, a smoke damper is required.

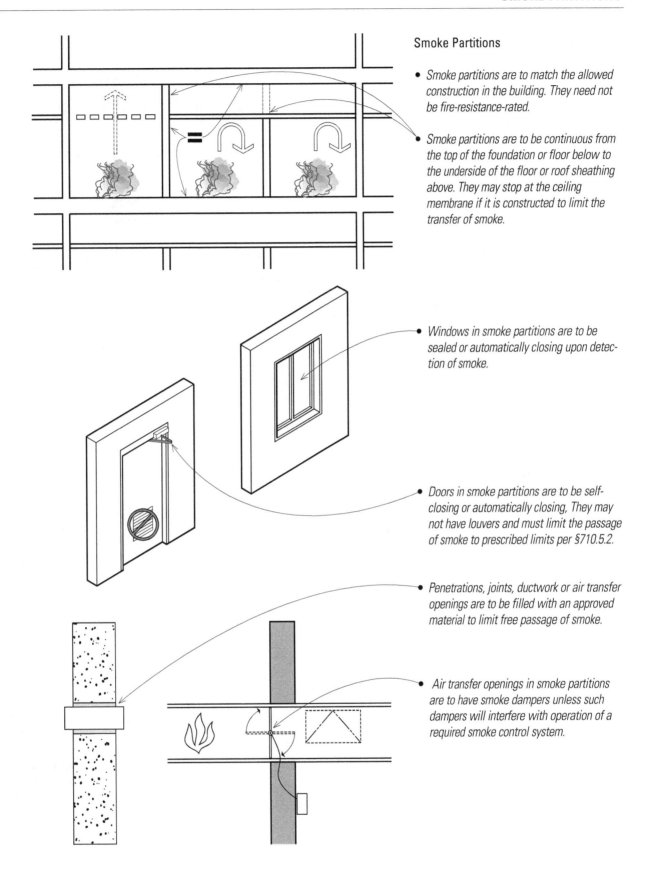

Smoke Partitions

- *Smoke partitions are to match the allowed construction in the building. They need not be fire-resistance-rated.*

- *Smoke partitions are to be continuous from the top of the foundation or floor below to the underside of the floor or roof sheathing above. They may stop at the ceiling membrane if it is constructed to limit the transfer of smoke.*

- *Windows in smoke partitions are to be sealed or automatically closing upon detection of smoke.*

- *Doors in smoke partitions are to be self-closing or automatically closing, They may not have louvers and must limit the passage of smoke to prescribed limits per §710.5.2.*

- *Penetrations, joints, ductwork or air transfer openings are to be filled with an approved material to limit free passage of smoke.*

- *Air transfer openings in smoke partitions are to have smoke dampers unless such dampers will interfere with operation of a required smoke control system.*

Horizontal Assemblies

§711 defines horizontal assemblies as floor/ceiling assemblies and roof/ceiling assemblies that require a fire-resistance rating. Their required fire-resistance rating is determined by their use. The rating is primarily determined by the fire-resistance rating based on type of construction as dictated by Table 601. When separating occupancies, the fire-resistance rating of the horizontal assembly must also be examined against the requirements of Table 508.3.3. For fire areas use the fire-resistance ratings from Table 706.3.9 per §901.7.

The criteria for penetrations of horizontal assemblies are more stringent than those for vertical assemblies, as the passage of smoke and gases vertically between floors of a building is of great concern and is facilitated by natural convective forces.

Where ducts are not required to have fire or smoke dampers complying with §716 they are required to be treated as penetrations, and the spaces around the penetrations to be sealed per this section.

Horizontal assemblies have the same continuity requirements as for vertical assemblies. Penetrations, joints and ducts are to be protected as for vertical assemblies; refer to §707.2, §712.4 and §713.

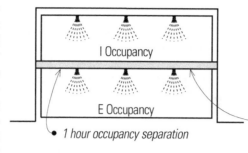

1 hour occupancy separation

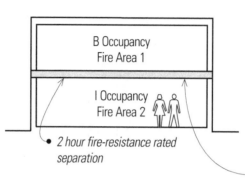

2 hour fire-resistance rated separation

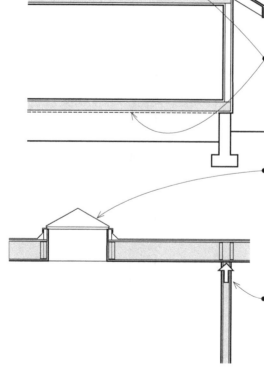

- *Horizontal smoke assemblies are to have a fire-resistance-rating equal to that required for the building type of construction.*

- *In horizontal smoke assemblies separating mixed occupancies the fire-resistance-rating is to be that required by §508.3.3 based on the occupancies being separated.*

- *Floor assemblies separating dwelling units in the same building or sleeping units in R-2 and I-1 occupancies shall be of minimum 1-hour fire-resistance-rated construction. There is an exception that such separations in Type IIB, IIIB or VB construction where there is an automatic sprinkler systems need only be of minimum ½-hour fire-resistance-rated construction.*

- *Where the floor assembly separates single or multiple occupancies into separate fire areas the assembly is to be rated per §706.3.9.*

- *Smoke resistant assemblies need to be continuous through concealed spaces unless they abut assemblies that will resist the passage of smoke, such as walls meeting fire-rated ceiling assemblies.*

- *In 1-hour fire-resistance-rated floor construction, §711.3.3 permits lower membranes to be omitted over unusable crawl spaces and upper membranes to be omitted below unusable attic spaces.*

- *Unprotected skylight openings are allowed when separation requirements to adjacent structures are met per §704.10 and the integrity of the roof construction around the skylight openings is maintained and protected as for the rest of the roof.*

- *Any structural members or walls supporting a horizontal assembly must have at least the same fire-resistance rating as the horizontal assembly.*

Penetrations

§712 requires that penetrations be protected to maintain the fire-resistive integrity of the assembly being penetrated.

The governing criteria for penetration protection systems are that they prevent the passage of flame and hot gasses into or through the assembly. Penetration treatment requirements are based on the size and quantity of penetrations. The requirements for through-penetration fire stops call for tested assemblies meeting minimum criteria for resistance to the passage of flame and hot gases. The basic criteria are that the required fire-resistance of the penetrated assembly not be compromised or reduced. Certain limited penetrations by small pipes or electrical components of specified sizes are allowed by exceptions.

Penetration firestop systems are given "F" and "T" ratings. An "F" rating is related to the time period that a through-penetration firestop system limits the movement of fire through a rated assembly. This rating should be nominally equal to the rating of the building assembly where the penetration is located. A "T" rating applies to any penetration, not just through penetrations, and measures how long it takes to raise the temperature of an assembly by 325 degrees F (163 degrees C) on the opposite side of the assembly from a fire of specified size.

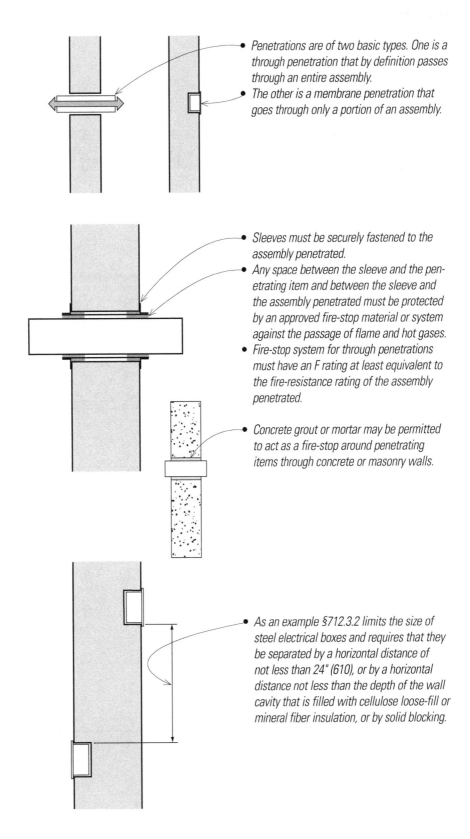

- Penetrations are of two basic types. One is a through penetration that by definition passes through an entire assembly.
- The other is a membrane penetration that goes through only a portion of an assembly.

- Sleeves must be securely fastened to the assembly penetrated.
- Any space between the sleeve and the penetrating item and between the sleeve and the assembly penetrated must be protected by an approved fire-stop material or system against the passage of flame and hot gases.
- Fire-stop system for through penetrations must have an F rating at least equivalent to the fire-resistance rating of the assembly penetrated.

- Concrete grout or mortar may be permitted to act as a fire-stop around penetrating items through concrete or masonry walls.

- As an example §712.3.2 limits the size of steel electrical boxes and requires that they be separated by a horizontal distance of not less than 24" (610), or by a horizontal distance not less than the depth of the wall cavity that is filled with cellulose loose-fill or mineral fiber insulation, or by solid blocking.

Fire-Resistant Joint Systems

§713 requires expansion, seismic movement and construction control joints to be treated in a similar manner to penetrations. The goal is that joints do not compromise the fire-resistive capabilities of the horizontal or vertical assemblies where they occur. Joints in fire-rated assemblies are to be fire tested to match the assembly rating. Such joint assemblies are tested before and after movement to verify that the joints maintain their fire-resistance after anticipated movements occur.

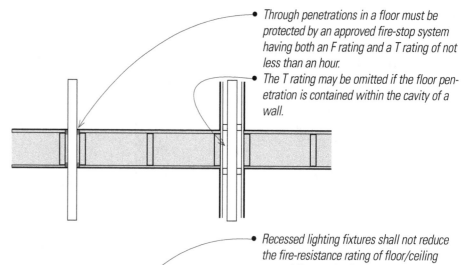

- Through penetrations in a floor must be protected by an approved fire-stop system having both an F rating and a T rating of not less than an hour.
- The T rating may be omitted if the floor penetration is contained within the cavity of a wall.

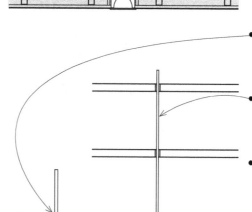

- Recessed lighting fixtures shall not reduce the fire-resistance rating of floor/ceiling assemblies required to have a minimum 1-hour fire-resistance rating.

- The annular space around a noncombustible item that penetrates only a single fire-resistance-rated floor need only be filled with an approved material.
- The noncombustible penetrating item may connect multiple fire-rated floors if limited in size per §712.4.1.

- §712.4.2 requires that even non-rated horizontal assemblies have annular spaces filled to resist the free passage of flame and products of combustion.
- Penetrating items may connect no more than two stories if so treated.
- Noncombustible items may penetrate no more than three stories.

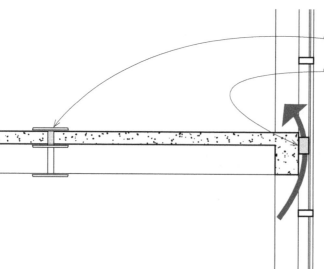

- Expansion joints are to have fire-resistance as for other penetrations.
- Joints at the intersection of floor assemblies and exterior curtain walls are included in §713.4. These intersections are to be sealed with approved materials tested for such applications. The materials are to be installed to prevent the spread of flame and hot gases at the intersection.

§714 requires that the fire-resistance ratings of structural members and assemblies comply with the requirements for type of construction as set forth in Chapter 6 and Table 601. The ratings should be complementary in that the structure supporting a fire-resistance-rated assembly should have at least the rating of the assembly supported.

§714.3 prohibits inclusion of service elements such as pipes or conduits into the fire-protection covering. This recognizes that such elements can conduct heat through the fire protection to the structural member and thus potentially compromise the time rating for fire-resistance.

§714.5 addresses the protection of structural members located on the exterior of a building. The protection of such members is set forth in Table 601 and must be as required for exterior bearing walls or for the structural frame, whichever is greater.

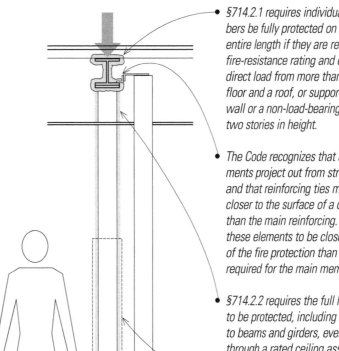

- §714.2.1 requires individual structural members be fully protected on all sides for their entire length if they are required to have a fire-resistance rating and either support a direct load from more than two floors or a floor and a roof, or support a load-bearing wall or a non-load-bearing wall more than two stories in height.

- The Code recognizes that attachment elements project out from structural members and that reinforcing ties may be located closer to the surface of a concrete element than the main reinforcing. §714.2.4 allows these elements to be closer to the surface of the fire protection than the thickness required for the main members.

- §714.2.2 requires the full height of a column to be protected, including its connection to beams and girders, even if it extends through a rated ceiling assembly.

- Fire-protection materials, especially spray-applied fire protection, are subject to impact damage that could knock off the fire protection and compromise the fire-resistance of the structural member. §714.4 requires impact protection for elements subject to damage. The protection is to extend to a height of at least 5' (1524) above the floor.

- §714.2.1 includes requirements for structural members that have less critical structural functions. These members may be protected by encasement or by membrane or ceiling protection.
- §714.2.3 permits the encapsulation of an entire truss assembly with fire-resistive materials.
- §714.2.4 permits attachments to structural members, such as lugs and brackets, to extend to within 1" (25.4 mm) of the surface of fire protection.
- §714.2.5 allows stirrups and spiral reinforcement in a masonry or concrete structural member to project no more than ½" (12.7 mm) into the protection.
- §714.6 allows the bottom flanges of lintels and shelf angles to be left unprotected when they span 6' (1829) or less or if they are not part of the structural frame.

The requirements of §715 for the protection of openings in fire-resistive construction allow the opening protectives to have different ratings than those for the wall where they are located. Table 715.4 lists the opening protection ratings for various assemblies such as fire doors and fire shutters. For example, a 2-hour-rated fire wall requires a 1-½ hour-rated opening protection assembly. Opening protectives are typically tested assemblies, so the designer can select compliant protection assemblies based on their tested performance. Test criteria for acceptable performance under given conditions in given occupancies is noted in this section. §715.3 allows alternative methods to be used to determine fire protection ratings for opening protectives.

Fire doors are to be self-closing. They typically require a latch as well. These requirements are based on maintaining the integrity of the assembly where the fire door is located. The doors should be closed under most circumstances and should be latched to prevent them being forced open by air currents generated by fire.

Fire assemblies in corridors and smoke barriers required by Table 715.4 to have a fire-resistance rating are to have a minimum fire protection rating of 20 minutes.

- Exception 1 to §715.4.3 allow viewports of less than 1 inch (25) diameter in metal frames.
- Exception 2 applies to I-2 occupancies and refers to §407.3.1 which allows doors in typical hospital corridors to not have a fire protection rating and are not required to be self closing or automatic closing. The doors are to provide a smoke barrier and are to have a positive self latching mechanism.

Glazing in opening protectives is limited by the Code based on tested performance and the area of the glazed openings. §715.4.3.2 notes that glazing in 20-minute rated fire doors must also have a 20-minute rating, but is exempt from the fire hose stream test which subjects glazing materials to a high pressure cold water stream under fire test conditions.

Wire glass is to be installed in steel frames of specified minimum material thickness. High-performance glazings tested under fire conditions may be installed in accordance with their rating based on tested sizes and durations of protection. Mullions in fire-protection-rated glazing over 12' (3658) tall must be protected as for the assembly wall construction.

Fire ratings of window protection at exterior walls requiring more than a 1-hour rating per Table 602 must be at least 1-½ hour. Other window openings where a 1-hour rating is required may have a ¾-hour rating. Where window-opening protection is required in otherwise unrated walls by proximity of building elements or exposure per §704.9 or 704.10, the windows must have a fire-resistance rating of at least ¾-hour.

- The Code typically requires that opening protectives, including fire doors and glazing materials, be labeled for test compliance and for their fire rating.

- Specialized fire-resistance-rated (as opposed to fire-protective-rated) glazing tested per ASTM E-119 in a fire-rated wall assembly may be used in walls with greater than a 1-hour rating without the area limitations of §715.5.7 when the entire wall and glazing assemblies have been tested.

- Table 715.4 specifies the minimum opening protective fire-protection ratings, in hours, based on:
  - type of assembly
  - required assembly rating

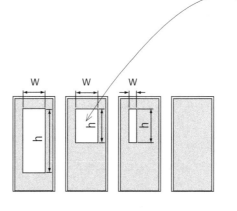

- Wire glass may be acceptable for use as glazing in opening protectives but is limited in maximum area (w × h) as well as maximum height (h) and width (w) according to the opening fire-protection rating per Table 715.5.3. Per the 2006 IBC, wire glass must now meet the requirements of CPSC 16 CFR Part 1201. Wire glass has traditionally been manufactured and tested to meet ANSI Z97.1. Wire glass is not specifically prohibited in areas requiring safety glazing, but typical wire glass, manufactured to the old standards, does not comply with the 2006 IBC.

- Per §715.5.7, fire-protection-rated glazing in fire-window assemblies may be used in fire partitions or fire barriers with a maximum fire rating of 1 hour. The total area of such windows cannot exceed 25% of the area of a common wall with any room.

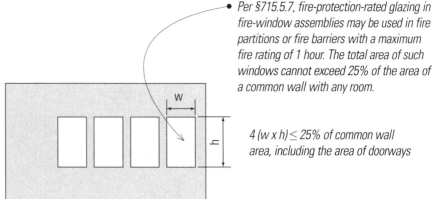

4 (w x h) ≤ 25% of common wall area, including the area of doorways

Ducts and air-transfer openings going through fire-rated assemblies must be treated in some way to prevent compromising the fire-resistance of the assemblies that the ducts pass through. Ducts that are allowed to be without dampers are treated as penetrations per §712. Where ducts must remain operational during a fire or smoke emergency—for example, in an atrium smoke-control system—they cannot have dampers. Alternate protection must be provided to maintain the integrity of fire protectives. Fire and smoke dampers must be tested and listed assemblies. Dampers must be provided with access for maintenance, testing and resetting the assemblies when they close. These access panels must not compromise the fire rating of the fire assembly where they are located.

The fire-resistive requirements for fire and smoke dampers are based on the type of assembly penetrated. Requirements are stated in the familiar pattern of basic requirements followed by exceptions. The assembly types are organized in the same hierarchy as in other parts of this chapter.

- *Fire dampers are listed mechanical devices installed in ducts and air-transfer openings and designed to close automatically upon detection of heat to restrict the passage of flame.*
- *Smoke dampers are similar to fire dampers but are intended to resist the passage of air and smoke. They are controlled either by a smoke-detection system or by a remote-control station.*
- *Both fire and smoke dampers require access for inspection and maintenance. The access panels must not compromise the integrity of the fire-resistant assembly.*

- *Fire Walls: §716.5.1 requires ducts that penetrate fire walls be provided with fire dampers located per the exception to §705.11.*

- *Fire Barriers: §716.5.2 requires fire dampers where ducts or air transfer openings penetrate fire barriers. The exceptions to this requirement are:*
  1. *The penetrations are listed in accordance with ASTM E119 as part of the fire-resistance-rated assembly.*
  2. *The ducts are part of a smoke-control system.*
  3. *The HVAC system is ducted, the wall is rated at 1 hour or less, the building is not an H occupancy and is fully sprinklered.*

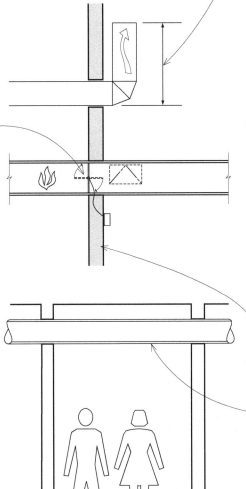

- *The opening language of §716.5.3 states that shaft enclosures that are permitted to be penetrated by ducts and air transfer openings are to be protected with fire and smoke dampers.*
- *The exceptions then go on to state that fire and smoke dampers are not required where:*
  1.1 *The duct turns up 22" (559) in exhaust ducts in the direction of a continuous upward air flow.*
  1.2 *The penetrations are listed in accordance with ASTM E119 as part of the fire-resistance-rated assembly.*
  1.3 *The ducts are part of a smoke-control system.*
  1.4 *The penetrations are in a parking garage shaft separated from other shafts by not less than 2-hour construction.*
  2. *In Group B and R occupancies with automatic sprinkler systems for kitchen, bath and toilet room exhaust enclosed in steel ducts that extend vertically 22" (559) in the shaft, have an exhaust fan at the upper termination of the shaft to maintain a continuous upward air flow*

- *Fire Partitions: §716.5.4 requires duct penetrations in fire partitions to have fire dampers except in the following conditions:*
  1. *The partitions penetrated are tenant separation or corridor walls in buildings with sprinkler systems.*
  2. *The ducts are steel, less than 100 square inches (0.06 m²) in cross-section, do not connect the corridor with other spaces, are above a ceiling and do not have a wall register in the fire-resistance-rated wall.*

- *Smoke Barriers: §716.5.5 requires that smoke dampers be provided when a duct penetrates a smoke barrier except when the duct is steel and serves only a single smoke compartment.*

- *Corridors: §716.5.4.1 requires smoke dampers at corridors except where:*
  1. *There is a smoke-control system.*
  2. *The duct is steel and passes through the corridor with no openings into it.*

HORIZONTAL ASSEMBLY PENETRATIONS

As noted above convective forces make penetrations of horizontal assemblies susceptible to the passage of heat, smoke or fire. Penetrations by ducts or air transfer openings through a floor, floor/ceiling assembly or the ceiling membrane of a roof/ceiling assembly are to be protected by a shaft enclosure per §707 or shall comply with the provisions of §716.6.1 through §716.6.3.

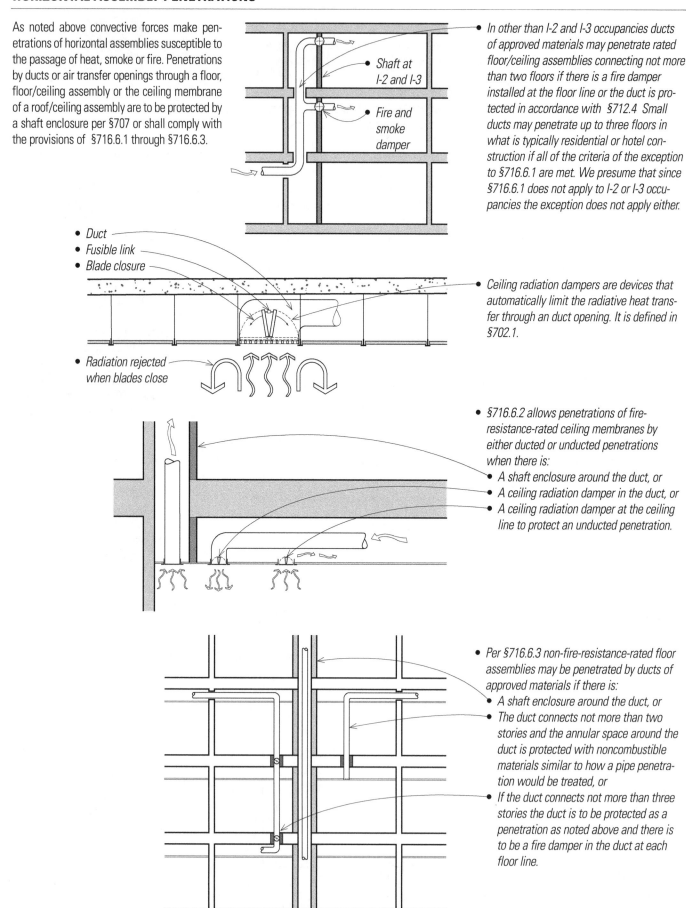

- *Shaft at I-2 and I-3*

- *Fire and smoke damper*

- In other than I-2 and I-3 occupancies ducts of approved materials may penetrate rated floor/ceiling assemblies connecting not more than two floors if there is a fire damper installed at the floor line or the duct is protected in accordance with §712.4 Small ducts may penetrate up to three floors in what is typically residential or hotel construction if all of the criteria of the exception to §716.6.1 are met. We presume that since §716.6.1 does not apply to I-2 or I-3 occupancies the exception does not apply either.

- *Duct*
- *Fusible link*
- *Blade closure*

- Ceiling radiation dampers are devices that automatically limit the radiative heat transfer through an duct opening. It is defined in §702.1.

- *Radiation rejected when blades close*

- §716.6.2 allows penetrations of fire-resistance-rated ceiling membranes by either ducted or unducted penetrations when there is:
  - A shaft enclosure around the duct, or
  - A ceiling radiation damper in the duct, or
  - A ceiling radiation damper at the ceiling line to protect an unducted penetration.

- Per §716.6.3 non-fire-resistance-rated floor assemblies may be penetrated by ducts of approved materials if there is:
  - A shaft enclosure around the duct, or
  - The duct connects not more than two stories and the annular space around the duct is protected with noncombustible materials similar to how a pipe penetration would be treated, or
  - If the duct connects not more than three stories the duct is to be protected as a penetration as noted above and there is to be a fire damper in the duct at each floor line.

Fire can spread rapidly inside concealed spaces in combustible construction if the spread of fire or movement of hot gases is not restricted. §717 sets out requirements for fire-blocking and draft-stopping.

The purpose of each requirement is to restrict or eliminate the spread of fire or the movement of hot gases in order to prevent the spread of fire within concealed spaces. These criteria typically apply to buildings with combustible construction, but they also apply where combustible decorative materials or flooring is installed in buildings of noncombustible construction.

Sprinklers are often a mitigation measure for draft-stopping of concealed spaces. However, it should be noted that in such circumstances the concealed spaces must usually be sprinklered as well as the occupied spaces.

Combustible construction materials are not allowed in concealed spaces in Type I or II buildings except when permitted by §603, when Class A finish materials are used, or when combustible piping is installed in accordance with the International Mechanical and Plumbing Codes.

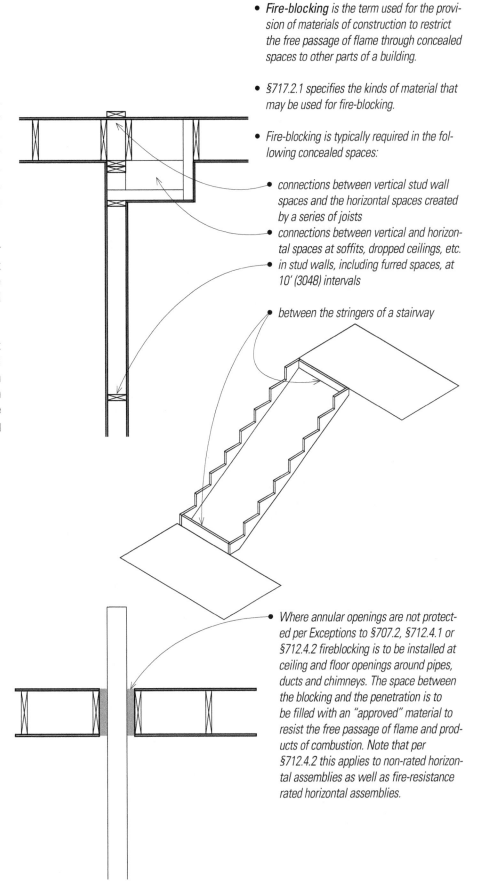

- *Fire-blocking is the term used for the provision of materials of construction to restrict the free passage of flame through concealed spaces to other parts of a building.*

- *§717.2.1 specifies the kinds of material that may be used for fire-blocking.*

- *Fire-blocking is typically required in the following concealed spaces:*

  - *connections between vertical stud wall spaces and the horizontal spaces created by a series of joists*
  - *connections between vertical and horizontal spaces at soffits, dropped ceilings, etc.*
  - *in stud walls, including furred spaces, at 10' (3048) intervals*

  - *between the stringers of a stairway*

- *Where annular openings are not protected per Exceptions to §707.2, §712.4.1 or §712.4.2 fireblocking is to be installed at ceiling and floor openings around pipes, ducts and chimneys. The space between the blocking and the penetration is to be filled with an "approved" material to resist the free passage of flame and products of combustion. Note that per §712.4.2 this applies to non-rated horizontal assemblies as well as fire-resistance rated horizontal assemblies.*

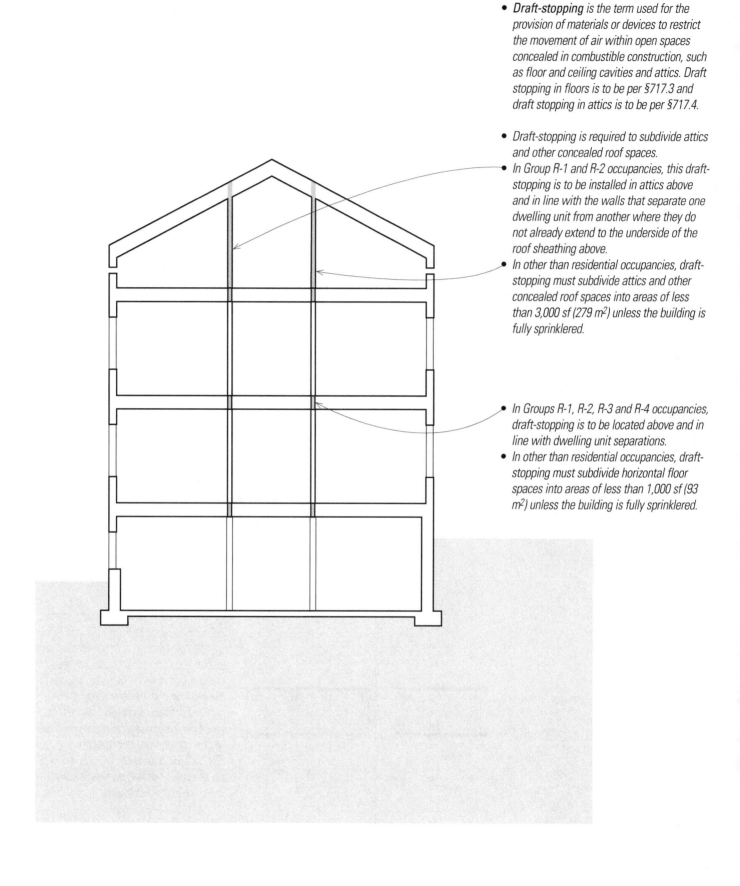

- *Draft-stopping* is the term used for the provision of materials or devices to restrict the movement of air within open spaces concealed in combustible construction, such as floor and ceiling cavities and attics. Draft stopping in floors is to be per §717.3 and draft stopping in attics is to be per §717.4.

- Draft-stopping is required to subdivide attics and other concealed roof spaces.
- In Group R-1 and R-2 occupancies, this draft-stopping is to be installed in attics above and in line with the walls that separate one dwelling unit from another where they do not already extend to the underside of the roof sheathing above.
- In other than residential occupancies, draft-stopping must subdivide attics and other concealed roof spaces into areas of less than 3,000 sf (279 m²) unless the building is fully sprinklered.

- In Groups R-1, R-2, R-3 and R-4 occupancies, draft-stopping is to be located above and in line with dwelling unit separations.
- In other than residential occupancies, draft-stopping must subdivide horizontal floor spaces into areas of less than 1,000 sf (93 m²) unless the building is fully sprinklered.

Plaster

Plaster is accepted as a fire-resistance-rated material when applied as prescribed in §718. Plaster assemblies must be based on tested assemblies. Plaster may be used to substitute for $1/2$" (12.7 mm) of the required overall assembly thickness when applied over concrete. Minimum concrete cover of $3/8$" (9.5 mm) at floors and 1" (25.4 mm) at reinforced columns must still be maintained.

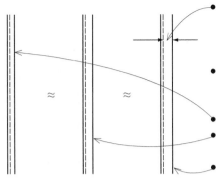

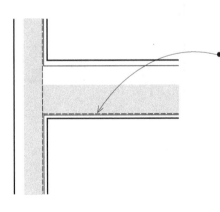

- The required plaster thickness is measured from the face of the lath when applied over metal or gypsum lath.

- §718.2 notes the following equivalencies for fire-resistance purposes:

- $1/2$" (12.7 mm) unsanded gypsum plaster
- $3/4$" (19.1 mm) one-to-three gypsum sand plaster
- 1" (25.4 mm) portland cement plaster

- An additional layer of lath embedded at least $3/4$" (19.1 mm) from the outer surface is required for plaster protection more than 1" (25.4 mm) thick.

Thermal- and Sound-Insulating Materials

§719 recognizes that thermal and acoustical insulating materials often have paper facings or contain combustible materials. When installed in concealed spaces, the materials must have a flame-spread index of not more than 25 and a smoke-developed index of not more than 450.

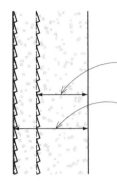

- Per §719.2.1 flame spread and smoke developed limitations do not apply in Type III, IV or V construction when the facing is turned toward and is in substantial contact with the unexposed surface of the wall, ceiling or floor.

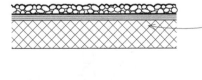

- Roof insulation may be combustible and need not meet the flame-spread and smoke-generation limits when it is covered with approved roof coverings.

PRESCRIPTIVE FIRE-RESISTANCES

The tables in §720 provide a laundry list of assemblies deemed to comply with fire-resistance requirements for the times noted when installed at the thickness indicated. There are tables for various elements:

- Table 720.1(1) for Structural Elements
- Table 720.1(2) for Wall and Partition Assemblies
- Table 720.1(3) for Floor and Roof Systems

The assemblies listed are by no means an exhaustive list. The designer will often refer to other testing agencies that are acceptable to the Code, such as Underwriters Laboratory or Factory Mutual, to find assemblies that meet the needs of the project. The designer must use these assemblies with care and identify where they are used in a project. Modifications to assembly designs should be done with caution as this may negate their approval and necessitate fire testing to prove their efficacy.

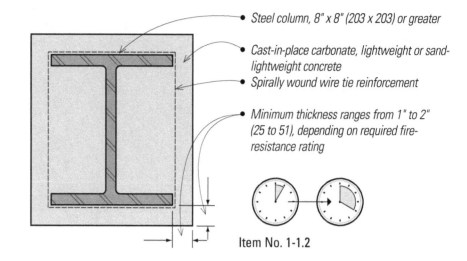

- Steel column, 8" x 8" (203 x 203) or greater
- Cast-in-place carbonate, lightweight or sand-lightweight concrete
- Spirally wound wire tie reinforcement
- Minimum thickness ranges from 1" to 2" (25 to 51), depending on required fire-resistance rating

Item No. 1-1.2

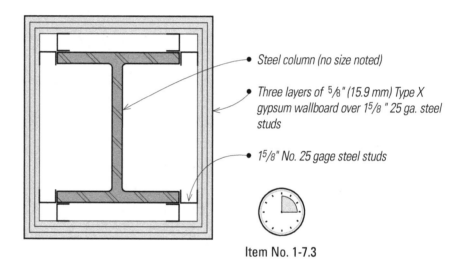

- Steel column (no size noted)
- Three layers of $^5/_8$" (15.9 mm) Type X gypsum wallboard over $1^5/_8$" 25 ga. steel studs
- $1^5/_8$" No. 25 gage steel studs

Item No. 1-7.3

Structural Elements

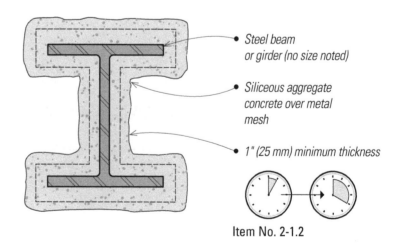

- Steel beam or girder (no size noted)
- Siliceous aggregate concrete over metal mesh
- 1" (25 mm) minimum thickness

Item No. 2-1.2

Structural Elements

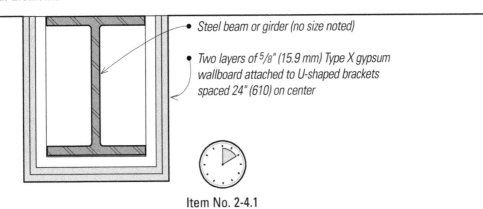

- Steel beam or girder (no size noted)

- Two layers of $5/8$" (15.9 mm) Type X gypsum wallboard attached to U-shaped brackets spaced 24" (610) on center

Item No. 2-4.1

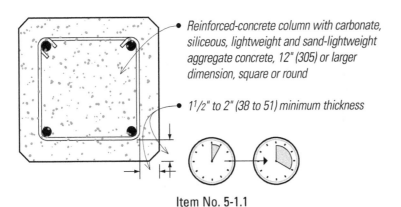

- Reinforced-concrete column with carbonate, siliceous, lightweight and sand-lightweight aggregate concrete, 12" (305) or larger dimension, square or round

- $1^1/2$" to 2" (38 to 51) minimum thickness

Item No. 5-1.1

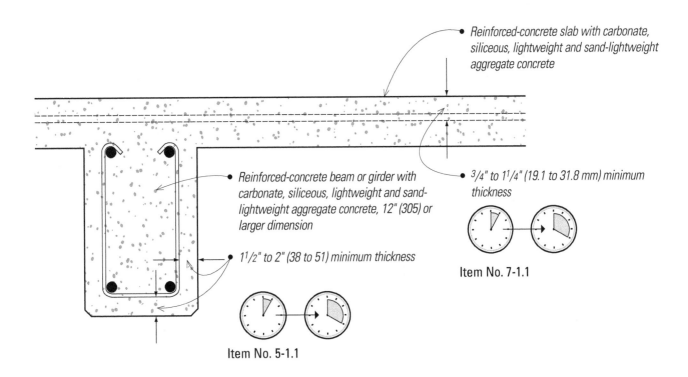

- Reinforced-concrete slab with carbonate, siliceous, lightweight and sand-lightweight aggregate concrete

- Reinforced-concrete beam or girder with carbonate, siliceous, lightweight and sand-lightweight aggregate concrete, 12" (305) or larger dimension

- $1^1/2$" to 2" (38 to 51) minimum thickness

- $3/4$" to $1^1/4$" (19.1 to 31.8 mm) minimum thickness

Item No. 7-1.1

Item No. 5-1.1

Wall and Partition Assemblies

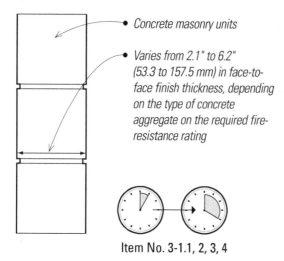

- Concrete masonry units

- Varies from 2.1" to 6.2" (53.3 to 157.5 mm) in face-to-face finish thickness, depending on the type of concrete aggregate on the required fire-resistance rating

Item No. 3-1.1, 2, 3, 4

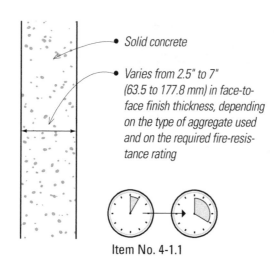

- Solid concrete

- Varies from 2.5" to 7" (63.5 to 177.8 mm) in face-to-face finish thickness, depending on the type of aggregate used and on the required fire-resistance rating

Item No. 4-1.1

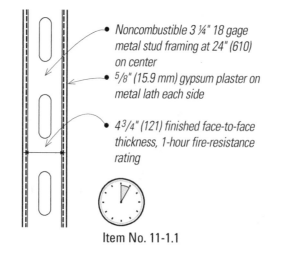

- Noncombustible 3 ¼" 18 gage metal stud framing at 24" (610) on center
- ⁵/₈" (15.9 mm) gypsum plaster on metal lath each side
- 4³/₄" (121) finished face-to-face thickness, 1-hour fire-resistance rating

Item No. 11-1.1

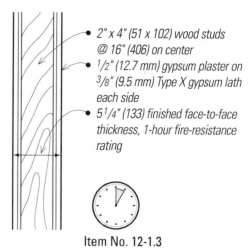

- 2" x 4" (51 x 102) wood studs @ 16" (406) on center
- ¹/₂" (12.7 mm) gypsum plaster on ³/₈" (9.5 mm) Type X gypsum lath each side
- 5¹/₄" (133) finished face-to-face thickness, 1-hour fire-resistance rating

Item No. 12-1.3

Wall and Partition Assemblies

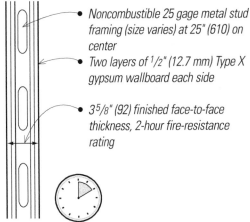

- Noncombustible 25 gage metal stud framing (size varies) at 25" (610) on center
- Two layers of $^1/_2$" (12.7 mm) Type X gypsum wallboard each side

- $3^5/_8$" (92) finished face-to-face thickness, 2-hour fire-resistance rating

Item No. 13-1.2

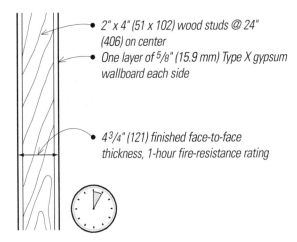

- 2" x 4" (51 x 102) wood studs @ 24" (406) on center
- One layer of $^5/_8$" (15.9 mm) Type X gypsum wallboard each side

- $4^3/_4$" (121) finished face-to-face thickness, 1-hour fire-resistance rating

Item No. 14-1.3

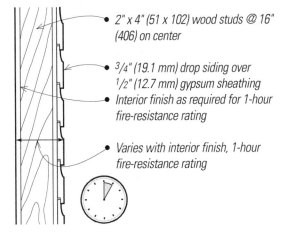

- 2" x 4" (51 x 102) wood studs @ 16" (406) on center

- $^3/_4$" (19.1 mm) drop siding over $^1/_2$" (12.7 mm) gypsum sheathing
- Interior finish as required for 1-hour fire-resistance rating

- Varies with interior finish, 1-hour fire-resistance rating

Item No. 15-1.1

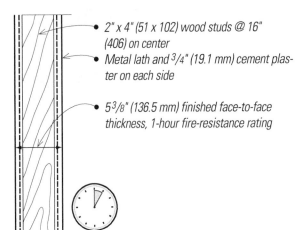

- 2" x 4" (51 x 102) wood studs @ 16" (406) on center
- Metal lath and $^3/_4$" (19.1 mm) cement plaster on each side

- $5^3/_8$" (136.5 mm) finished face-to-face thickness, 1-hour fire-resistance rating

Item No. 15-1.2

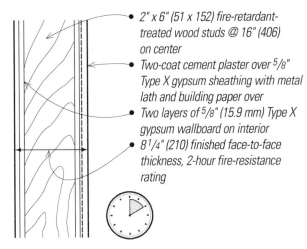

- 2" x 6" (51 x 152) fire-retardant-treated wood studs @ 16" (406) on center
- Two-coat cement plaster over $^5/_8$" Type X gypsum sheathing with metal lath and building paper over
- Two layers of $^5/_8$" (15.9 mm) Type X gypsum wallboard on interior
- $8^1/_4$" (210) finished face-to-face thickness, 2-hour fire-resistance rating

Item No. 15-1.6

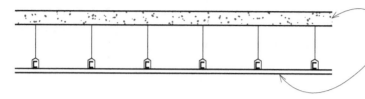

- 3" (76) reinforced-concrete floor or roof slab

- 1" (25.4 mm) thick ceiling consists of vermiculite gypsum plaster over metal lath attached to ³/₄" (19.1 mm) channels @ 12" (305) on center

- 4-hour fire-resistance rating

Item No. 5-1.1

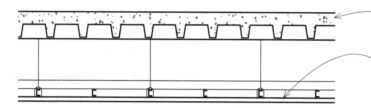

- 3" (76) deep cellular steel deck with 2¹/₂" (64) minimum thickness concrete floor or roof slab above top of deck flutes
- 1¹/₈" (28.6 mm) thick ceiling consists of vermiculite gypsum plaster base coat and vermiculite acoustical plaster on metal lath attached to ³/₄" (19.1 mm) channels @ 12" (305) on center attached to 1¹/₂" channels @ 36" (915) on center and suspended with wire hangers at 36" (915) on center
- 4-hour fire-resistance rating

Item No. 9-1.1

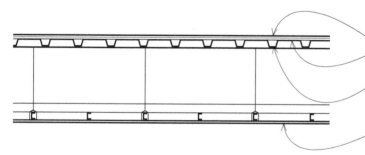

- Class A or B roof covering
- Insulation board of wood fibers with cement binders bonded to deck
- 1¹/₂" (38.1 mm) deep steel roof deck on steel framing

- ³/₄" (19.1 mm) thick ceiling of gypsum plaster over metal lath attached to 2" (50 mm) channels @ 36" (915) on center and suspended with wire hangers at 36" (915) on center
- 2-hour fire-resistance rating

Item No. 10-1.1

Floor and Roof Systems

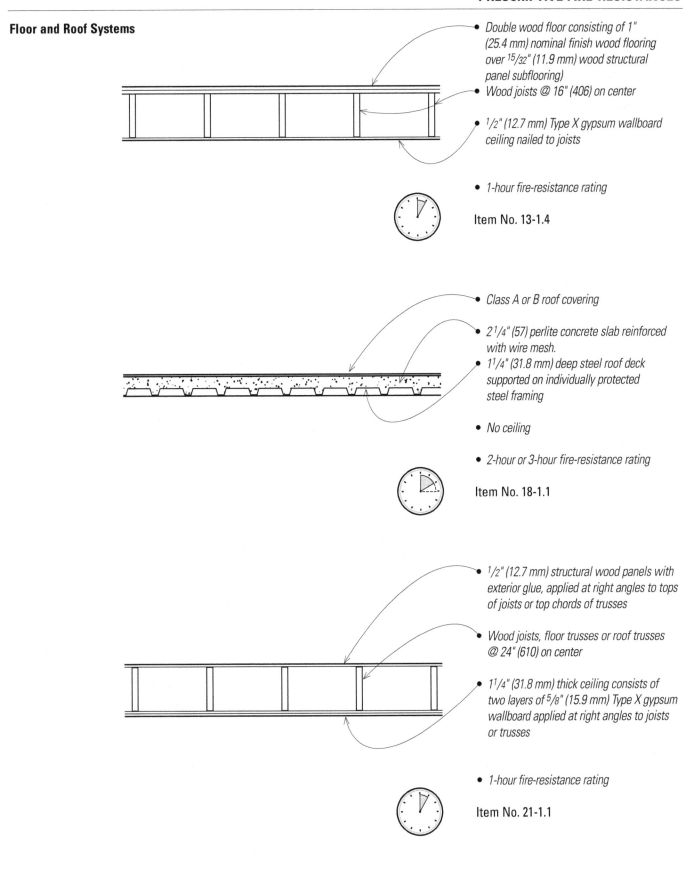

- Double wood floor consisting of 1" (25.4 mm) nominal finish wood flooring over $^{15}/_{32}$" (11.9 mm) wood structural panel subflooring)
- Wood joists @ 16" (406) on center

- $^1/_2$" (12.7 mm) Type X gypsum wallboard ceiling nailed to joists

- 1-hour fire-resistance rating

Item No. 13-1.4

- Class A or B roof covering

- $2^1/_4$" (57) perlite concrete slab reinforced with wire mesh.
- $1^1/_4$" (31.8 mm) deep steel roof deck supported on individually protected steel framing

- No ceiling

- 2-hour or 3-hour fire-resistance rating

Item No. 18-1.1

- $^1/_2$" (12.7 mm) structural wood panels with exterior glue, applied at right angles to tops of joists or top chords of trusses

- Wood joists, floor trusses or roof trusses @ 24" (610) on center

- $1^1/_4$" (31.8 mm) thick ceiling consists of two layers of $^5/_8$" (15.9 mm) Type X gypsum wallboard applied at right angles to joists or trusses

- 1-hour fire-resistance rating

Item No. 21-1.1

CALCULATED FIRE-RESISTANCES

When project conditions cannot be met using prescriptive assemblies, §721 provides a methodology to calculate the fire-resistive performance of specific materials or combinations of materials. These calculations are developed for use in the Code and meant to apply only to the section in which they are contained. They are designed to facilitate design and documentation of fire assemblies so that the designer and the building official will have reasonable assurance of the performance of the calculated assembly under actual fire conditions. The formulas are based on data for heat transfer in structural members, thermal conductance of insulating materials, conduction in materials, fire-test data of the fire-resistance of various building materials; individually and in concert with each other, along with anticipated loads on assemblies in place.

The designer can use the data contained in this section to determine fire-resistance for assemblies that do not fit neatly into the prescriptive, pretested categories contained in §720.

On the following pages are illustrated several examples for both structural protection and assembly calculations to illustrate the principles at work in this section.

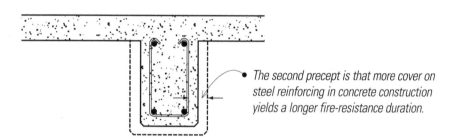

There are several fundamental precepts for the calculation of fire-resistance.

The first is that there is a quantifiable relationship between the thickness of fire-protection materials applied to building elements and the resulting time of fire-resistance. In short, thicker fire protection yields longer fire-resistance duration.

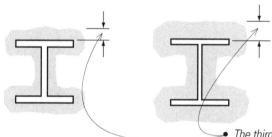

The second precept is that more cover on steel reinforcing in concrete construction yields a longer fire-resistance duration.

The third precept is that there is a relationship between the overall exterior surface dimension of a steel structural member (referred to as the "heated perimeter"), the thickness of the parts of the member and the thickness of the fire-protection materials necessary to achieve a certain fire rating. For example, compact steel wide flange sections with thick webs and flanges usually require thinner layers of fire-protective materials to achieve a given fire-resistance duration than a larger-dimension beam made up of thinner steel.

§721.2 Concrete Assemblies

Typical concrete has either siliceous or carbonate aggregate; for our example, we will assume the use of siliceous aggregate.

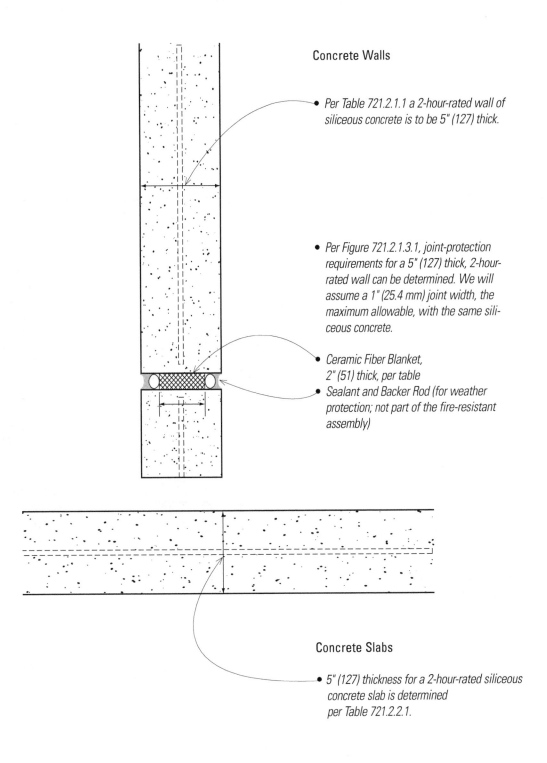

Concrete Walls

- *Per Table 721.2.1.1 a 2-hour-rated wall of siliceous concrete is to be 5" (127) thick.*

- *Per Figure 721.2.1.3.1, joint-protection requirements for a 5" (127) thick, 2-hour-rated wall can be determined. We will assume a 1" (25.4 mm) joint width, the maximum allowable, with the same siliceous concrete.*

- *Ceramic Fiber Blanket, 2" (51) thick, per table*
- *Sealant and Backer Rod (for weather protection; not part of the fire-resistant assembly)*

Concrete Slabs

- *5" (127) thickness for a 2-hour-rated siliceous concrete slab is determined per Table 721.2.2.1.*

CALCULATED FIRE-RESISTANCES

§721.5 Steel Assemblies

Fire protection for steel columns is dependent on the weight per lineal foot (W) of the column and the heated perimeter (D) of the column, which is related to the physical dimensions of the column.

Table 721.5.1(1) shows W/D ratios for typical columns. We will assume a W12 x 96 column.

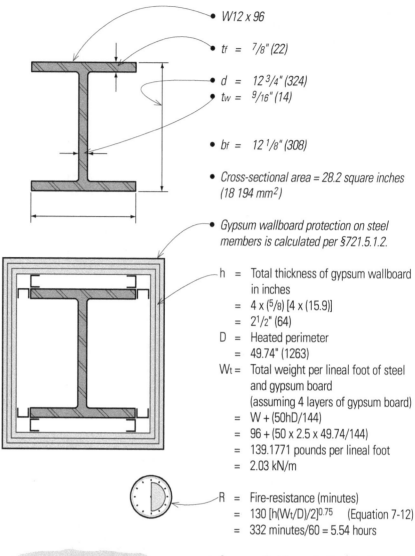

- W12 x 96

- t_f = $^7/_8$" (22)

- d = 12 $^3/_4$" (324)
- t_w = $^9/_{16}$" (14)

- b_f = 12 $^1/_8$" (308)

- Cross-sectional area = 28.2 square inches (18 194 mm²)

- Gypsum wallboard protection on steel members is calculated per §721.5.1.2.

- For box profile:
  D = $2(b_f + d)$
  = 2(12.16 + 12.71)
  = 49.74

  Per Table 721.5.1(1), W/D ratio (box)
  = 96/49.74
  = 1.93

h = Total thickness of gypsum wallboard in inches
  = 4 x ($^5/_8$) [4 x (15.9)]
  = 2$^1/_2$" (64)
D = Heated perimeter
  = 49.74" (1263)
W_t = Total weight per lineal foot of steel and gypsum board (assuming 4 layers of gypsum board)
  = W + (50hD/144)
  = 96 + (50 x 2.5 x 49.74/144)
  = 139.1771 pounds per lineal foot
  = 2.03 kN/m

R = Fire-resistance (minutes)
  = 130 $[h(W_t/D)/2]^{0.75}$ (Equation 7-12)
  = 332 minutes/60 = 5.54 hours

- For contour profile:
  D = $4b_f + 2d - 2t_w$
  = 4(12.16) + 2(12.71) - 2(0.55)
  = 72.96

  Per Table 721.5.1(1), W/D ratio (contour)
  = 96/72.96
  = 1.32

- Spray-applied fire protection of columns is based on the heated perimeter (usually of a contour profile), the thickness of the material and material-dependent fire-resistance constants. For our W12x96 column the formula is:

C_1 and C_2 are material-dependent constants obtained from material manufacturers. Assume both to be 45 for our example.

W/D ratio= 1.32
h = Thickness of spray-applied fire-resistant material in inches
  = 2" (51)

R = fire-resistance
  = $[C_1 (W/D) + C_2]$ h   (Equation 7-13)
  = [45 (1.32) + 45] 2
  = 209 minutes/60 = 3.48 hours

§721.6 Wood Assemblies

These calculations may be used only for 1-hour rated assemblies. We will assume a wall located more than 5' (1524) from the property line. Thus the fire side is presumed to be on the interior of the building, per §721.6.2.3.

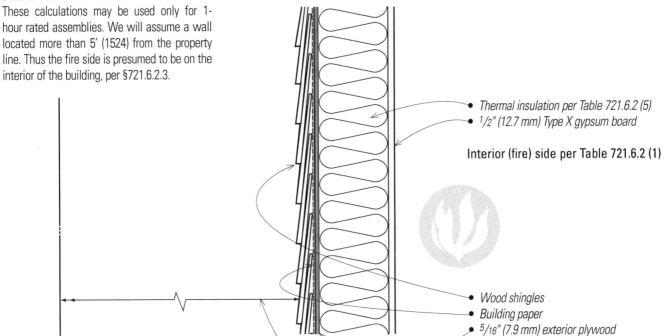

- Thermal insulation per Table 721.6.2 (5)
- 1/2" (12.7 mm) Type X gypsum board

Interior (fire) side per Table 721.6.2 (1)

- Wood shingles
- Building paper
- 5/16" (7.9 mm) exterior plywood

Exterior side per Table 721.6.3 (3)

- 75' (22 860) to property line

Tabulation

Material	Time (minutes)	Alternate assembly*	Reference
Exterior	0		Excluded; not on fire-exposed side; see §721.6.2.1.
Insulation	15	0	Table 721.6.2 (5)
Wood studs @ 16" (406) o.c.	20	20	Table 721.6.2 (2)
1/2" (12.7 mm) Type X gypsum board	25	40	Table 721.6.2 (1)
Total	60	60	

* Alternatively a 1-hour rating may be achieved without insulation by using two layers of 1/2" (12.7 mm) Type X gypsum wallboard (25 minutes x 2 = 50 minutes of fire-resistance assigned) or one layer of 5/8" (15.9 mm) Type X gypsum wallboard (40 minutes of fire-resistance assigned). Compliance may thus be achieved in several different ways.

8

Interior Finishes

Chapter 8 of the Code governs the use of materials for interior finishes, trim and decorative materials. The primary consideration of this chapter is the flame-spread and smoke-generation characteristics of materials when they are applied to underlying surfaces. These regulations are meant to govern those decorative finishes that are exposed to view. The regulations include all interior surfaces; floors, walls and ceilings. Trim applied to finish surfaces is also included in these regulations. Note that §801.1.3 requires finishes to be flood-damage-resistant when located below design flood elevation in buildings in flood hazard areas.

Classification

Materials are classified by their flame spread and smoke generation as tested on the test protocols of ASTM E 84. This test sets a standard for the surface-burning characteristics of building materials. The materials are placed in a test furnace and exposed to a flame of a calibrated size and the ignition, spread of flame, and generation of smoke are noted and assigned values based on a standard scale. The standard material against which other materials are measured for flame spread and smoke generation is Red Oak.

Typically the lower the number for each measurement, the slower the flames spread and the less smoke is generated. Once materials are tested, they are classified as A, B or C. The classifications, per §803.1 are:

The exception allows compliance with NFPA 286 instead of using ASTM E 84. This test is known as a "room corner" fire test. Many materials are tested using this standard test, which probably gives a better indication of actual performance in the field than does the ASTM test. §803.2.1 lists the acceptance criteria for using NFPA 286 data.

The required classes of finish materials are determined by occupancy. The requirements are further subdivided into sprinklered and unsprinklered occupancies. The final consideration is whether the finish occurs in exits, in exit-access areas or in rooms. Per Table 803.5, unsprinklered buildings require higher classifications of finish materials. Also, egress paths require higher finishes than rooms or occupied spaces. This is consistent with the general philosophy of the Code to offset active fire-protection measures with passive measures and to hold egress paths to higher standards than for occupied rooms.

Per Table 803.5 the most restrictive flame-spread and smoke-developed classifications are required for:
- *Exits, exit passageways, exit-access corridors and other components of emergency egress*
- *Group A (assembly), H (high-hazard) and I (institutional) occupancies*

For healthcare occupancies the requirements are shown in the table to the right:

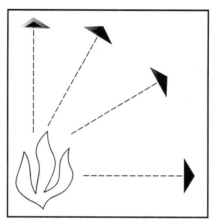

	Flame-Spread Index	Smoke-Developed Index
Class A	0–25	0–450
Class B	26–75	0–450
Class C	76–200	0–450

Interior Wall and Ceiling Finish Requirements for Healthcare Occupancies

	Sprinklered			Nonsprinklered		
Group	Exit Enclosures and Exit Passageways	Corridors	Rooms and Enclosed Spaces	Exit Enclosures and Exit Passageways	Corridors	Rooms and Enclosed Spaces
B Clinic	B	C	C	A	B	C
I-1 Res. for >16	B	C	C	A	B	B
I-2 Hospital	B	B	B (C at Admin. and with fewer than 4 occupants/room)	A	A	B
I-3 Detention	A	A (Cl. B wainscot OK to 48")	C	A	A	B
R-3 Congregate 5 < occ. 16	C	C	C	C	C	C
R-4 Assisted 5 < occ. < 16	B	C	C	A	B	C
R-3 IRC < 5 occ.	C	C	C	C	C	C

Application

§803.4 governs the application of materials over walls or ceilings required to be fire-resistance rated or of noncombustible construction. Where materials are set off from fire-resistive construction by more than the distances noted, the finish materials must be Class A except where sprinklers protect both the visible and the concealed side of the finish materials.

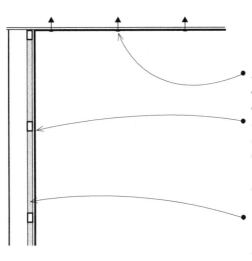

• *§803.3 requires finish materials be securely attached so that they will remain in place for at least 30 minutes at 200°F (93°C).*

• *§803.4 states that wall and ceiling finishes in fire-resistance-rated or noncombustible construction must be placed directly against the backing material, or the spaces behind must be furred with furring strips less than 1-¾" (44.5 mm) deep.*

• *The spaces between the furring strips filled with fire-resistive inorganic or Class A materials, or the furring must be fire-blocked every 8' (2438).*

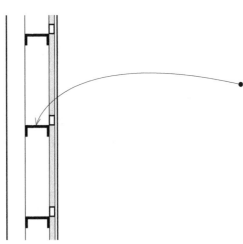

• *Alternately, where stand-off finishes are desired without sprinklers, Class "A" materials must be used, or a second layer of noncombustible construction may be used to build out the walls and then use furring as noted above when it is not desirable or feasible to sprinkler both sides of the set-out.*

Acoustical Ceiling Systems

§803.9 regulates the quality, design, fabrication and installation of acoustical tile and lay-in panel ceiling systems. The suspension systems must meet general engineering principles for vertical and lateral load capabilities. The acoustic materials must comply with the classification ratings noted for the occupancy and configuration in relation to fire-resistance rated construction per §803.4.2. Where the ceiling suspension system is part of the fire-resistance rated construction the assembly must be installed consistent with the tested assemblies and must conform to the fire-resistance-rating requirements of Chapter 7.

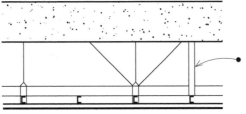

• *In areas prone to seismic activity provisions must be made to resist uplift as well as lateral motion in suspended ceilings.*

§804 deals primarily with carpets or similar fibrous flooring materials. Traditional-type flooring—such as wood, vinyl, linoleum or terrazzo—are excepted. Flooring is to be classified into two categories, Class I or II, per the results of fire testing done per NFPA 253, with Class I materials being the more resistant to flame spread than Class II materials. This test records the fire and smoke response of flooring assemblies to a radiant heat source. Carpets are to be tested in configurations similar to how they are to be installed, including pads and adhesives or fasteners. Flooring materials are to be tagged with test results.

§804.4 governs the interior floor finish in vertical exits, exit passageways, exit access corridors, and rooms not separated from exit access corridors by full-height partitions extending from the underside of the floor to the underside of the ceiling. The finishes in these conditions are to withstand a minimum critical radiant flux as specified in §804.4.1.

Note that this section applies to all occupancies and introduces one additional test criteria, the Department of Commerce FF-1 Pill Test (per Consumer Product Safety Commission 16, Code of Federal Regulations 1630). This criteria is similar to, but less stringent than the NFPA test.

§805 addresses the use of combustible materials on floors in Type I and II construction. The requirements for interior floor finishes are thus based on occupancy, construction type and sprinklering, and can be summarized as follows:

- There are provisions allowing use of wood flooring in Type I or II buildings. In these conditions wood floors are to be installed over fire-blocked sleepers with blocking under walls.
- §805.1.1 regulates subfloor construction and prohibits combustible sleepers and nailing blocks unless the space between the fire-resistance-rated floor construction and the subfloor is solidly filled with noncombustible material or fire-blocked per §717.

- §805.1.2 permits wood flooring to be attached directly to embedded or fire-blocked wood sleepers, or be cemented directly to the top surface of fire-resistance-rated construction or to a wood subfloor attached to sleepers per §805.1.1.
- Another stipulation of §805.1.1 prohibits such spaces from extending under or through permanent walls and partitions.

	Unsprinklered		Sprinklered	
	Vertical Exits Exit Passageways Exit Access Corridors	All Other Areas	Vertical Exits Exit Passageways Exit Access Corridors	All Other Areas
Occupancy				
I-2, I-3	Class I	FF-1	Class II	FF-1
A, B, E, H, I-4 M, R-1, R-2, S	Class II	FF-1	FF-1	FF-1

§804.4 states that all other areas are to comply with FF-1. Thus this criteria would apply to I-1 occupancies which are not specifically mentioned in the list of I occupancies.

§806 regulates the flame-resistance of decorative materials suspended from ceilings or walls in occupancies A, E, I, R-1 and R-2 dormitories. These decorative materials, such as curtains, draperies and hangings, must be flame-resistant per NFPA 701 or be noncombustible. In I-1 and 1-2 occupancies, higher standards apply, even including such items as paintings or photographs unless they are in limited quantities. If large paintings or photos are contemplated for use by the designer in I-1 and I-2 occupancies this should be reviewed with the AHJ. In 1-3 occupancies, all combustible decorations are prohibited.

Fixed or movable walls and partitions, wall and crash pads, and acoustical panels are to be considered interior finish if they cover more than 10% of the wall or ceiling area and are not to be considered as decorative materials or furnishings. Thus the interior finish rules apply for such areas.

§806.1 also requires that fabric partitions suspended from the ceiling and not supported by the floor in Group B or M occupancies meet the flame propagation performance criteria of NFPA 701 or be noncombustible.

§806.5 requires all interior trim other than foam plastic to have minimum Class C flame-spread and smoke-developed indices (FS = 76-200 and S-D = 0-450 per §803.1). The amount of combustible trim, excluding handrails and guardrails, may not exceed 10% of the aggregate wall or ceiling area where it is located.

- *Even though the connecting word in this clause is* or *instead of* and *note that in §806.1.2, the word* aggregate *appears, which implies that the standard applies to the areas of walls and ceilings when taken together. Thus the area of trim should not exceed 10% of the total areas of both walls and ceilings when added together.*

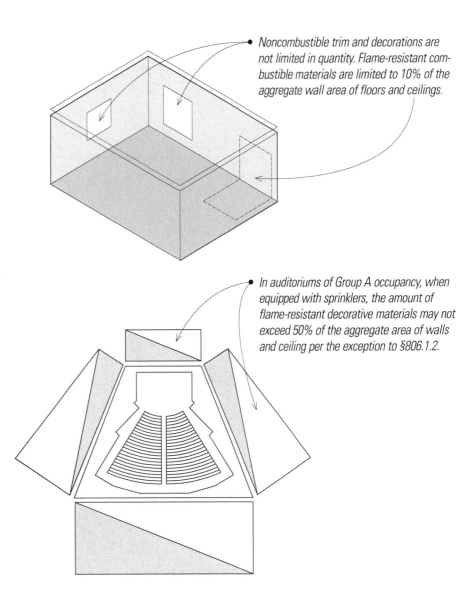

- *Noncombustible trim and decorations are not limited in quantity. Flame-resistant combustible materials are limited to 10% of the aggregate wall area of floors and ceilings.*

- *In auditoriums of Group A occupancy, when equipped with sprinklers, the amount of flame-resistant decorative materials may not exceed 50% of the aggregate area of walls and ceiling per the exception to §806.1.2.*

9
Fire-Protection Systems

As we have seen in Chapter 7, the Code recognizes the effectiveness of active fire-protection systems. The provision of an automatic sprinkler system allows trade-offs with passive fire-resistance in numerous sections of the Code. The Code also requires that active systems be provided in buildings above a certain height, above certain occupant loads and in certain occupancies deemed to required greater protection for occupants. Thus the Code recognizes the efficacy of active systems in concert with passive fire-protection to provide a balanced approach to fire and life safety for building occupants.

Active systems, especially fire sprinklers, are very effective but subject to interruption of water supplies unless emergency water sources with backup pressure systems are provided. At the same time, passive systems are subject to failure due to construction defects, poor maintenance, remodeling, failure of attachments, damage during seismic or weather events and damage from building use. The designer should consider the integration of both passive and active systems while developing the building design. The best designs do not rely on just one set of systems to provide fire protection but use an overall holistic approach to provide maximum protection.

We will touch on those active systems that integrate most effectively with passive fire-resistive strategies. The primary systems considered are fire sprinklers, fire alarms and detectors, smoke-control systems, and smoke and heat vents. Some systems respond to hazards with defined actions designed to suppress fires or provide for occupant safety and egress. Other systems—such as fire alarms, fire extinguishers and standpipe systems—provide notification for egress or auxiliary fire-fighting capabilities for the fire service. The Code requires their presence for certain occupancies but typically does not allow balancing their use with alternate design considerations.

Fire Sprinklers

The design and installation of automatic sprinkler systems is typically governed by rules developed by the National Fire Protection Association and promulgated as NFPA 13. An automatic sprinkler system is designed to respond to a fire when the sprinkler heads are activated by a fire or by heat. The systems are designed to function without intervention by the building occupants to cause their operation. The history of fire sprinklers as fire-suppression devices has been a good one, and they have been required under a greater range of circumstances in each new edition of model codes.

Residential sprinkler systems may be installed under the provisions of NFPA 13D or 13R, which are subsets of the provisions of NFPA 13. These may be acceptable, or even required as sprinkler systems in specific instances, but the Code does not recognize these as fully equivalent to NFPA 13 systems for the trade-offs contained in exceptions or reductions unless they are specifically allowed or required. See the requirements for I and R occupancies for allowable uses for these sprinkler-system classes.

Fire-protection systems consist of:

- *Detection systems that sense heat, fire or smoke (particles of combustion) and activate an appropriate alarm.*

- *Alarm systems that alert occupants of an emergency in a building by the sense of hearing, sight or, in come cases, touch.*

- *Automatic fire-extinguishing or sprinkler systems that are activated by heat from a fire and discharge either an approved fire-extinguishing agent or water over the fire area in order to extinguish or control a fire.*

- *Automatic sprinkler systems consist of underground and overhead piping from a suitable water supply, to which automatic sprinklers are attached in a regular pattern designed to provide even coverage for water discharge. Upon activation by heat from a fire, the systems discharge water over the fire area.*

- *A fire-protection system may also include equipment to control or manage smoke and combustion products of a fire.*

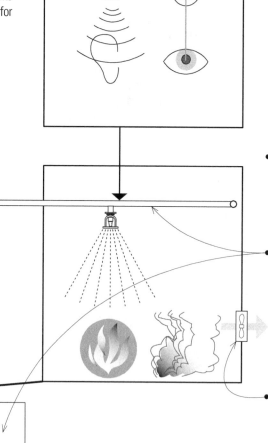

NFPA
13

NFPA
13D
≠
13

NFPA
13R
≠
13

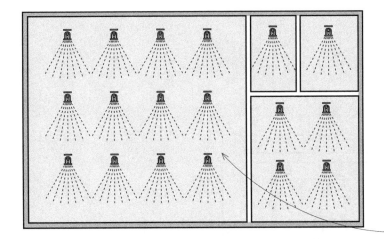

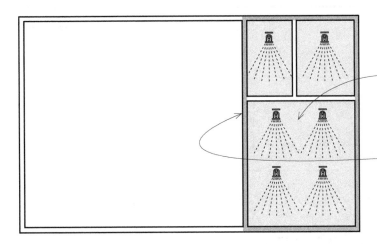

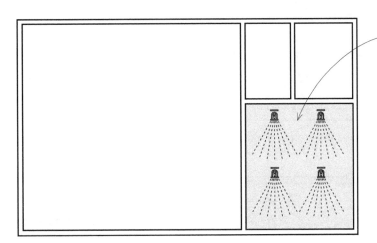

Various occupancies are required to have automatic sprinkler systems per this section. These requirements are based on several factors:

- Occupancy type
- Occupant load
- Area of occupied space
- Locations not providing ready egress or ready fire-department access

The requirements are listed by occupancy group. Carefully note that an automatic sprinkler system may be required:

- *Throughout an entire building* because of a code requirement or because the sprinkler system is used as a substitute for other passive fire-protection features

- *Throughout a fire area* that may exceed a certain size or occupant load, or one that is located in a specific portion of a building

- *Per the definitions contained in §702.1, fire areas are enclosed and bounded by fire walls, fire barriers, exterior walls or fire-resistance-rated horizontal assemblies.*

- *In specific rooms or areas to protect against a specific hazard*

Note also that the increases for building heights and areas allowed in Chapter 5 for provision of sprinkler systems apply only when those systems are automatic and are installed throughout a building.

In addition to the requirements contained in §903.2, Table 903.2.13 is a useful cross reference for sprinkler requirements contained in other code chapters.

AUTOMATIC SPRINKLER SYSTEMS

Group A: §903.2.1 requires that when any of
the conditions listed below exist, an automatic
sprinkler system be installed:

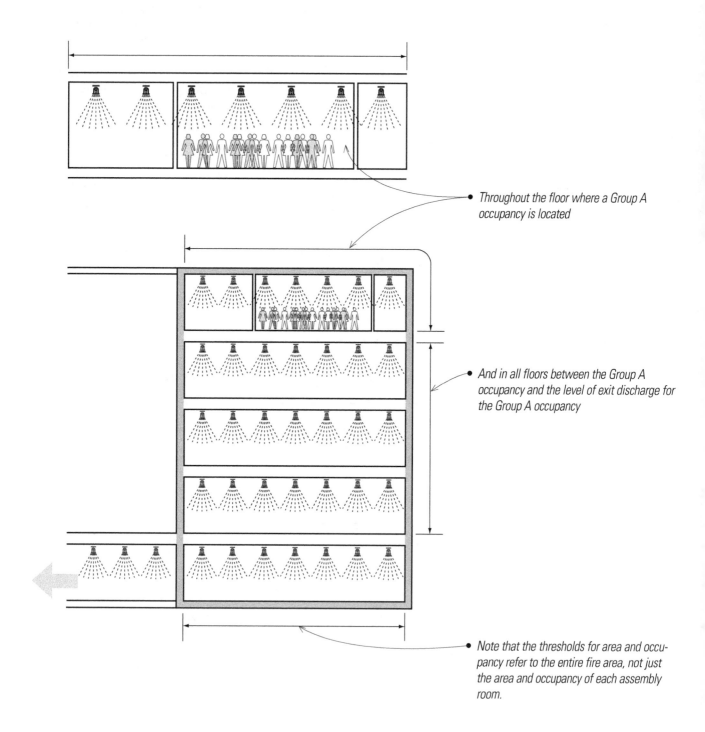

• Throughout the floor where a Group A
  occupancy is located

• And in all floors between the Group A
  occupancy and the level of exit discharge for
  the Group A occupancy

• Note that the thresholds for area and occupancy refer to the entire fire area, not just
  the area and occupancy of each assembly
  room.

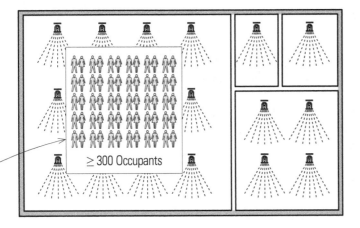

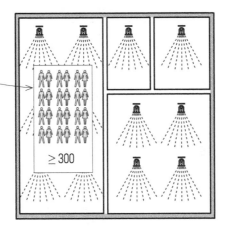

• Fire area exceeds 12,000 sf (1115 m²) for Group A-1, A-3, and A-4 occupancies, or 5,000 sf (465 m²) for Group A-2 occupancies.

• The sprinkler threshold for overall fire area occupant load in Group A occupancies is 100 for A-2 occupancies and 300 for A-1, A-3 and A-4 occupancies.

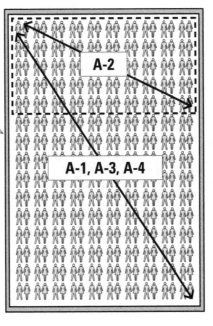

Note that for any of the following Group A occupancies, any one of the specified conditions triggers the requirement for sprinklers. The requirements are:

• A-1: Throughout a fire area containing an A-1 occupancy when one of the following conditions exist:
1. The fire area exceeds 12,000 sf (1115 m²).
2. The fire area has an occupant load of 300 or more.
3. The fire area is located on a floor other than the level of exit discharge.
4. The fire area contains a multitheater complex.

• A-2: Throughout a fire area containing an A-2 occupancy when one of the following conditions exist:
1. The fire area exceeds 5,000 sf (465 m²).
2. The fire area has an occupant load of 300 or more.
3. The fire area is located on a floor other than the level of exit discharge.

• A-3: Throughout a fire area containing an A-3 occupancy when one of the following conditions exist:
1. The fire area exceeds 12,000 sf (1115 m²).
2. The fire area has an occupant load of 300 or more.
3. The fire area is located on a floor other than the level of exit discharge.

• A-4: Throughout a fire area containing an A-4 occupancy when one of the following conditions exist:
1. The fire area exceeds 12,000 sf (1115 m²).
2. The fire area has an occupant load of 300 or more.
3. The fire area is located on a floor other than the level of exit discharge.

• A-5: A sprinkler system is to be provided in concession stands, retail areas, press boxes and other accessory use areas larger than 1,000 sf (93 m²) in A-5 occupancies, such as outdoor stadiums.

• *Group H: §903.2.4 indicates that Group H occupancies should be assumed to require sprinkler systems as a basic design premise.*

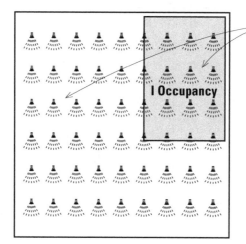

• *Group I: §903.2.5 requires fire-area sprinklers throughout the building wherever there is a Group I occupancy fire area. Note that the sprinklered area may extend beyond the area of the Group I occupancy, depending upon the building configuration.*

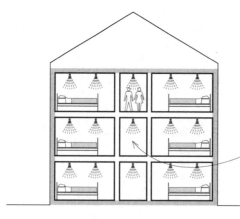

• *Group R-1: §903.2.7 requires an automatic sprinkler system throughout buildings with a Group R fire area.*

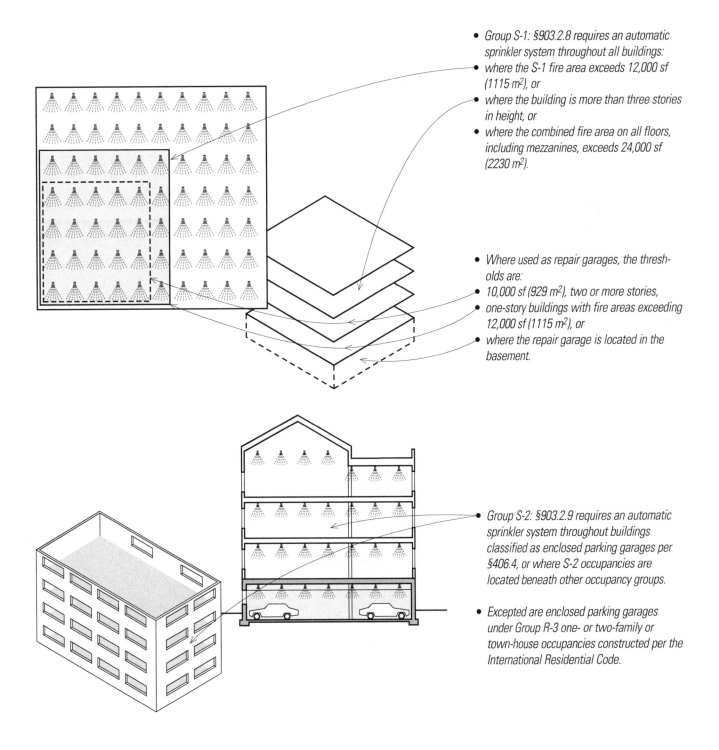

- Group S-1: §903.2.8 requires an automatic sprinkler system throughout all buildings:
- where the S-1 fire area exceeds 12,000 sf (1115 m²), or
- where the building is more than three stories in height, or
- where the combined fire area on all floors, including mezzanines, exceeds 24,000 sf (2230 m²).

- Where used as repair garages, the thresholds are:
- 10,000 sf (929 m²), two or more stories,
- one-story buildings with fire areas exceeding 12,000 sf (1115 m²), or
- where the repair garage is located in the basement.

- Group S-2: §903.2.9 requires an automatic sprinkler system throughout buildings classified as enclosed parking garages per §406.4, or where S-2 occupancies are located beneath other occupancy groups.

- Excepted are enclosed parking garages under Group R-3 one- or two-family or town-house occupancies constructed per the International Residential Code.

AUTOMATIC SPRINKLER SYSTEMS

§903.2.10.1 requires that, in buildings other than R-3 and U occupancies, sprinklers are required at floors below grade or with limited fire-fighting access. Any floor area exceeding 1,500 sf (139.4 m²) must be provided with one type or the other of exterior wall openings noted below, or automatic sprinklers must be provided for the story or basement.

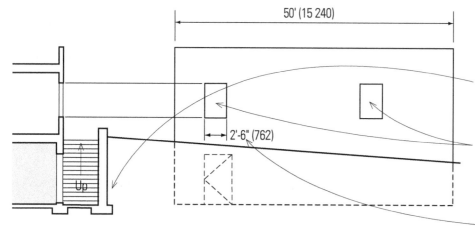

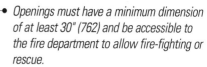

1. Openings below grade must lead directly to the ground floor via a stair or ramp. Openings must be no more than 50' (15 240) apart on at least one side of the space.

2. Openings that are entirely above the adjoining ground level must total at least 20 sf (1.86 m²) for every 50 lineal feet (15 240) of exterior wall.

• Openings must have a minimum dimension of at least 30" (762) and be accessible to the fire department to allow fire-fighting or rescue.

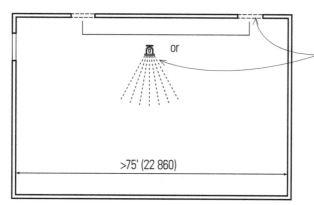

• Where openings in a story are provided on one side only and the opposite wall is more than 75' (22 860) from such openings, the story must be equipped with an automatic sprinkler system, or openings as described in 1 and 2 above must be provided on at least two sides of the story.

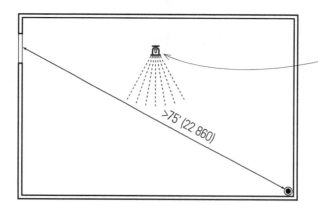

• Where any portion of a basement (even with openings as noted above) is more than 75' (22 860) from openings required by §903.2.10.1, the basement must be equipped with an automatic sprinkler system.

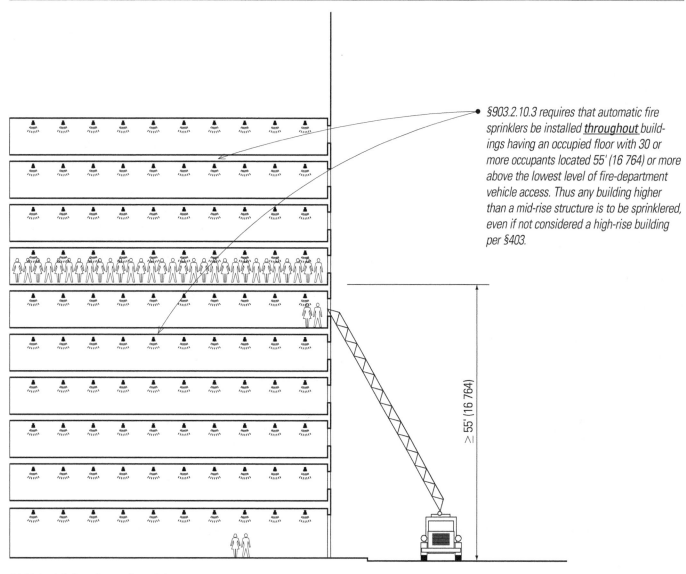

§903.2.10.3 requires that automatic fire sprinklers be installed __throughout__ buildings having an occupied floor with 30 or more occupants located 55' (16 764) or more above the lowest level of fire-department vehicle access. Thus any building higher than a mid-rise structure is to be sprinklered, even if not considered a high-rise building per §403.

≥ 55' (16 764)

§903.2.13 Other Areas Requiring Supression Systems

§903.2.13 refers to Table §903.2.13. This table refers back to §407.5 for sprinkler requirements in I-2 occupancies. This section requires that smoke compartments containing sleeping units are to be equipped with sprinklers per §903.3.1.1. It also requires that these areas have quick response or residential sprinklers in accordance with §903.3.2. These sections expand on the requirements of §407.5. Taken together these sections require quick response or residential sprinklers systems installed per NFPA 13 in accordance with §903..3.1.1 in Group I-1 and Group R dwelling units and in Group I-2 patient sleeping units.

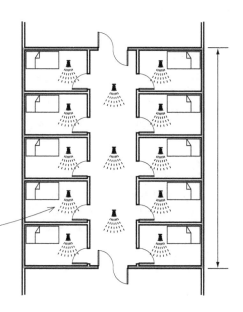

STANDPIPE SYSTEMS

§905 contains the requirements for the installation of standpipe systems. Standpipes are permanent pipes rising through a building that provide hose connections for use in interior fire fighting. Standpipes are classified as Class I, II or III in §902.1 based upon the hose connections they furnish.

- Class I standpipes provide large, 2½" (64) hose connections for use by fire fighters who are trained in the use of the heavy flow of water these connections provide.
- Class II standpipes provide 1½" (38) hose connections that are lower volume and pressure and can conceivably be used either by untrained building occupants or first responders to help fight a fire inside the building.
- Class III standpipe systems provide access to both sizes of connections to allow use by either building occupants or fire fighters with the choice of which connection to use based on the ability and training of the responders.
- Standpipes are to have approved standard threads for hose connections that are compatible with fire department hose threads. These connections should be verified and approved by the Authorities Having Jurisdiction.

Standpipe types are classified as either wet or dry and automatic or manual. Dry standpipes contain air that is displaced by water when they are put to use. Wet standpipes contain water at all times. The use of these is often determined by location. For example, dry standpipes can be on open landings in cold climates, where the water in wet standpipes would freeze.

Standpipe classifications for automatic standpipes presume that the water supply for the standpipe can meet the system demand. Manual standpipes depend on additional pressure and water supply to meet the demand. Hooking up a fire-department pumper truck to the standpipe typically provides such additional supply.

Standpipes are often referred to as fire risers, a term also applied to vertical fire-sprinkler supply lines. Avoid confusion in language by determining to which pipes the term applies.

Standpipes are permitted to be combined with automatic sprinkler systems. Thus a single riser pipe may supply both the sprinkler system and the hose connections. However, such a single pipe will be larger in diameter than two separate pipes, since it must meet the demand of both the sprinkler and hose connection systems. Typically the systems are combined as a single riser, even if somewhat larger in diameter, it is easier to locate in the building than two such risers.

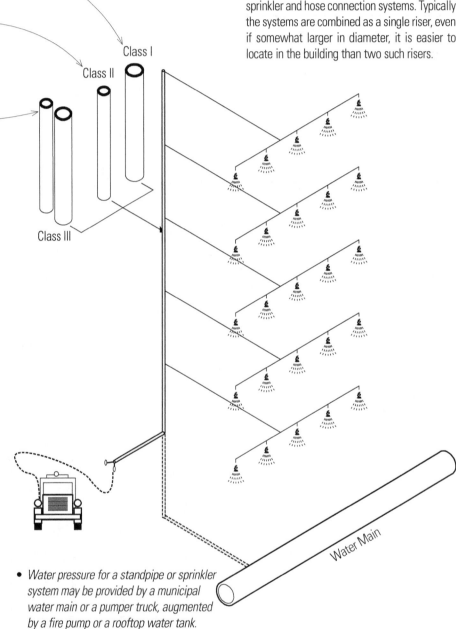

Class I

Class II

Class III

Water Main

- *Water pressure for a standpipe or sprinkler system may be provided by a municipal water main or a pumper truck, augmented by a fire pump or a rooftop water tank.*

Required Installations

Triggers for standpipe requirements have different thresholds depending upon building height, occupancy type, provision of sprinklers and building area.

§905.3.1 requires that buildings having a floor level 30' (9144) above or below the lowest level of fire vehicle access are to have Class III standpipes.

Exceptions to §905.3.1 are:
1. Class I standpipes are allowed in buildings that are fully sprinklered.
2. Class I manual standpipes are allowed in open parking garages where the highest floor is not more than 150' (45 720) above the lowest level of fire-department access.
3. Class I manual dry standpipes are allowed in open parking garages subject to freezing as long as the standpipes meet spacing and location criteria per §905.5.
4. Class I standpipes are allowed in basements that have automatic sprinklers.
5. Determining the lowest level of fire department access does not require consideration of recessed loading docks for four or fewer vehicles and topographic conditions that make fire department vehicle access impractical or impossible.

§905.3.5 requires Class I automatic wet or manual wet standpipes be provided in all underground buildings

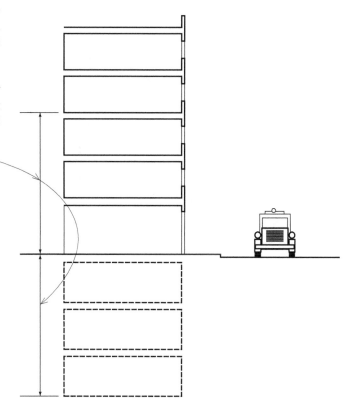

Standpipe Locations

Locations are described for hose connections for the various standpipe classes. Class III standpipes, having components of both other classes must comply with both sets of regulations.

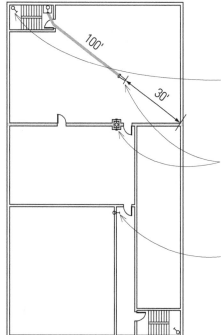

Class I Standpipe Hose Connections

§905.4 requires Class I standpipe hose connections be provided at the following locations:

1. A hose connection is to be located in every required stairway. The hose connections are to be at the intermediate level between floors at every floor level above or below grade.
2. At each side of the wall adjacent to an exit opening in a horizontal exit unless there is a 100' (30 480) hose with a 30-foot (9144) hose stream in an adjacent exit stair that can reach the floor areas adjacent to the horizontal exit.
3. In every exit passageway at the entrance from the exit passageway to other areas of a building.

(list of locations continued on page 130)

Class I Standpipe Hose Connections
(list of locations continued from page 129)

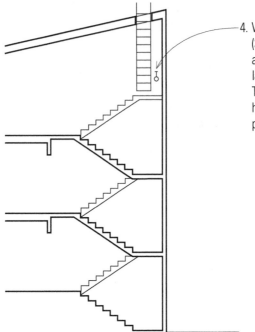

4. Where a roof has a slope of less than 4:12 (33.3% slope), each standpipe is to have a connection on the roof or at the highest landing of stairs with access to the roof. There is also to be a connection at the most hydraulically remote standpipe for testing purposes.

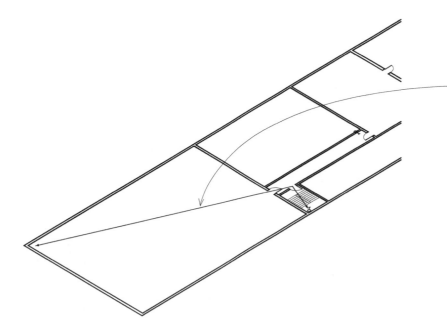

5. The building official is authorized to require additional hose connections where the most remote portion of a nonsprinklered building is more than 150' (45 720) from a hose connection, or the most remote part of a sprinklered building is more than 200' (60 960) from a hose connection.

When risers and laterals of Class I standpipes are not located in an enclosed stairway or pressurized enclosure, §905.4.1 requires the piping to be protected by fire-resistant materials equal to those required in the building for vertical enclosures. Protection is not required for laterals in fully sprinklered buildings.

Class II Standpipe Hose Connections

Where Class II standpipes are required, §905.5 requires the standpipe hose connections to be located so that all portions of the building are within 30' (9144) of a nozzle attached to a 100' (30 480) hose.

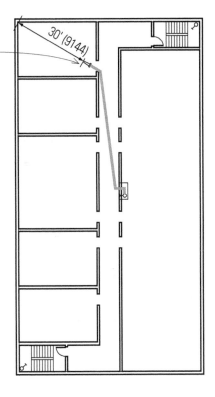

- *Class II standpipe laterals and risers do not require fire-resistance-rated protection.*
- *For light-hazard occupancies, the hose for Class II systems may be minimum 1" (25.4 mm) hose where approved by the building official.*

Class III Standpipe Hose Connections

Class III standpipe hose connections, by definition, have both Class I and II hose connections. Therefore, §905.6 refers back to §905.4 for the location of Class I connections and to §905.5 for the location of Class II hose connections. The laterals and risers are to be protected as for Class I systems per §905.4.1. Where there is more than one Class III standpipe, they are to be hydraulically interconnected at the bottom.

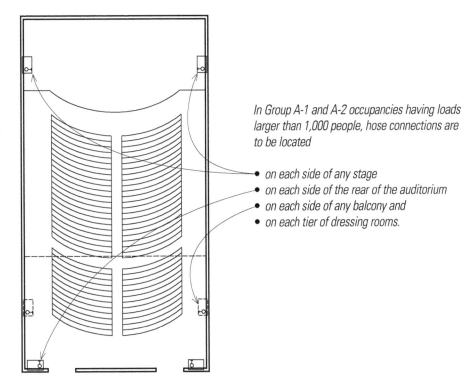

In Group A-1 and A-2 occupancies having loads larger than 1,000 people, hose connections are to be located

- *on each side of any stage*
- *on each side of the rear of the auditorium*
- *on each side of any balcony and*
- *on each tier of dressing rooms.*

FIRE-ALARM & DETECTION SYSTEMS

Fire alarms are audible and visual devices to alert occupants and responders of emergencies. Detectors are automatic devices that usually sound an audible and visual alarm to alert occupants and also trigger other responses in systems such as sprinklers, smoke-control systems or HVAC controls. Systems to notify building occupants of an emergency can be either manually or automatically actuated.

The Code requires differing types of alarms or detectors, and different levels of actuation in various occupancies and types of buildings. When buildings are sprinklered and the sprinklers connected to the alarm system, automatic heat detection is not required. Sprinklers are heat-actuated and thus are a type of heat detector themselves. Fire detectors are to be smoke detectors as well, except in areas like boiler rooms where products of combustion would tend to set off smoke detectors. In those rooms alternative types of fire detectors may be used. Alarms are to be per NFPA 72.

The Code offers trade-offs between manual alarms and automatic detection systems under certain conditions. There are instances where having manual alarms may result in a large number of false alarms, so providing automatic detection may be a desirable alternative.

Detailed requirements are grouped by occupancy or use. The table on the following pages gives general requirements; it does not provide for every nuance of conditions.

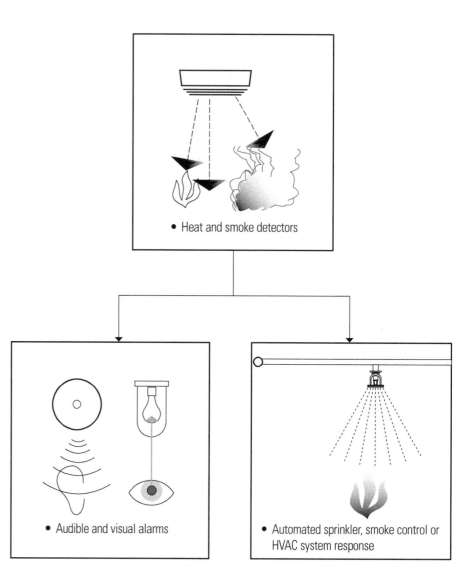

- Heat and smoke detectors

- Audible and visual alarms

- Automated sprinkler, smoke control or HVAC system response

Section 907.2.x	Occupancy or Building Type	Occupant Load or Condition	Device Type	Exceptions	Notes
.1	A	>300	Manual fire alarm	Not required if sprinklered and w/ water-flow alarm	E assembly to be per Group E requirements
.1.1	A	>1,000	Voice/alarm communication system		Alarm on emergency power
.2	B	>500, or >100 above/below exit discharge	Manual fire alarm	Not required if sprinklered and w/ water-flow alarm	
.3	E	>50	Manual fire alarm	Not required if meets long criteria list for detection or if the building is fully sprinklered with automatic notification to a normally occupied location	When installed, sprinklers and smoke detectors must interconnect to alarm system
.4	F	>2 stories >500 above/below exit discharge	Manual fire alarm	Not required if sprinklered and w/ water-flow alarm	
.6	I-1, I-2	All	Manual fire alarm	Not required in resident or patient sleeping rooms if at constantly attended, visible and accessible staff stations	
.6.1	I-1	Corridors, habitable spaces open to corridor	Automatic smoke detection	Not required if fully sprinklered and not required for exterior balconies	
.6.2	I-2, nursing homes	Corridors in nursing homes	Automatic fire detection		Nursing home includes both intermediate care and skilled nursing
.6.2	I-2 hospital		Smoke detection per §407.2	Not required if smoke compartments containing rooms have smoke detectors or auto closing doors	
.6.3	I-3	All	Manual and auto fire alarm for alerting staff		Alarms to alert staff
.6.3.3	I-3	Housing area	Smoke detectors	Alternate detector location and size limits for small groups	Not required for sleeping units with 4 or fewer occupants in sprinklered smoke compartments

FIRE-ALARM & DETECTION SYSTEMS

Section 907.2.×	Occupancy or Building Type	Occupant Load or Condition	Device Type	Exceptions	Notes
.8	R-1	All	Manual fire alarm and auto fire detection	1. No alarm if < 2 stories with 1-hour separations and direct exit 2. No alarm if sprinklered and local alarm and one manual pull box at approved location 3. Automatic alarm at interior corridors serving guest rooms 4. Guest room smoke alarms	
.9	R-2	1. 3 or more stories 2. Dwelling units >1 level below exit discharge 3. > 16 Dwelling units	Fire-alarm system (not clear if manual or automatic)	1. Not required if < 2 stories with 1-hour separations and direct exit 2. No alarm if sprinklered and local alarm	
.10	R Occupancies		Smoke alarms		Where indicated below
.10.1.1	R-1	1. Sleeping areas 2. Every room in means of egress path 3. Each story in multistory units			
.10.1.2	R-2, R-3, R-4, I-1, regardless of occupant load	1. Near each separate sleeping area 2. In each sleeping room 3. In each story of dwelling units			
.10.1.3	I-1	At sleeping areas	Single or multiple station smoke alarms		Smoke alarms not required when building has automatic fire detection system throughout per §907.2.6
.10.2			Smoke-alarm power		From electrical source w/battery backup in new construction; battery power OK in retrofit
.10.3	R-2, R-3, R-4, single guestroom in R-1		Smoke alarm interconnection	Not required for solely battery-powered alarms	Interconnect so one alarm activation will sound all alarms in the unit; audible with doors closed

Section 907.2.×	Occupancy or Building Type	Occupant Load or Condition	Device Type	Exceptions	Notes
.12.1	High-rise buildings		Automatic alarm, smoke detectors	In I-1 and I-2 occupancies the alarm is to sound in a constantly attended area and a general occupant notification delivered over a voice paging system	Voice/alarm system and smoke detectors at: 1. Mechanical, electrical, elevator equipment rooms, elevator lobbies 2. Main return air and exhaust air at HVAC 3. Vertical duct risers > 2 stories
.12.2	High-rise buildings		Emergency voice/alarm communication system		Activation of detector, fire flow or manual pull station will sound tone and activate alert instructions per International Fire Code at: 1. Elevator lobbies 2. Corridors 3. Rooms/tenant spaces > 1,000 sf (93 m²) 4. Dwelling units in R-2 occupancies 5. Guest rooms in R-1 occupancies 6. Areas of refuge per this Code
.12.3	High-rise buildings		Fire-department communication	Fire-department radios may be used in lieu of two-way communication when approved by the fire department	Two-way fire-department communication per NFPA 72 from central command to elevators, elevator lobbies, emergency power rooms, fire-pump rooms, areas of refuge and on each floor inside enclosed stairways
.13	Atrium connecting 2 or more stories			Fire alarm activated per §907.6 by automatic alarm, fire flow or manual alarm	Emergency voice/alarm system in Group A, E or M occupancies per §907.2.12.2

FIRE-ALARM & DETECTION SYSTEMS

Section 907.2.×	Occupancy or Building Type	Occupant Load or Condition	Device Type	Exceptions	Notes
.14		High-piled combustible storage	Automatic fire-detection system		International Fire Code
.15		Delayed egress locks	Automatic smoke- or heat-detection system		When devices are installed per Chapter 10
.18		Underground buildings with smoke exhaust	Automatic		Similar conditions to high-rise buildings; also activation of smoke exhaust activates audible alarm in attended location.
.19	Underground buildings > 60 below lowest exit discharge		Manual fire alarm		Emergency voice/alarm system per §907.2.12.2.
.20	Covered mall buildings > 50,000 sf (4645 m²)		Emergency voice/alarm system per §907.2.12.2		System to be accessible to fire department
.23	Battery rooms, lead-acid batteries, capacity > 50 gal. [189.3 L]		Automatic smoke-detection system		System to be supervised by central, proprietary or remote station with alarm at attended location

Manual Alarm Boxes

§907.3 requires manual alarm boxes be located no more than 5' (1524) from the entrance to each exit. Travel distance to the nearest box is not to exceed 200' (60 960).

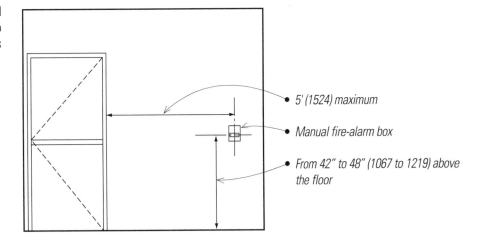

- 5' (1524) maximum
- Manual fire-alarm box
- From 42" to 48" (1067 to 1219) above the floor

Visual Alarms

§907.9.1 requires visual alarms be provided to notify persons with hearing impairments of alarm conditions. These are white strobe lights mounted on the wall of certain areas. Accessibility criteria require they be located at least 80" (2032) but no more than 96" (2438) above the floor.

Alarms having both visible and audible functions are to be provided in the following locations:
1. Public areas and common areas
2. Make provisions for visual alarms to be added in employee work areas
3. I-1 and R-1 sleeping accommodations per the quantities noted in Table 907.9.1.3.
4. R-2 occupancies that are required by §907 to have a fire-alarm system. (All dwelling units are required to be adaptable to accommodate visual alarm appliances per accessibility requirements of ICC/ANSI A117.1.)

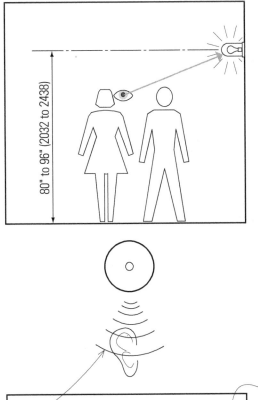

80" to 96" (2032 to 2438)

Audible Alarms

§907.9.2 requires audible alarms to have a distinctive sound not used for any other purpose. They are to be loud enough to provide a sound-pressure level at least 15 decibels (dBA) above the ambient sound or 5 dBA above the maximum sound level. The minimum sound-pressure level in R-1 and I-1 occupancies is to be 70 dBA.

Per §907.14 fire alarm systems are to be monitored by an approved supervising station in accordance with NFPA 72., Smoke detectors in I-1 occupancies where §907.2.10 is met, or those in I-3 occupancies, are not required to be supervised.

- Visible alarm notification appliances may be used in lieu of audible alarms in critical-care areas of Group I-2 occupancies. This recognizes that having an audible alarm going off in a critical-care area could have severe detrimental effects on patients.

SMOKE-CONTROL SYSTEMS

The purpose of smoke-control systems as stated in §909 is to provide a tenable environment for the evacuation or relocation of occupants. The provisions are not intended for the preservation of contents, the timely restoration of operations, or for assistance in fire suppression or overhaul activities.

Smoke-control systems are classified as active fire-resistance systems for our discussion in that they actively perform their basic function for life safety. They respond to fire not by their presence as barriers to fire or heat but by activating a sequence of operations to safeguard the building's occupants. These systems are referred to in the code as either of the active or passive type. This refers to whether the systems exhaust smoke through natural convection or by the use of mechanical ventilation. All smoke-control systems rely on automatic activation, whether the exhaust mechanisms are passive or active.

The systems are provided in certain building types such as malls or atriums to contain or evacuate smoke to allow building occupants to leave areas where smoke might hinder their egress. Buildings with smoke-control systems are typically those having large areas with interconnected air spaces where smoke cannot be contained by barriers but must be moved or exhausted for occupant protection. The design criteria for smoke-control systems require detailed calculations and modeling. They are almost invariably designed with the aid of a consultant experienced in design and construction of such systems.

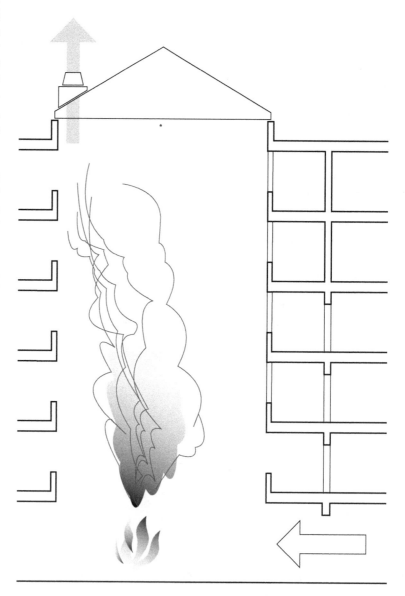

Smoke Barriers

Per §909.5 smoke barriers are to comply with §709. Openings in smoke barriers are to have automatic closing devices actuated by the smoke control system detectors. Doors are to be protected per §715.4.3 which typically requires 20-minute rated doors in smoke-barrier walls. An exception to §909.5.2 allows the use of opposite swinging doors as are often found in I-2 hospital occupancies. There is also an exception for doors in smoke barriers in I-3 occupancies recognizing that these are highly supervised occupancies where staff typically performs the functions of automatic systems when building occupants are under restraint.

§909.20 requires that smoke-proof enclosures required by §1020.1.7 consist of an enclosed interior stairway accessed by way of an outside balcony or a ventilated vestibule. This requirement refers back to the requirements of §403 and 405 for high-rise buildings and underground buildings.

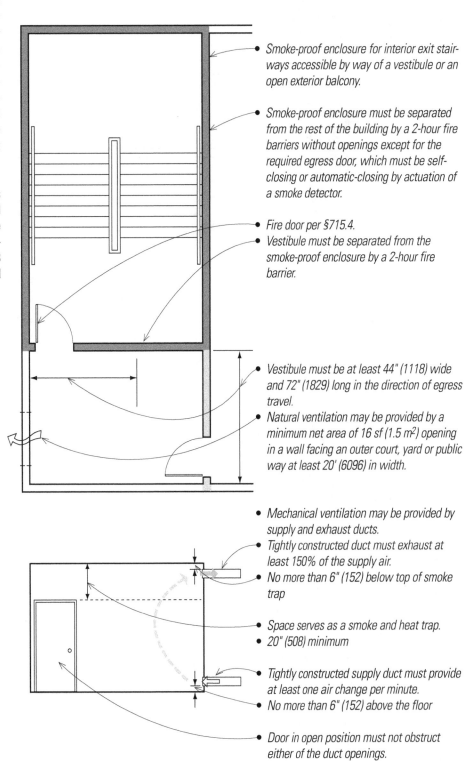

- *Smoke-proof enclosure for interior exit stairways accessible by way of a vestibule or an open exterior balcony.*

- *Smoke-proof enclosure must be separated from the rest of the building by a 2-hour fire barriers without openings except for the required egress door, which must be self-closing or automatic-closing by actuation of a smoke detector.*

- *Fire door per §715.4.*
- *Vestibule must be separated from the smoke-proof enclosure by a 2-hour fire barrier.*

- *Vestibule must be at least 44" (1118) wide and 72" (1829) long in the direction of egress travel.*
- *Natural ventilation may be provided by a minimum net area of 16 sf (1.5 m2) opening in a wall facing an outer court, yard or public way at least 20' (6096) in width.*

- *Mechanical ventilation may be provided by supply and exhaust ducts.*
- *Tightly constructed duct must exhaust at least 150% of the supply air.*
- *No more than 6" (152) below top of smoke trap*

- *Space serves as a smoke and heat trap.*
- *20" (508) minimum*

- *Tightly constructed supply duct must provide at least one air change per minute.*
- *No more than 6" (152) above the floor*

- *Door in open position must not obstruct either of the duct openings.*

§910 covers the requirements for smoke and heat vents, which have functions similar to smoke-control systems. They typically use simpler technology to actively respond to fire conditions. Smoke and heat vents allow products of combustion or explosion to vent from the building and minimize the damage they can cause.

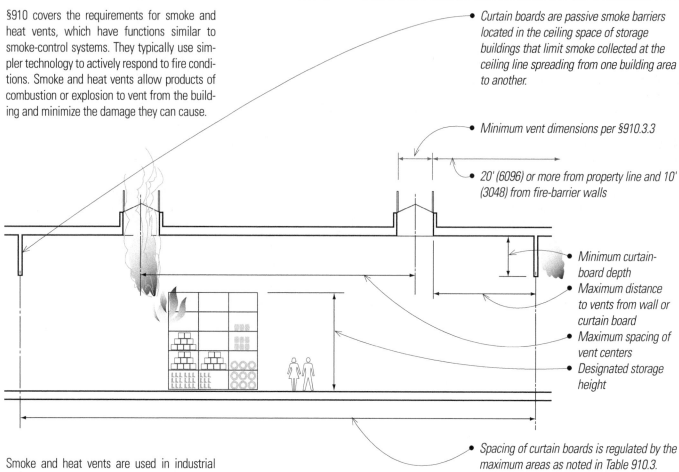

- Curtain boards are passive smoke barriers located in the ceiling space of storage buildings that limit smoke collected at the ceiling line spreading from one building area to another.

- Minimum vent dimensions per §910.3.3

- 20' (6096) or more from property line and 10' (3048) from fire-barrier walls

- Minimum curtain-board depth
- Maximum distance to vents from wall or curtain board
- Maximum spacing of vent centers
- Designated storage height

- Spacing of curtain boards is regulated by the maximum areas as noted in Table 910.3.

Smoke and heat vents are used in industrial and commercial occupancies where high-piled storage, manufacturing or warehousing activities have high fuel loads or large quantities of hazardous materials. There is a potential for large catastrophic fires in such occupancies. The occupancy groups covered by this section are Group F-1 and S-1 occupancies larger than 50,000 sf (4645 m²) in undivided area as well as in buildings with high-piled combustible storage per §413 of the International Fire Code. These uses are unlikely to occur in healthcare occupancies other than in very large facilities which incorporate central storage. This reference material is included for general information.

Smoke and heat vents are to operate automatically, either tied to sprinkler activation in sprinklered buildings or by response to heat in unsprinklered buildings. Engineered smoke-exhaust systems, similar in principle to a smoke-control system may be substituted for prescriptive systems if approved by the building official.

Table 910.3 shows the requirements for curtain board spacing and depth. The table also shows detailed requirements for vent size and spacing. The requirements are based on the hazard of the materials stored or used in the building as determined by the Commodity Classification per the International Fire Code. The areas formed by curtain boards, or by fire-resistive walls, are a factor in the requirements. Also considered is the height of the material-storage configuration. Thus higher-hazard materials or higher storage racks will generate requirements for more vents and/or closer spacing of curtain boards.

10
Means of Egress

Chapter 10 of the International Building Code contains the requirements for designing exiting systems known as "means of egress." The fundamental purpose for a means of egress is to get all of the occupants out of a building in a safe and expeditious manner during a fire or other emergency. A means of egress must, therefore, provide a continuous and unobstructed path of exit travel from any occupied point in a building to a public way, which is a space such as a street or alley, permanently dedicated for public use.

The term *means of egress* is a general one describing ways of getting out of a building. Websters Third New International Dictionary of the English Language defines egress as "1. the act or right of going or coming out." This term has been chosen to encompass all system components used for getting people out of a building in an emergency. The emphasis is on emergency egress, not convenience or function. However, access for persons with disabilities is also a consideration in designing a means of egress.

In the definitions contained in this chapter, such common terms as *exit access*, *exit*, and *exit discharge* take on more specific, code-defined meanings as components of a means of egress, which is the complete system for escape. Pay close attention to code-specific nomenclature as you familiarize yourself with this chapter.

MEANS OF EGRESS

§1002 defines "means of egress" as a continuous, unobstructed path of vertical and horizontal exit travel from any occupied portion of a building or structure to a public way. The means of egress for every occupancy in a building must meet the provisions of Chapter 10. The requirements are driven by the occupancy type of the area to be provided with egress and the occupant load calculated per Table 1004.1.1.

A building may contain several different types of occupancies. The designer must consider all of the uses, present or contemplated, so that adequate numbers and widths of exits are provided. The occupant load is typically calculated as if each area is occupied at the same time to determine the maximum possible number of occupants to be accommodated by the means of egress. The code presumes that means of egress components themselves are generally not occupied spaces but serve the occupants of the building. There are specific and narrow exceptions to the requirements for calculating occupant load that will be discussed later.

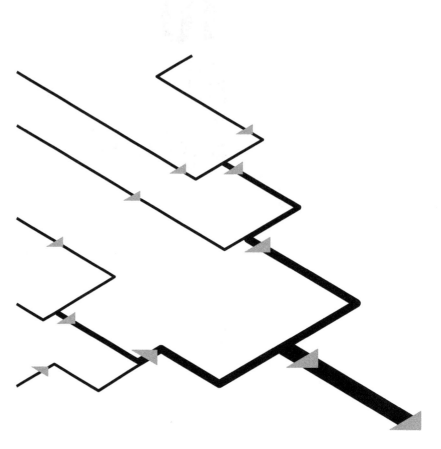

Egress Concepts

The general objective of egress, that of providing a continuous, unobstructed, and protected path to the outside of a building, is fundamental to the Code and must be thoroughly understood. It must also be understood that when the egress path from a major occupancy of the building passes through an area containing another occupancy, possibly a public assemblage or a garage area, the egress path retains the occupancy-class designation and fire-protection requirements of the main use. Protection should never be reduced as the occupants proceed downstream in the means-of-egress system.

Flow

- There is a direction of flow to every means of egress, leading from an occupied space to a final safe outside area, away from any hazard or emergency in the building. The best analogy for means-of-egress design is that of watercourses. Small streams lead to creeks, which feed rivers, which then lead to the sea. As rivers get larger and carry more water as they go downstream as tributaries feed into them, so too should the conceptual design of a means of egress. Exit paths must remain the same size or get larger as the building occupants proceed downstream from areas of lesser safety to areas of greater safety and ultimately out of the building through the exit discharge.

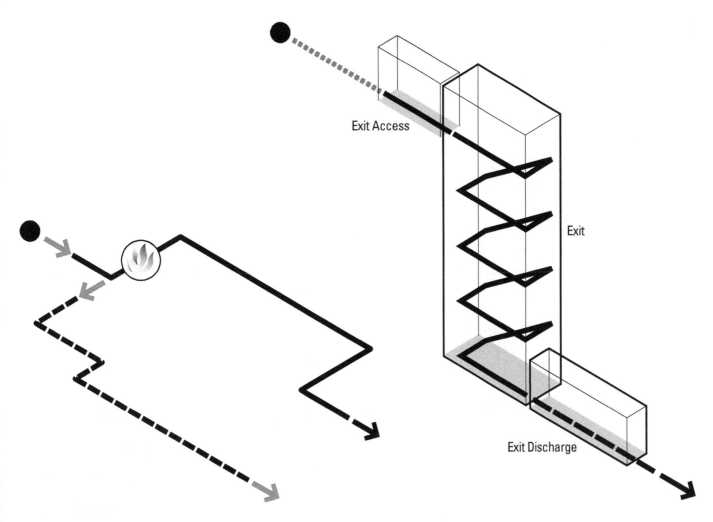

Exit Access

Exit

Exit Discharge

Alternative Paths

- Another basic exiting principle is that at least two different egress paths should lead from the interior of a building to the outside at ground level. The rationale is that if one path is blocked or endangered by a hazard, then the other path will be available. There are exceptions to this rule for situations involving relatively small numbers of occupants, but the designer should always begin with the basic design concept that two ways out of each space should typically be provided in case of an emergency.

Protection

- Various sections of the egress path are required to provide fire-resistive protection to the occupants. These components begin at the access door to a corridor, then lead to an exit, extend in the exit enclosure down through the building, and then to the exit discharge leading to the exterior. Note that often corridors will not be required to be of fire-resistive construction.

  Once in a rated corridor or exit stairway, the occupant is to remain protected until the building exterior is reached. In the process of exiting, the occupant should not be required to move out of a protected environment into adjacent occupied spaces or unprotected egress paths to get to the next portion of the protected egress path or to the exterior, because doing so would necessitate removing the protection the exit enclosure is required to provide.

EGRESS COMPONENTS

There are three basic components to a means of egress: the exit access, the exit and the exit discharge. It is important to clearly understand how the Code defines these components so that one can apply the requirements of Chapter 10.

It is important that components of a means of egress be integrated with the requirements for accessible means of egress contained in Chapter 11. Some provisions of Chapter 11 or of the Americans with Disabilities Act (ADA) may supersede those of Chapter 10. The egress requirements of Chapter 10 must always be considered together along with the ADA and local accessibility regulations.

1. EXIT ACCESS is that portion of the means-of-egress system that leads from any occupied portion in a building or structure to an exit.

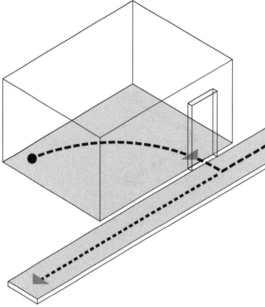

Exit access is the part of the building where the occupants are engaged in the functions for which the building was designed and are thus the spaces to be escaped from in an emergency. Although the exit access allows the occupants the most freedom of movement, it also offers the lowest level of life-safety protection of any of the components of the means of egress.

In the typical exit path from a room or space one may encounter aisles, passages, corridors or other rooms before gaining access to a protected fire-rated enclosure called the exit. The distance to be traversed is limited by the Code as will be covered in this analysis.

- *It is important to remember that the distance one may travel in an exit access, from the most remote point in the room or space to the door of an exit, is regulated by the Code. It may be necessary to provide a fire-rated passage, such as an exit passageway, for larger floor areas that may exceed the travel distances for various other means-of-egress components.*
- *Travel distances are not restricted in exits or in the portion of the exit discharge located at grade.*

2. EXIT is that portion of the means-of-egress system between the exit access and the exit discharge or the public way.

The exit portion of the means of egress allows the occupants of a building to move through a protected enclosure in a protected environment from the area where a hazardous event is occurring to a place where they may finally escape the building. Exits are therefore separated from other interior spaces of a building or structure by fire-resistance-rated construction and opening protectives as required to provide a protected path-of-egress travel from the exit access to the exit discharge.

Exits include exit enclosures, exit passageways, exterior exit stairs, exterior exit ramps, horizontal exits and exterior exit doors at ground level. The distance one may travel within a protected exit enclosure is not limited by the Code. This part of the egress path can be very long, as in the case of the stairways in a high-rise building.

3. EXIT DISCHARGE is that portion of the means-of-egress system between the termination of an exit and a public way. The exit discharge may include exterior exit balconies, exterior stairs, exterior exit ramps, exit courts and yards. Many of these are building-related components, but an exit discharge may also include site elements.

This portion of the means of egress is basically assumed to be at or near grade and open to the atmosphere. The occupants are able to clearly see where an area of safety outside the building lies and are able to move toward it.

The Code defines "public way" as any street, alley or other parcel of land open to the outside air leading to a street that has been deeded, dedicated or otherwise permanently appropriated to the public for public use and having a clear width and height of not less than 10' (3048).

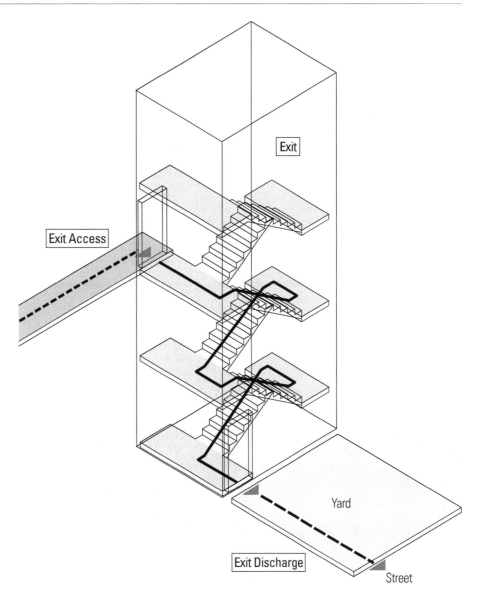

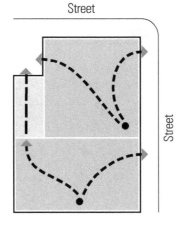

- *The means of egress for a small single-story building is usually simple, because the second and third components of the exit path are often combined.*
- *In many one-story buildings, such as retail stores and banks, only the first portion of the means of egress exists. A corridor may extend to the exterior wall and open onto a street, yard or other public space. This simultaneously provides the exit access, the exit and the exit discharge to the exterior public way of the building at ground level. The room or space opens directly to the building's exterior without the need of protected corridors or stairways.*

GENERAL EGRESS REQUIREMENTS

§1003 through §1013 specify the requirements that apply to all three portions of the means of egress. Thus, requirements for doors, stairs, ramps and similar egress components apply to all three parts of the means of egress when included in these sections. Note also that there may be additional modifications to conditions in this section contained in the specific sections applicable to the three parts of a means of egress.

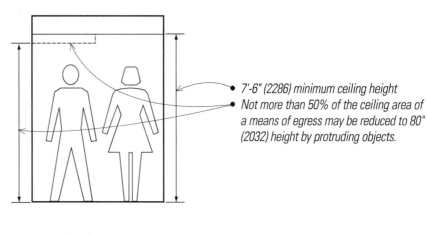

- *7'-6" (2286) minimum ceiling height*
- *Not more than 50% of the ceiling area of a means of egress may be reduced to 80" (2032) height by protruding objects.*

Ceiling Height

§1003.2 requires that egress paths have a ceiling height of not less than 7'-6" (2286), although it is acceptable to have some projections that reduce the minimum headroom to 80" (2032) for any walking surface.

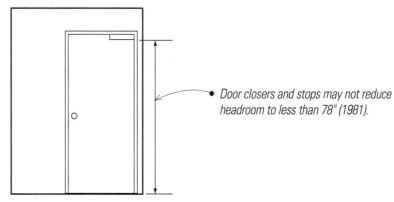

- *Door closers and stops may not reduce headroom to less than 78" (1981).*

Protruding Objects

§1003.3 governs how much objects may project into the required ceiling height and required egress widths.

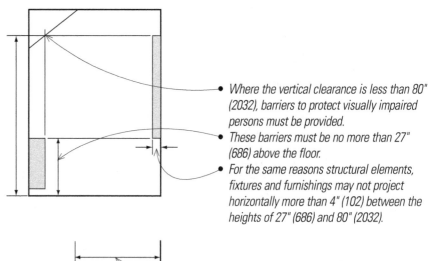

- *Where the vertical clearance is less than 80" (2032), barriers to protect visually impaired persons must be provided.*
- *These barriers must be no more than 27" (686) above the floor.*
- *For the same reasons structural elements, fixtures and furnishings may not project horizontally more than 4" (102) between the heights of 27" (686) and 80" (2032).*

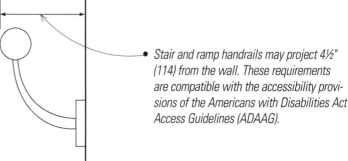

- *Stair and ramp handrails may project 4½" (114) from the wall. These requirements are compatible with the accessibility provisions of the Americans with Disabilities Act Access Guidelines (ADAAG).*

Elevation Changes

§1003.5 restricts elevation changes in the means of egress. If the elevation change is less than 12" (305), it should be sloped.

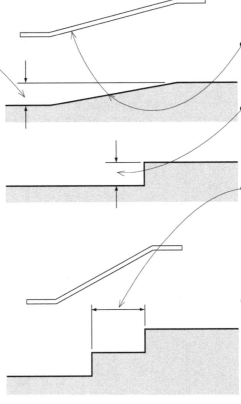

Maintaining Width of the Means of Egress

§1003.6 emphasizes that when objects, such as brackets or columns, occur in a means of egress, they shall not decrease the required width of the means of egress, Where these obstructions or projections do occur, additional width is needed to maintain the required egress width.

Elevators, Escalators and Moving Walks

§1003.7 does not permit any of these modes of transportation to be used as components of a means of egress. The only exception is for elevators per §1007.4, where they are provided with emergency power and also with operation and signal devices per §2.27 of ASME A17.1.

- *When the slope exceeds 1 in 20, then the transition should be made by an accessible ramp compliant with the provisions of Chapter 11.*
- *Slopes rising less than 6" (152) must be equipped with handrails or a floor finish that contrasts with adjacent flooring.*

- *In aisles serving seating areas not required to be accessible by Chapter 11, exceptions allow for single steps with a maximum riser height of 8" (178) per §1025.11 and provision of a handrail per §1025.13 or,*
- *a stair with up to two risers, complying with §1009.3, having a minimum tread depth of 13" (330) and one handrail complying with §1012, and within 30" (762) of the centerline of the normal path-of-egress travel on the stair.*

- *This section requires that grade changes in any I-2 occupancy serving non-ambulatory persons must be made by means of a ramp (slope between 1:20 and 1:12) or sloped walkway (slope greater than 1:20). It seems prudent that designers of I-2 occupancies assume that this criteria would apply in essentially all parts of such occupancies.*

- *It is not recommended to use single steps or short stairways in any passageway, as such steps may be overlooked by occupants and present a severe tripping hazard in daily use. They are to be avoided whenever possible.*

- *Maintain minimum width of egress travel around obstructions.*

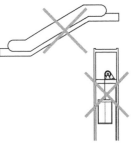

Design Occupant Load

§1004.1 specifies that means-of-egress facilities for a building are to be designed to accommodate the number of occupants as computed in accordance with §1004.1.1 and Table 1004.1.1 The occupant load is to be based on actual occupant loads when it can be determined, as for the number of seats in a theater, or by using Table 1004.1.1, where occupant-load factors are assigned to determine occupant loads based upon use. These factors are based on past history of anticipated occupant loads for various uses. The code does not state requirements regarding rounding. We suggest rounding all fractional calculations up to the next whole number in typical cases.

The floor-area allowances in the table are based not on the actual physical situation within any building but for code purposes, on probabilities and observation. The values are codified so that they can be used uniformly by everyone in determining the exit details; i.e., the number and width of the exit components. They are not intended as design program guidelines. It is not necessary that every office worker have a 100 sf (9.3 m2) office even though the occupant load factor is 100.

Per §1004.2 occupant loads may be increased beyond those listed in Table 1004.1.1 as long as the means of egress system in sized to accommodate the occupant load. In no case may the occupant load calculated under this provision exceed one occupant per 7 sf (0.65 m2) of occupiable floor area.

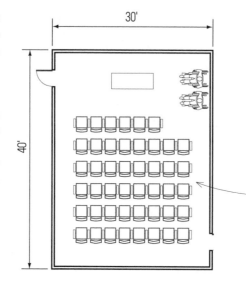

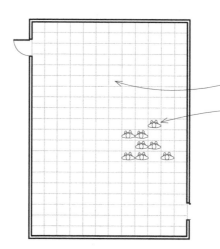

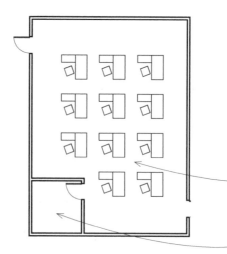

Actual Number

- *§1004.1.1 states that the design occupant load shall be based on the number of occupants determined by assigning one occupant per unit of area from Table 1004.1.1. For areas with fixed seating the actual number of fixed seats in a theater will determine the occupant load for that space, per §1004.7.*
- *Per §1004.7 occupant loads for areas such as waiting areas or wheelchair areas for persons with disabilities are to be calculated using Table 1004.1.1.*
  - *30' × 40' (9144 × 12 192) room*
  - *Fixed seating for 46 occupants*
  - *No standing room, fixed walls*
  - *Two wheelchair locations for accessibility*
  - *Occupant load = 48*
  - *This example is based upon §1004.7 as referenced in Table 1004.1.1.*

Number by Table 1004.1.1

- *§1004.1 uses a table to prescribe the design occupant load, computed on the basis of one occupant per unit of area.*
  - *30' × 40' (9144 × 12 192) room*
  - *Standing assembly area*
  - *From Table 1004.1.1, allow 5 sf (0.46 m2) of floor area per occupant.*
  - *Occupant load = 1200/5 = 240*

- *Note that the words **net** and **gross** used in the table are defined in terms of code usage in §1002. It is expected that the interpretation of these terms will be that **net** refers to the actual area where occupants may stand, less the space occupied by building elements and not including accessory areas such as corridors, stairways or toilet rooms. **Gross** would be expected to apply to large areas defined by the perimeter walls, excluding vent shafts.*
- *Note also that there are no occupant-load factors for such service spaces as restrooms where it can be anticipated that the users of those spaces are already occupants of the building. It is anticipated that these spaces will be considered a part of the gross area per the discussion above.*
  - *30' × 40' (9144 × 12 192) room*
  - *Business area*
  - *From Table 1004.1.1, allow 100 sf (9.29 m2) of floor area per occupant.*
  - *Deduct store room area 10' x 10' = 100 sf (9.29 m2)*
  - *Occupant load = (1200-100)/100 = 11*

Number By Combination

§1004.1 states that occupants from accessory spaces who exit through a primary area are to be added to the design occupant load. This section addresses such areas as conference rooms or alcoves, and adheres to the principle that each tributary area adds to the occupant load "downstream" of the areas being served by a means of egress.

Example One

Seating Area
- *Assembly without fixed seats Concentrated; chairs only*
- *15' × 15' (4572 × 4572)*
- *From Table 1004.1.1, allow 7 sf (0.65 m^2) of floor area per occupant*
- *Occupant load = 225/7 = 32*

Conference Room
- *Assembly without fixed seats Unconcentrated; table and chairs*
- *15' × 20' (4572 × 6096)*
- *From Table 1004.1.1. allow 15 sf (1.39 m^2) of floor area per occupant.*
- *Occupant load = 300/15 = 20*
- *Because the occupant load is less than 50, only one egress door is required per Table 1015.1.*

Business Area
- *From Table 1004.1.1 allow 100 gross sf (9.29 m^2) of floor area per occupant.*
- *Floor area = (30 × 40) − (15 × 20) = 900*
- *Occupant load = 900/100 = 9*

- *Assuming a Group B occupancy. Table 1014.1 requires at least two means of egress because the occupant load is greater than 50.*
- *The egress width for each door is based on an occupant load of 31. [Rounding up 61 ÷ 2 = 30.5 to 31, which should always be done.]*

Total Design Occupant Load = 32 + 20 + 9 = 61

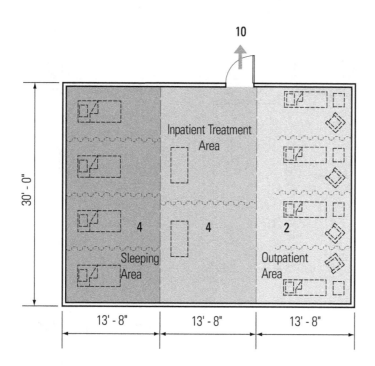

Example Two, Healthcare

Healthcare occupancies are often arranged in suites of rooms. Table 1004.1.1 calls out different floor areas for various uses for "Institutional areas":

- Inpatient treatment rooms: 240 gross square feet
- Outpatient areas: 100 gross square feet
- Sleeping areas: 120 gross square feet

The occupant load, and subsequently the egress widths necessary for a hypothetical suite of rooms in a 30' x 40' space (9144 x 12 192) are:

Sleeping Area

- *30' x 13"-8" (9144 x 4064) = 400 sf (37.16 m²)*
- *From Table 1004.1.1, allow 120 sf (11.15 m²) of floor area per occupant*
- *Occupant load = 400/120 = 3.33, assume 4 occupants (Occupant loads should be whole numbers for design. Occupant loads should be rounded up on a room-by-room basis, not at the end of a summary. Generally it is good practice to round up fractions to the next highest integer to be conservative and to allow user flexibility over the life of the space without exceeding design occupant load.)*

Inpatient Treatment Area

- *30' x 13"-8" (9144 x 4064) = 400 sf (37.16 m²)*
- *From Table 1004.1.1, allow 1000 sf (92.9 m²) of floor area per occupant*
- *Occupant load = 400/100 = 4 occupants*

Outpatient Area

- *30' x 13"-8" (9144 x 4064) = 400 sf (37.16 m²)*
- *From Table 1004.1.1, allow 240 sf (22.3 m²) of floor area per occupant*
- *Occupant load = 400/240 = 1.6, assume 2 occupants*

The total design occupant load for the space is thus:

- *Sleeping area:*   4
- *Inpatient treatment:*   2
- *Outpatient area:*   4
- *Total design occupant load:*   10

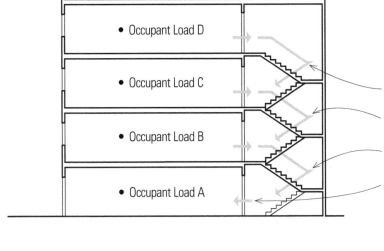

Maximum Occupant Load

75
persons

Sign to remain posted

- *§1004.3 requires assembly occupancies to have signs posted in a conspicuous place showing the occupant load. This is to prevent overloading of the spaces, which could impair exiting in a panic situation.*

- *§1004.4 states that the accumulation required by §1004.1 does not apply to occupant loads from levels above or below the level in question as long as the exit capacity does not decrease in the direction of egress travel. This is based upon observations of the time factor, along with their directional nature, in determining how people move through egress systems. The occupants of the floor below are assumed to exit that level before those behind them get to the same point in the egress system.*

- *Assuming A, B, C and D are approximately of equal size and occupant load:*

- Occupant Load D

- Occupant Load C

- Occupant Load B

- Occupant Load A

Egress Capacity D

Egress Capacity C

Egress Capacity B

Egress Capacity A

- Egress width A + B

- Egress width A

- Egress width B

- *Note also that §1004.4 states that egress paths shall not decrease in exit capacity in the direction of egress travel. This corresponds to the watercourse analogy as well. This is further supported by §1004.5, which requires that when egress paths merge, the capacity of the egress path serve both tributary areas.*

GENERAL EGRESS REQUIREMENTS

§1004.6 applies the same principle of convergence as §1004.5 to the means of egress from a mezzanine, based on the assumption that the occupants of the mezzanine must pass through the floor below to get to a common set of exit paths and thus are added to the occupant load.

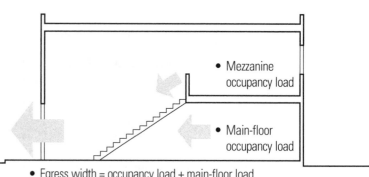

- Mezzanine occupancy load
- Main-floor occupancy load
- Egress width = occupancy load + main-floor load

§1004.8 states that outdoor areas, such as a dining patio outside of a hospital cafeteria, are to be included in occupant-load calculations and provided with exits. When it can be anticipated that the occupants of an outdoor area would be in addition to those occupants inside the building, this occupant load must be added to the egress-capacity calculations. When the occupants can be expected to be either inside or outside, but not both, the capacity need not be additive.

The designer should make assumptions about uses of spaces with great caution. For example, even when a courtyard serves an office area for the office occupants only, it is conceivable that there could be a public event in the office that would have guests and employees occupying both spaces. Such special events with crowded conditions can be the most conducive to panic in an emergency, and egress systems should take such possibilities into account.

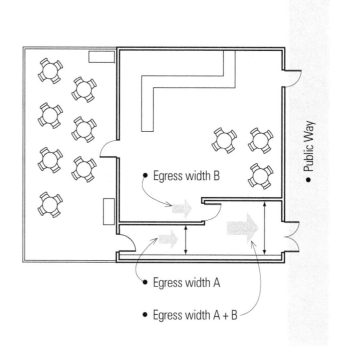

- Egress width B
- Public Way
- Egress width A
- Egress width A + B

Multiple Occupancies

§1004.9 covers the means-of-egress requirements that apply to buildings housing multiple occupancies. When there are special egress requirements based upon occupancy, they apply in that portion of the building housing that occupancy. When different occupancies share common egress paths, the most stringent requirements for each occupancy will govern the design of the means-of-egress system.

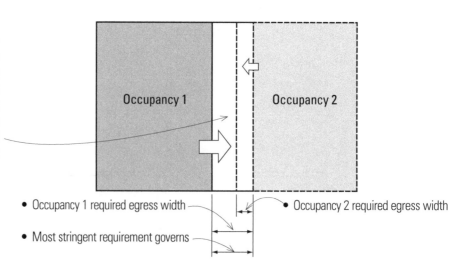

Occupancy 1   Occupancy 2

- Occupancy 1 required egress width
- Occupancy 2 required egress width
- Most stringent requirement governs

Egress Width

§1005 specifies that the width of exit pathways be governed by: the occupant load, the hazard of the occupancy, whether the building is sprinklered, and whether the path is a stair or other component of a means of egress.

Table 1005.1 specifies the required egress width for various occupancies in terms of inches per occupant served.

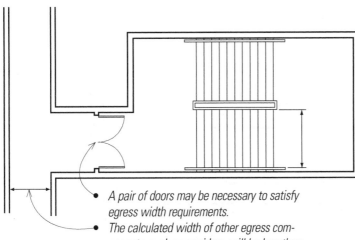

- *Stairs are required to provide more width than corridors since people move more slowly in stairways than in corridors or passages. Thus a corridor with a given occupant capacity will be narrower than a stair in the same egress system.*
- *The table also requires greater egress widths for hazardous occupancies, recognizing the increased hazards in such occupancies.*
- *It also requires wider egress paths in institutional occupancies, recognizing the need for moving bedridden patients and the fact that occupants are often not fully capable of mobility or cognizant enough for self-preservation without assistance.*
- *The table also gives credit for provision of sprinklers by reducing the egress widths in sprinklered buildings.*
- *Note that the table highlights the fact that I-2 Occupancies must be sprinklered by not specify egress widths for non-sprinklered I-2 Occupancies.*

- *A pair of doors may be necessary to satisfy egress width requirements.*
- *The calculated width of other egress components, such as corridors, will be less than the required width calculated for egress stairways because of the egress width factors from Table 1005.1.*

- *Based on the principle of maintaining capacity, not width, it should be permissible to have a narrower egress passage width downstream of a stairway as long as the required capacity is provided. This may be subject to interpretation and negotiation between the building official and the designer.*

- *Because exit doors typically swing in the direction of exit travel, doors from rooms will often swing into paths of egress travel, such as corridors. §1005.2 requires that in such situations:*
- *the door should project a maximum of 7" (178) into the required width when fully opened against the wall of the passage, and*
- *the opening of the door should not reduce the required width by more than one-half.*

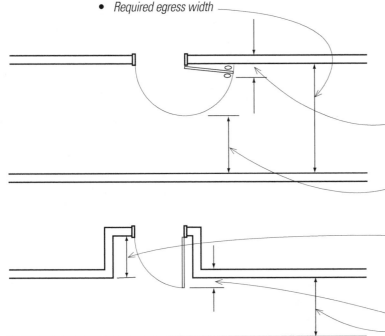

- *Required egress width*

- *Along narrow corridors, doors should be recessed. Minimum recess for a 36" (914) door would be:*

  36"–7" = 29" (914–178 = 736)

- *7" (178) maximum projection*
- *Required egress width*

Means-of-Egress Illumination

§1006.2 requires that the illumination level for a means of egress be not less than 1 foot-candle (11 lux) at the walking surface level. The areas to be illuminated should include all three parts of the means-of-egress system to be certain that occupants can safely exit the building to the public way. In assembly areas such as movie theaters, the illumination may be dimmed during a performance but should be automatically restored in the event of activation of the premises' fire-alarm system. Exceptions are also allowed at sleeping areas in I and R occupancies. Emergency-power requirements are intended to apply to those buildings that require two or more exits, thus exempting small structures and residences.

Emergency-lighting requirements for means of egress require an average of 1 foot-candle (11 lux) of illumination within means of egress with a minimum of 0.1 foot-candle (1 lux) upon initial operation. Such systems are allowed to dim slightly over the 90-minute life of the emergency operations but must not fall below 0.6 foot-candle (6 lux) average and 0.06 foot-candle (0.6 lux) minimum.

Accessible Means of Egress

§1007 addresses the egress requirements for people with disabilities. The basic requirement is that accessible spaces in a building must have an accessible means of egress. Where more than one means of egress is required, based on occupancy or occupant load, then at least two accessible means of egress are to be provided. Note that this requirement for two accessible means of egress applies even when three or more exits may be required based on occupant load. This provision applies only to new construction, it is not required for alterations to existing buildings, although work in altered areas must comply with code requirements for the altered area.

For buildings with four or more stories, one accessible means of egress is to be provided by elevator with standby power and signal devices per §1007.4.

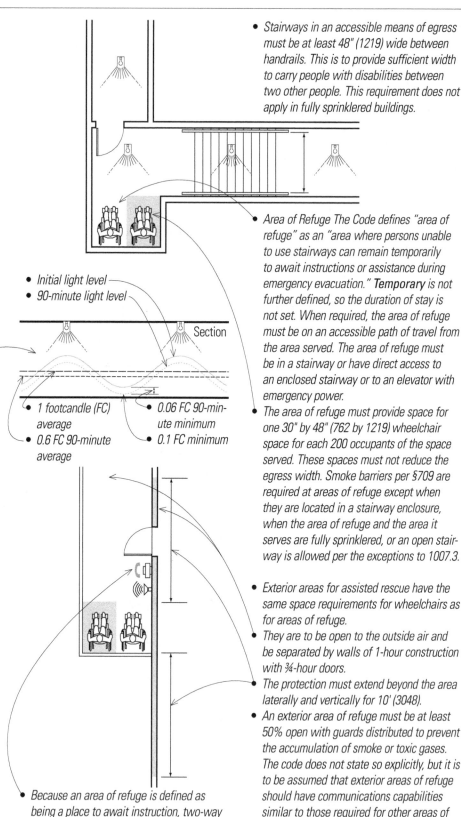

- Initial light level
- 90-minute light level

Section

- 1 footcandle (FC) average
- 0.6 FC 90-minute average
- 0.06 FC 90-minute minimum
- 0.1 FC minimum

- Because an area of refuge is defined as being a place to await instruction, two-way communications are required between the area of refuge and a central control point, or via a public telephone if the central control is not continuously attended. The communications shall be both visual and audible.

- Stairways in an accessible means of egress must be at least 48" (1219) wide between handrails. This is to provide sufficient width to carry people with disabilities between two other people. This requirement does not apply in fully sprinklered buildings.

- Area of Refuge The Code defines "area of refuge" as an "area where persons unable to use stairways can remain temporarily to await instructions or assistance during emergency evacuation." *Temporary* is not further defined, so the duration of stay is not set. When required, the area of refuge must be on an accessible path of travel from the area served. The area of refuge must be in a stairway or have direct access to an enclosed stairway or to an elevator with emergency power.

- The area of refuge must provide space for one 30" by 48" (762 by 1219) wheelchair space for each 200 occupants of the space served. These spaces must not reduce the egress width. Smoke barriers per §709 are required at areas of refuge except when they are located in a stairway enclosure, when the area of refuge and the area it serves are fully sprinklered, or an open stairway is allowed per the exceptions to 1007.3.

- Exterior areas for assisted rescue have the same space requirements for wheelchairs as for areas of refuge.

- They are to be open to the outside air and be separated by walls of 1-hour construction with ¾-hour doors.

- The protection must extend beyond the area laterally and vertically for 10' (3048).

- An exterior area of refuge must be at least 50% open with guards distributed to prevent the accumulation of smoke or toxic gases. The code does not state so explicitly, but it is to be assumed that exterior areas of refuge should have communications capabilities similar to those required for other areas of refuge. This will be subject to interpretation and should be reviewed carefully with the AHJ.

Doors, Gates and Turnstiles

§1008 is a general section that applies to all doors, gates and turnstiles that are part of any of the three parts of the means of egress.

Egress Doors

§1008.1 requires that egress doors be readily distinguishable from adjacent construction. They should never be blind doors, hidden by decorative materials.

Egress doors should typically be side-hinged swinging doors. They must swing in the direction of exit travel when serving an occupant load of more than 50 in a typical occupancy, or when serving any occupant load in high-hazard occupancy. There are exceptions to §1008.1.2:

- *Exception 2 allows the use of other types of doors in I-3 occupancies used for detention. This allows the use of such doors as remotely operated sliding doors for cells.*
- *Exception 8 allows non-swinging doors serving bathrooms within individual sleeping units in R-1 occupancies.*

Revolving Doors

Where egress doors are power-operated, they shall be openable in the event of a power failure. Doors with access controls must also have a failsafe mechanism to allow egress in the event of power failure. The operation of any egress door shall be obvious and not require special knowledge or effort on the part of the user.

§1008.1.3.1. Revolving doors may be used as egress doors when the leaves of the door collapse in the direction of egress and provide an aggregate width of egress of at least 36" (914). For practical purposes, it is desirable that each opening be at least the 32" (813) minimum required for typical doors.

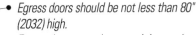

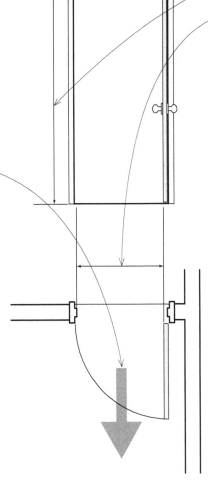

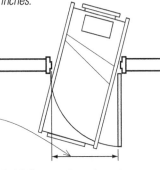

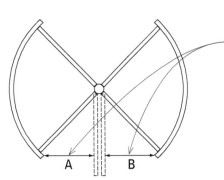

- *Egress doors should be not less than 80" (2032) high.*
- *Egress doors must have a minimum clear width of 32" (813), measured from the face of the door to the stop when the door is open 90° (1.57 rad).*
- *Practically speaking this means that egress doors should always be 3'-0" × 6'-8" (914 × 2032) doors to provide the required minimum opening clearances. Maximum individual door widths for swinging egress doors is 48 inches.*

- *In I-2 Occupancies where doors must allow for passage of beds the doors must provide a clear width not less than 41.5 inches (1054)*
- *Exception 2 to §1008.1.1 allows the door openings of I-3 occupancies to be 28 inches (711). We do not recommend using this smaller dimension unless dictated by security considerations, as this could be in conflict with accessibility requirements.*
- *There are other minor exceptions for doors in small closets or in residences, but these should be avoid by the designer. We recommend that the typical 3 foot wide door width criteria be applied to doors in residential occupancies, even when not strictly required by access standards.*
- *Access criteria limit the force that must be exerted on doors to open them. The maximum force is 5 pounds (22 N) at interior doors and 15 pounds (67 N) at exterior doors.*

- *The capacity and use of revolving doors is limited. Revolving doors may be credited for no more than 50% of the required egress capacity, each door may have no more than a 50-occupant capacity, and the collapsing force can be no greater than 130 pounds (578 N). A plus B must be at least 36 inches (914.4 mm).*

Landings

§1008.1.4. There should be a landing or floor on each side of a door, and the elevation of the floor or landing should be the same. There are exceptions in residential occupancies for screen doors and at interior stairways to allow doors to swing over landings.

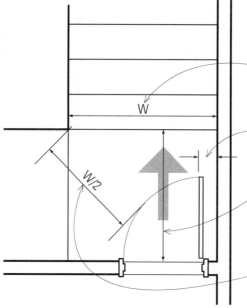

§1008.1.5. Landings are to have a width not less than that of the stairway or the door, <u>whichever is greater</u>.

Doors in the open position should not reduce the required width or depth of a landing by more than 7" (178).

Landings are to be a minimum of 44" (1118) in length in the direction of travel, except they may be 36" (914) long in residential occupancies.

For occupant loads of 50 or more, the door in any position may not reduce the width of the landing to less than half of its required width.

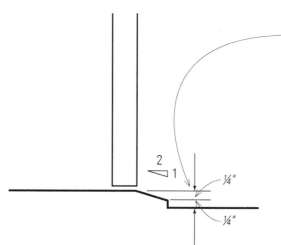

§1008.1.6. Door thresholds should be level whenever possible to allow passage by people with disabilities. When thresholds are necessary at exterior doors or at level changes for materials, they may be no higher than ¾" (19.1 mm) at sliding doors in dwelling units and no higher than ½" (12.7 mm) for other doors. Floor level changes greater than ¼" (6.4 mm) must be beveled with a slope not greater than 2:1 (50% slope). These requirements should be compared with ADAAG provisions as the ADA may have more stringent requirements for lower thresholds.

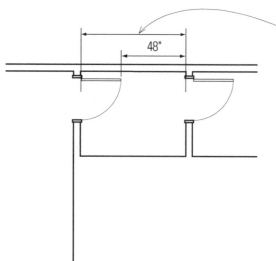

§1008.1.7. Doors in series must be at least 48" (1219) apart, not including the swing of each door. Thus, where a 3' (914) wide door swings toward another door, the space between them must be at least 7' (2134) long.

Exceptions allow power doors operating in tandem and storm or screen doors in residential occupancies to be closer together.

Locks and Latches

§1008.1.8 requires egress doors to be readily openable from the egress side without the use of a key or special knowledge or effort. The unlatching of any door or leaf should not require more than one operation. This recognizes the directional nature of the egress path and that the path serves to make exiting clear and simple for the least familiar building occupant.

Exceptions to the basic rule recognize the special nature of detention centers and of time-limited occupancies in such facilities as stores, assembly areas and churches with fewer than 300 occupants. Key lock devices may be used when occupancies have exits that are unlocked during times of use, and signs are posted requiring that the exit doors are to remain unlocked when the building is occupied.

Power-operated locks are permissible under certain conditions. The space layout must not require an occupant to pass through more than one access-controlled door before entering an exit. Door locks must be deactivated upon actuation of the sprinkler system or automatic detection system. Also, locks must have a fail-safe mechanism to allow egress in the event of power failure. Mechanisms that open the door upon application of a constant force to the egress mechanism for a maximum of 15 seconds are also allowed.

Panic Hardware

§1008.1.9 contains the provisions for panic and fire exit hardware. Panic hardware is typically bars that open a door when depressed. It is named panic hardware based on experience in assembly occupancies where people may crush against doors in panic situations such that normal hardware is jammed by pressure, even if the door opens outward. Panic hardware is designed to open the door if a person is pressed against it. This hardware is always used in conjunction with doors opening in the direction of egress.

Panic hardware is required in Group A or E occupancies having occupant loads of 100 or more and in Group H occupancies with any occupant load. Panic hardware is not required in very large assembly occupancies such as stadiums, where the gates are under constant immediate supervision, as is usually the case.

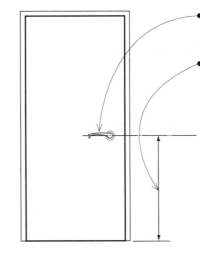

- *§1008.1.8.1 Doors need to be accessible without pinching or grasping the hardware. Lever handles meet these criteria.*
- *§1008.1.8.2. Door handles, pulls, locks and other operating hardware are to be installed from 34" to 48" (864 to 1219) above the finished floor.*

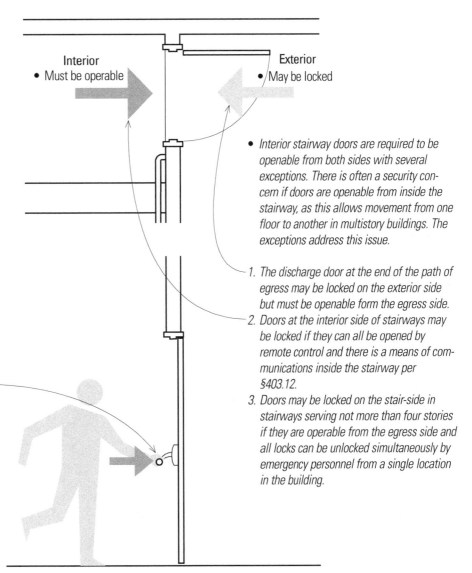

Interior
- Must be operable

Exterior
- May be locked

- *Interior stairway doors are required to be openable from both sides with several exceptions. There is often a security concern if doors are openable from inside the stairway, as this allows movement from one floor to another in multistory buildings. The exceptions address this issue.*

1. *The discharge door at the end of the path of egress may be locked on the exterior side but must be openable form the egress side.*
2. *Doors at the interior side of stairways may be locked if they can all be opened by remote control and there is a means of communications inside the stairway per §403.12.*
3. *Doors may be locked on the stair-side in stairways serving not more than four stories if they are operable from the egress side and all locks can be unlocked simultaneously by emergency personnel from a single location in the building.*

Stairways

§1009 covers requirements for the design of stairs and stairways.

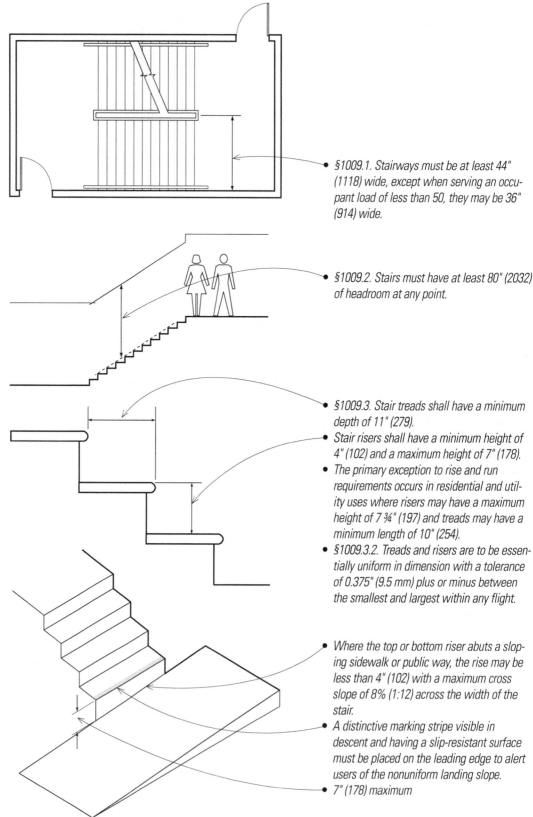

§1009.1. Stairways must be at least 44" (1118) wide, except when serving an occupant load of less than 50, they may be 36" (914) wide.

§1009.2. Stairs must have at least 80" (2032) of headroom at any point.

§1009.3. Stair treads shall have a minimum depth of 11" (279).

Stair risers shall have a minimum height of 4" (102) and a maximum height of 7" (178).

The primary exception to rise and run requirements occurs in residential and utility uses where risers may have a maximum height of 7 ¾" (197) and treads may have a minimum length of 10" (254).

§1009.3.2. Treads and risers are to be essentially uniform in dimension with a tolerance of 0.375" (9.5 mm) plus or minus between the smallest and largest within any flight.

Where the top or bottom riser abuts a sloping sidewalk or public way, the rise may be less than 4" (102) with a maximum cross slope of 8% (1:12) across the width of the stair.

A distinctive marking stripe visible in descent and having a slip-resistant surface must be placed on the leading edge to alert users of the nonuniform landing slope.

7" (178) maximum

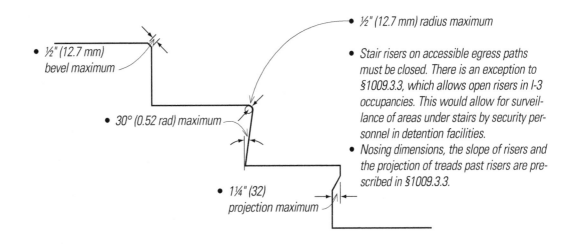

- ½" (12.7 mm) bevel maximum
- 30° (0.52 rad) maximum
- 1¼" (32) projection maximum
- ½" (12.7 mm) radius maximum
- Stair risers on accessible egress paths must be closed. There is an exception to §1009.3.3, which allows open risers in I-3 occupancies. This would allow for surveillance of areas under stairs by security personnel in detention facilities.
- Nosing dimensions, the slope of risers and the projection of treads past risers are prescribed in §1009.3.3.

Enclosures under Stairs

§1009.5.3 requires that enclosed usable space under stairs be protected with 1-hour fire-resistance-rated construction. The only exception to this is for stairs in R-2 or R-3 occupancies.

This requirement applies to similar spaces under exterior stairs. The usable space is not to be accessed from inside the stair enclosure.

Note that these requirements do not require the underside of a stairway to have a fire-rated enclosure merely because there is usable space under the stairs. These requirements apply only when the usable space itself is enclosed. However, note that open space under exterior stairs is not to be used for any purpose.

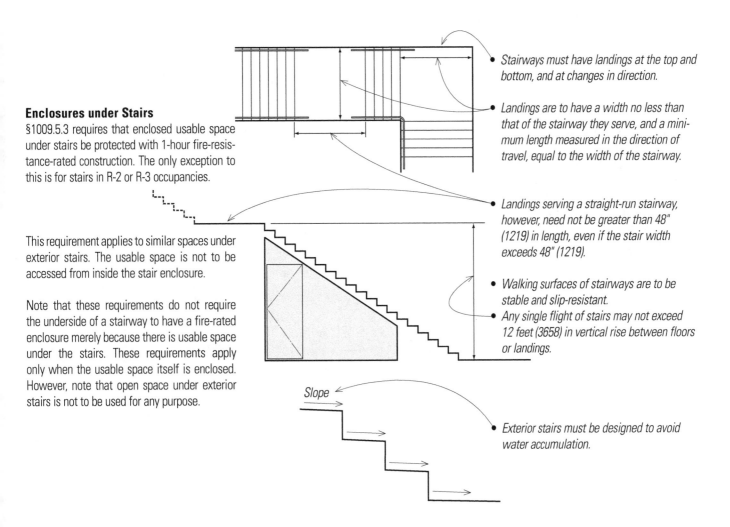

- Stairways must have landings at the top and bottom, and at changes in direction.
- Landings are to have a width no less than that of the stairway they serve, and a minimum length measured in the direction of travel, equal to the width of the stairway.
- Landings serving a straight-run stairway, however, need not be greater than 48" (1219) in length, even if the stair width exceeds 48" (1219).
- Walking surfaces of stairways are to be stable and slip-resistant.
- Any single flight of stairs may not exceed 12 feet (3658) in vertical rise between floors or landings.

Slope

- Exterior stairs must be designed to avoid water accumulation.

GENERAL EGRESS REQUIREMENTS

Curved stairs, winding stairs and spiral stairs
are allowable as egress elements in limited use,
primarily in residential occupancies.

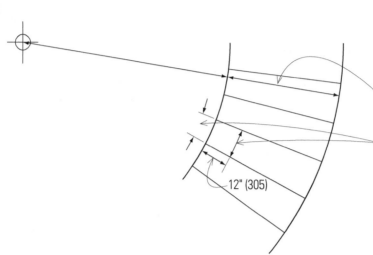

Circular stairs may be part of a means-of-egress
system when they comply with the following
dimensional requirements.

- The treads and risers of curved stairs must
  comply with §1009.3.
- The smaller radius of a curved stairway must
  be no less than twice the required width of
  the stairway.
- The treads of a circular stairway shall be no
  less than 10" (254) at the narrow end, and/
  not less than 11" (279) when measured at a
  point 12" (305) from the narrower end of the
  tread.

12" (305)

Alternating tread devices are allowable as
egress for very limited uses in I-3 occupancies
for guard towers and control rooms, or in F, H
or S occupancies. All of these interior uses are
limited to areas of less than 250 sf (23 m²).
Such stairs may also be used for access to
unoccupied roofs.

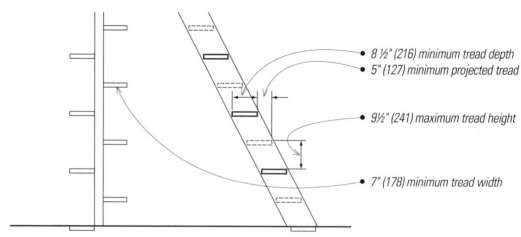

- 8 ½" (216) minimum tread depth
- 5" (127) minimum projected tread

- 9½" (241) maximum tread height

- 7" (178) minimum tread width

Roof Access

§1009.11 requires that, in buildings four of more stories in height, at least one stairway must extend to the roof, unless the roof is sloped at more than a 4 in 12 pitch. An exception allows access to unoccupied roofs by a hatch not less than 16 sf (1.5 m²) in area and having a minimum dimension of 2' (610).

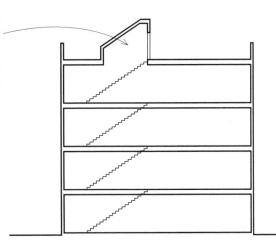

Ramps

§1010 contains provisions that apply to all ramps that are part of a means of egress, except when amended by other provisions where indicated. For example. §1025.11 governs ramps in assembly areas.

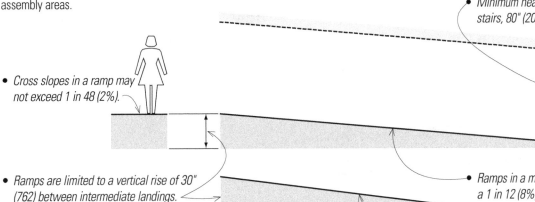

- Minimum headroom is the same as for stairs, 80" (2032).

- Cross slopes in a ramp may not exceed 1 in 48 (2%).

- Ramps are limited to a vertical rise of 30" (762) between intermediate landings.

- Ramps in a means of egress may not exceed a 1 in 12 (8%) slope.

- Other ramps may not exceed a 1 in 8 (12.5%) slope. It is recommended that the designer never use ramps steeper than 1 in 12, even in nonaccessible paths of travel. The use of 1 in 12 ramps makes those paths of travel accessible and safer for all building users.

- The width of ramps in a means of egress shall not be less than the width of corridors as required by §1017.2; this width is typically 44" (1118).
- Other ramps may have a minimum clear width, between handrails, if provided, of 36" (914). Once the width of an egress ramp is established, it should not be reduced in the direction of egress.

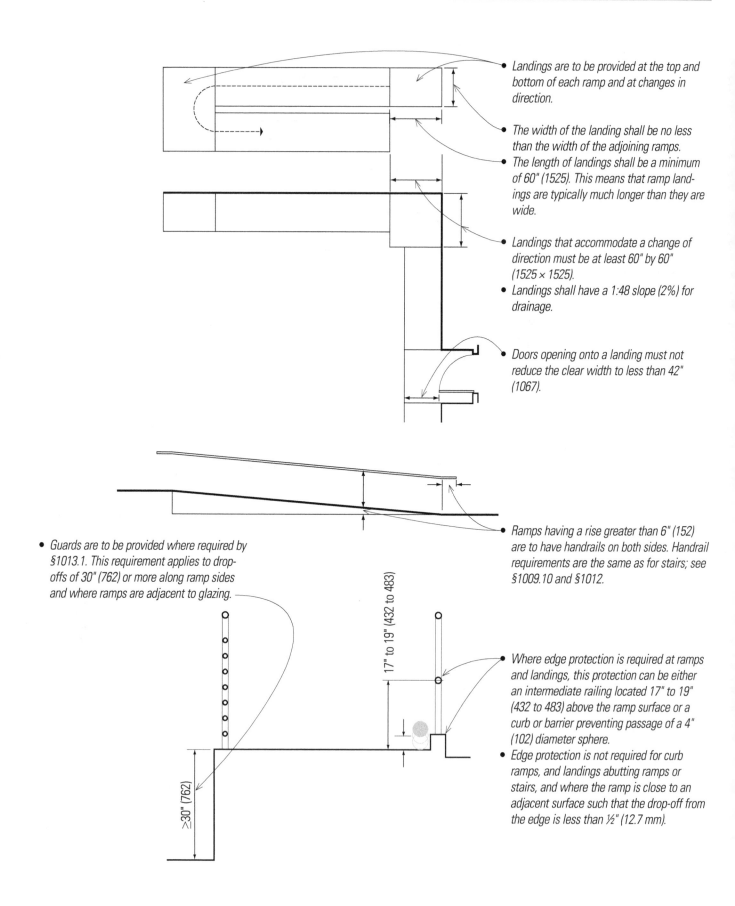

- Landings are to be provided at the top and bottom of each ramp and at changes in direction.

- The width of the landing shall be no less than the width of the adjoining ramps.

- The length of landings shall be a minimum of 60" (1525). This means that ramp landings are typically much longer than they are wide.

- Landings that accommodate a change of direction must be at least 60" by 60" (1525 × 1525).

- Landings shall have a 1:48 slope (2%) for drainage.

- Doors opening onto a landing must not reduce the clear width to less than 42" (1067).

- Ramps having a rise greater than 6" (152) are to have handrails on both sides. Handrail requirements are the same as for stairs; see §1009.10 and §1012.

- Guards are to be provided where required by §1013.1. This requirement applies to drop-offs of 30" (762) or more along ramp sides and where ramps are adjacent to glazing.

17" to 19" (432 to 483)

≥30" (762)

- Where edge protection is required at ramps and landings, this protection can be either an intermediate railing located 17" to 19" (432 to 483) above the ramp surface or a curb or barrier preventing passage of a 4" (102) diameter sphere.

- Edge protection is not required for curb ramps, and landings abutting ramps or stairs, and where the ramp is close to an adjacent surface such that the drop-off from the edge is less than ½" (12.7 mm).

Exit Signs

§1011.1 calls for exit signs to be provided at exits and exit-access doors. Exit signs may be omitted in some residential occupancies and in cases where the exit pathway is obvious to occupants. They can be omitted in the sleeping areas of I-3, R-1, R-2 and R-3 occupancies. Such conditions where signs are to be omitted should be reviewed and approved by the building official.

Exit sign with directional arrow

100' (30 480)

100' (30 480)

EXIT

Exit sign

60" (1524)

- Exit signs must be of an approved design, be illuminated by internal or external means, and have the capability to remain illuminated for up to 90 minutes either by battery, internal illumination or connection to an emergency power source.
- Exit signs must be clearly visible and be not more than 100' (30 480) from any point in an exit-access corridor.
- Tactile exit signs accessible for persons with disabilities are to be provided at doors to egress stairways, exit passageways and the exit discharge.

- Handrails must extend horizontally for 12" (305) beyond the top riser of a stairway.
- Handrails are to be between 34" and 38" (864 and 965) above the stair-tread nosing.
- Handrails must also continue their slope for the depth of one tread beyond the bottom riser.

- Note that ADAAG requires an additional 12" (305) horizontal extension at the bottom of a stairway. In no case should the designer use less than the ADAAG dimensions, except where the stairway is in a residence and not on an accessible path.

Handrails

§1009.10 specifies that stairways are to have handrails on each side complying with §1012, except in aisle stairs, where a center rail is provided, or in dwellings. Handrails are not required on decks having a single level change between two areas that are equal to or greater than a landing dimension and in residences where there is only one riser.

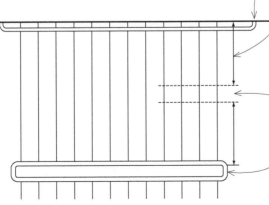

- When handrails do not continue to the handrail of an adjacent flight, they are required to return to a wall or to the walking surface.

- Only portions of a stairway width within 30" (762) of a handrail may count toward the width required for egress capacity. This means that intermediate handrails may be required for stairways that are required to be more than 60" (1524) wide.
- Stair width more than 30" (762) from handrails does not count toward required egress capacity.
- Railings are to be continuous except in residences where newel posts and turnouts are acceptable.
- Handrail extensions are not required where the handrails are continuous between flights.

Guards

§1013 requires that railings or similar protective elements be provided where any grade change of 30" (762) or more occurs in a means of egress. This also applies when a means of egress is adjacent to glazing elements that do not comply with the strength requirements for railings and guards per §1607.7.

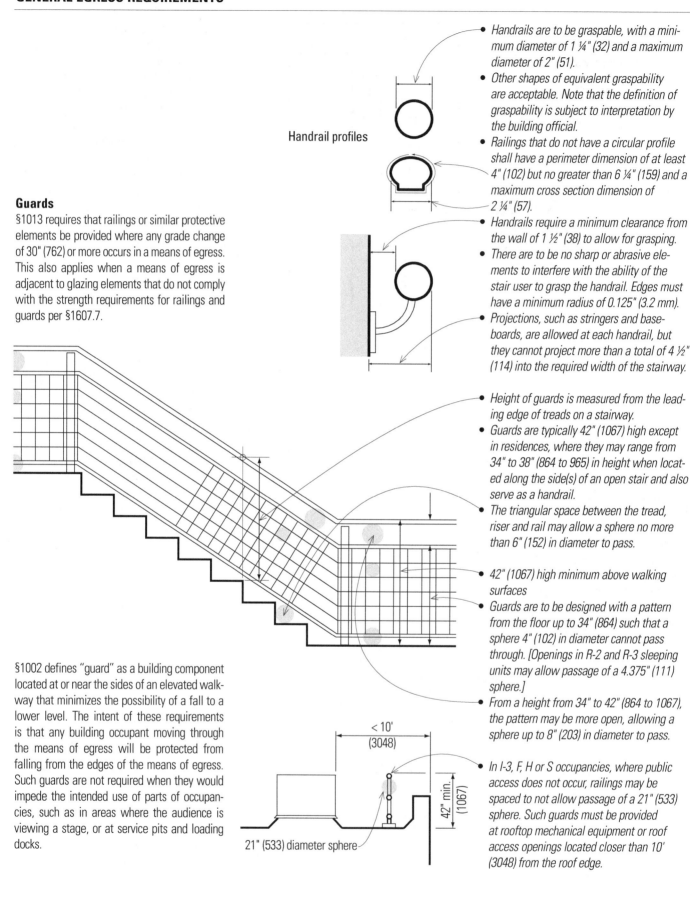

Handrail profiles

- Handrails are to be graspable, with a minimum diameter of 1 ¼" (32) and a maximum diameter of 2" (51).
- Other shapes of equivalent graspability are acceptable. Note that the definition of graspability is subject to interpretation by the building official.
- Railings that do not have a circular profile shall have a perimeter dimension of at least 4" (102) but no greater than 6 ¼" (159) and a maximum cross section dimension of 2 ¼" (57).
- Handrails require a minimum clearance from the wall of 1 ½" (38) to allow for grasping.
- There are to be no sharp or abrasive elements to interfere with the ability of the stair user to grasp the handrail. Edges must have a minimum radius of 0.125" (3.2 mm).
- Projections, such as stringers and baseboards, are allowed at each handrail, but they cannot project more than a total of 4 ½" (114) into the required width of the stairway.

- Height of guards is measured from the leading edge of treads on a stairway.
- Guards are typically 42" (1067) high except in residences, where they may range from 34" to 38" (864 to 965) in height when located along the side(s) of an open stair and also serve as a handrail.
- The triangular space between the tread, riser and rail may allow a sphere no more than 6" (152) in diameter to pass.

- 42" (1067) high minimum above walking surfaces
- Guards are to be designed with a pattern from the floor up to 34" (864) such that a sphere 4" (102) in diameter cannot pass through. [Openings in R-2 and R-3 sleeping units may allow passage of a 4.375" (111) sphere.]
- From a height from 34" to 42" (864 to 1067), the pattern may be more open, allowing a sphere up to 8" (203) in diameter to pass.

§1002 defines "guard" as a building component located at or near the sides of an elevated walkway that minimizes the possibility of a fall to a lower level. The intent of these requirements is that any building occupant moving through the means of egress will be protected from falling from the edges of the means of egress. Such guards are not required when they would impede the intended use of parts of occupancies, such as in areas where the audience is viewing a stage, or at service pits and loading docks.

< 10'
(3048)

42" min.
(1067)

21" (533) diameter sphere

- In I-3, F, H or S occupancies, where public access does not occur, railings may be spaced to not allow passage of a 21" (533) sphere. Such guards must be provided at rooftop mechanical equipment or roof access openings located closer than 10' (3048) from the roof edge.

§1014 through §1017 cover the exit-access portion of the means of egress that leads from any occupied portion of a building to any component of an exit—an exterior exit door at grade, exit enclosures and passageways, horizontal exits, and exterior exit stairways. Exit access therefore includes the functional areas of the building along with various levels of egress pathways that will be discussed in detail.

As will become clear as we proceed through the analysis of §1014 through §1017, there is a hierarchy of protection requirements for various elements of the exit access. As the number of occupants increases or as occupants get closer to an exit these requirements become more stringent.

Exit-Access Design Requirements
§1015 and §1016 contain the key design requirements for exits and exit-access doorways, addressing such issues as the number and arrangement of exit paths, travel distances, as well as travel through intervening rooms and spaces.

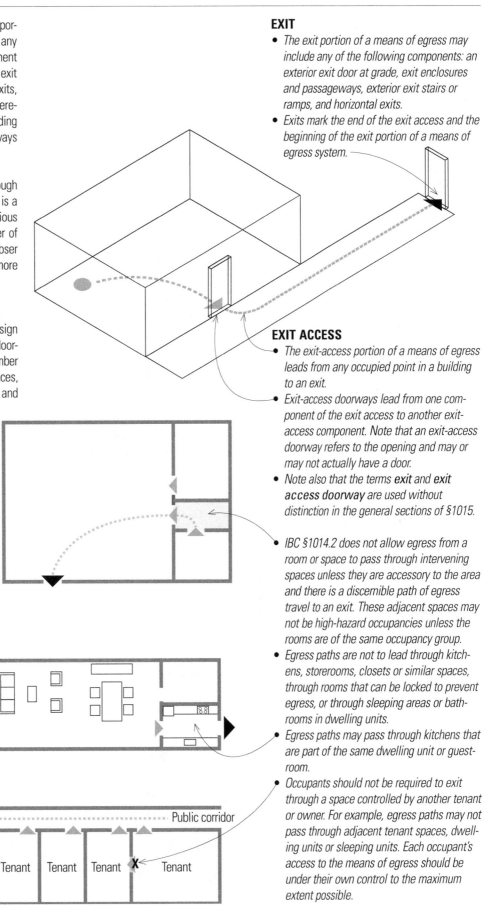

EXIT

- *The exit portion of a means of egress may include any of the following components: an exterior exit door at grade, exit enclosures and passageways, exterior exit stairs or ramps, and horizontal exits.*
- *Exits mark the end of the exit access and the beginning of the exit portion of a means of egress system.*

EXIT ACCESS

- *The exit-access portion of a means of egress leads from any occupied point in a building to an exit.*
- *Exit-access doorways lead from one component of the exit access to another exit-access component. Note that an exit-access doorway refers to the opening and may or may not actually have a door.*
- *Note also that the terms **exit** and **exit access** doorway are used without distinction in the general sections of §1015.*

- *IBC §1014.2 does not allow egress from a room or space to pass through intervening spaces unless they are accessory to the area and there is a discernible path of egress travel to an exit. These adjacent spaces may not be high-hazard occupancies unless the rooms are of the same occupancy group.*
- *Egress paths are not to lead through kitchens, storerooms, closets or similar spaces, through rooms that can be locked to prevent egress, or through sleeping areas or bathrooms in dwelling units.*
- *Egress paths may pass through kitchens that are part of the same dwelling unit or guest-room.*
- *Occupants should not be required to exit through a space controlled by another tenant or owner. For example, egress paths may not pass through adjacent tenant spaces, dwelling units or sleeping units. Each occupant's access to the means of egress should be under their own control to the maximum extent possible.*

I-2 Egress Requirements

Group I-2 occupancies are those where the occupants are not capable of self-preservation without assistance, such as in hospitals and nursing homes. In these occupancies, IBC §1014.2.2 requires that habitable rooms or suites have an exit-access door leading directly to an exit-access corridor. There are a number of exceptions. Several items in this section refer to "suites." This term is not specifically defined in the IBC. Suite is defined by the Merriam-Webster Dictionary as "a group of rooms occupied as a unit." The exceptions are:

1. Rooms that have exit doors opening directly to the outside at ground level.

2. Patient sleeping rooms may have one intervening room between them and a corridor when no more than 8 patient beds use the intervening space for exit access.

3. Special nursing suites may have one intervening room where there is direct and constant visual supervision by nursing personnel.

4. Rooms other than patient sleeping rooms located within a suite are allowed to have exit access travel through one intervening room where travel to the exit access door is not greater than 100 feet (30 480).

5. Rooms other than patient sleeping rooms located within a suite are allowed to have exit access travel through two intervening rooms where travel to the exit access door is not greater than 50 feet (15 240).

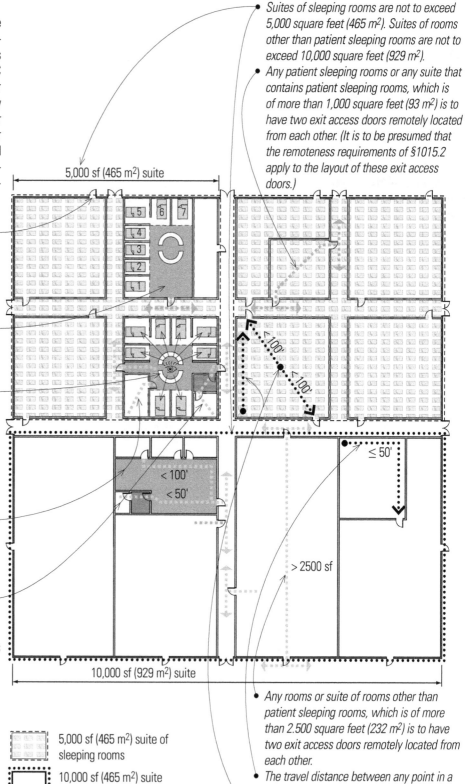

• Suites of sleeping rooms are not to exceed 5,000 square feet (465 m²). Suites of rooms other than patient sleeping rooms are not to exceed 10,000 square feet (929 m²).

• Any patient sleeping rooms or any suite that contains patient sleeping rooms, which is of more than 1,000 square feet (93 m²) is to have two exit access doors remotely located from each other. (It is to be presumed that the remoteness requirements of §1015.2 apply to the layout of these exit access doors.)

5,000 sf (465 m²) suite

10,000 sf (929 m²) suite

5,000 sf (465 m²) suite of sleeping rooms

10,000 sf (465 m²) suite of non-sleeping rooms

Intervening space

Second intervening space

• Any rooms or suite of rooms other than patient sleeping rooms, which is of more than 2.500 square feet (232 m²) is to have two exit access doors remotely located from each other.

• The travel distance between any point in a Group I-2 occupancy and an exit access door in the room is not to exceed 50 feet (15 240).

• The travel distance between any point in a suite of sleeping rooms and an exit access door of that suite is not to exceed 100 feet (30 480).

Common-Path-of-Egress Travel

§1002.1 defines a common-path-of-egress travel as that portion of an exit access that occupants are required to traverse before two separate and distinct paths of egress travel to two exits are available. As noted in the table on page 168, common-paths-of-egress travel are limited per §1014.3.

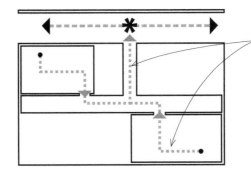

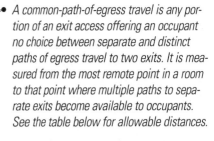

• A common-path-of-egress travel is any portion of an exit access offering an occupant no choice between separate and distinct paths of egress travel to two exits. It is measured from the most remote point in a room to that point where multiple paths to separate exits become available to occupants. See the table below for allowable distances.

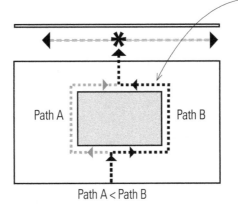

Path A   Path B

Path A < Path B

• Common-paths-of-egress travel include paths that split and merge within the exit access prior to the location where multiple paths lead to separate exits.

• While there are two choices within the space illustrated the two alternate paths still lead to a single means of egress from the space under consideration. The code is silent on how to measure the length of the common-path-of-egress travel in this condition, but the most conservative measurement of the length of the common-path-of-egress travel would be to measure along the longest of the two alternate egress pathways. The end of the common-path-of-egress travel is noted by the mark ✱

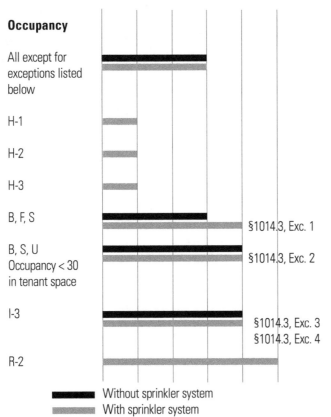

| | 0 | 25 | 50 | 75 | 100 | 125 | Feet |
| | 0 | 7.8 | 15.2 | 22.9 | 30.5 | 38.1 | Meters |

Occupancy

All except for exceptions listed below

H-1

H-2

H-3

B, F, S   §1014.3, Exc. 1

B, S, U
Occupancy < 30
in tenant space   §1014.3, Exc. 2

I-3   §1014.3, Exc. 3
                  §1014.3, Exc. 4

R-2

━━━━ Without sprinkler system
━━━━ With sprinkler system

Exit-Access Components

The criteria contained in §1014 through §1017 apply to those components that occur within the exit access only. Some common terminology may also occur in the exit and exit discharge portions of the means of egress, but their application must be viewed in the context of which parts of the means of egress are being described by the various code sections.

There is a definite hierarchy to the components of the exit access, as described below.

Aisles

§1014.4 applies not just to the conventional definition of aisles in assembly areas with fixed seats, but also to any occupied portions of the exit access. Components may be fixed or movable, such as tables, furnishings or chairs. Assembly-area aisles for fixed seats, grandstands or bleachers are governed by §1025.

The aisle is the simplest defined component in the exit access with the most flexibility. The intent of aisle requirements is to provide clear pathways so that building occupants can easily find egress pathways to exits in an emergency.

Aisle requirements relate to unobstructed widths; there are no fire-resistive construction requirements for aisles. Indeed, most aisle requirements apply to nonfixed building elements, such as tables, chairs or merchandise display racks. It should also be recognized that the designer does not have day-to-day control over how spaces with non-fixed elements are used by the building occupants. However, the means-of-egress design should be based on delineated design assumptions as to occupant load and furniture layout to justify the design of the means-of-egress system. Such design should be rationally based on quantified determinations of how the space is anticipated to be used.

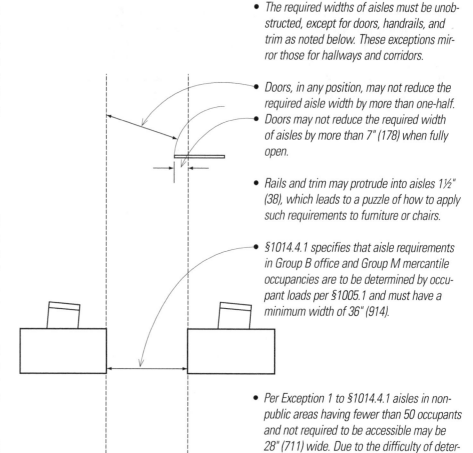

- The required widths of aisles must be unobstructed, except for doors, handrails, and trim as noted below. These exceptions mirror those for hallways and corridors.

- Doors, in any position, may not reduce the required aisle width by more than one-half.
- Doors may not reduce the required width of aisles by more than 7" (178) when fully open.

- Rails and trim may protrude into aisles 1½" (38), which leads to a puzzle of how to apply such requirements to furniture or chairs.

- §1014.4.1 specifies that aisle requirements in Group B office and Group M mercantile occupancies are to be determined by occupant loads per §1005.1 and must have a minimum width of 36" (914).

- Per Exception 1 to §1014.4.1 aisles in nonpublic areas having fewer than 50 occupants and not required to be accessible may be 28" (711) wide. Due to the difficulty of determining exactly what is nonpublic as well as what need not be accessible, especially in office uses, we recommend using the criteria for public aisles for design purposes. This is also more likely to be compliant with accessibility regulations that may apply in both public and employee areas.

§1014.4.3 addresses seating at tables and chairs, whether loose or fixed.

Aisle Accessways

§1014.4.3.1 divides areas with tables and chairs into aisles and aisle accessways.

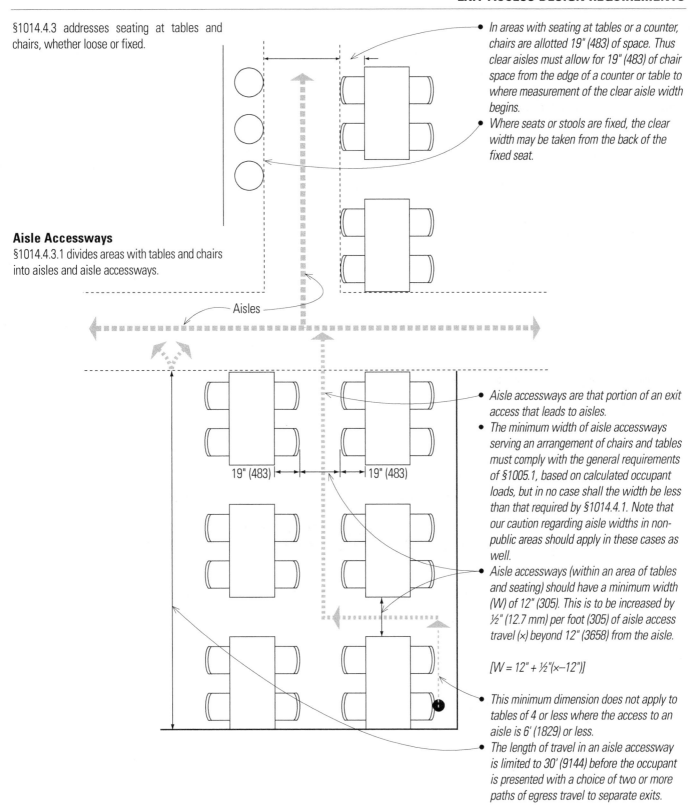

Aisles

19" (483)   19" (483)

• In areas with seating at tables or a counter, chairs are allotted 19" (483) of space. Thus clear aisles must allow for 19" (483) of chair space from the edge of a counter or table to where measurement of the clear aisle width begins.

• Where seats or stools are fixed, the clear width may be taken from the back of the fixed seat.

• Aisle accessways are that portion of an exit access that leads to aisles.

• The minimum width of aisle accessways serving an arrangement of chairs and tables must comply with the general requirements of §1005.1, based on calculated occupant loads, but in no case shall the width be less than that required by §1014.4.1. Note that our caution regarding aisle widths in non-public areas should apply in these cases as well.

• Aisle accessways (within an area of tables and seating) should have a minimum width (W) of 12" (305). This is to be increased by ½" (12.7 mm) per foot (305) of aisle access travel (x) beyond 12" (3658) from the aisle.

[W = 12" + ½"(x–12")]

• This minimum dimension does not apply to tables of 4 or less where the access to an aisle is 6' (1829) or less.

• The length of travel in an aisle accessway is limited to 30' (9144) before the occupant is presented with a choice of two or more paths of egress travel to separate exits.

EXIT-ACCESS DESIGN REQUIREMENTS

Number of Exits and Exit-Access Doorways

§1015.1 sets the requirements for the number of exit or exit-access doorways required from any space. Two exits or exit-access doorways are required when the occupant load exceeds the values in Table 1015.1 or when the conditions of the common path of exit travel exceed the requirements of §1014.3. We will address the common path of exit travel requirements in a later section. I-2 occupancies have very different and very specific requirements for when two exit access doors are required. I-2 occupancies are not addressed in Table 1015.1. An exception in the language of the table reference in §1015.1 refers the code user back to §1014.2.2 which is discussed above.

The table below combines the requirements from several code sections and compares interrelated requirements for occupant load, number of exits and lengths of common paths of egress travel as they relate to various occupancy groups.

- *Typically 49 or fewer occupants, or per Table 1015.1*

- *More than the maximum occupant load per Table 1015.1*

- *Note that §1015.1.1 references §1019.1 to determine when three or more exits must be provided for a floor area.*

Occupancy	Per Tables 1015.1 & 1019.1				Per Section 1014.3	
	Maximum Occupant Load per story w/1 Exit	Minimum No. of Occupants with 2 Exits	No. of Occupants Requiring 3 Exits	No. of Occupants Requiring 4 Exits	Length of Common Path-of-Egress Travel** before 2 Paths of Egress Travel are Required	
All	–	Up to 500 (Except as modified by §1015.1 or §1019.2)	501–1,000	>1,000	Nonsprinklered	Sprinklered
A, E	49	50–500	"	"	75(22 860)	75
B, F	49	50–500	"	"	75*	100
H-1, 2, 3	3	4–500	"	"	25 (7620)	25
H-4, 5	10	11–500	"	"	75	75
I-1	10	11–500	"	"	75	75
I-2	Per §1014.2.2.	1–500	"	"	75	75
I-3	10	11–500	"	"	100 (30 480)	100
I-4	10	11–500	"	"	75	75
M	49	50–500	"	"	75	75
R	10	11–500	"	"	75	75 (125 @ R-2)
S	20	30–500	"	"	75*	100
U	49	50–500	"	"	75*	75*

* Tenant spaces in occupancy groups B, S & U with an occupant load of not more than 30 may have a common path-of-egress travel up to 100 feet (30 480).

**See illustration on page 167.

Exits and Exit-Access Doorway Arrangement

§1015.2 requires that all exits be obvious to the occupants and be arranged for ease of use in a rational manner.

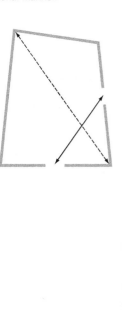
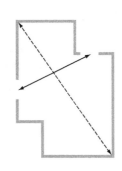
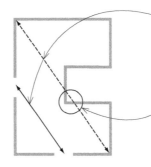

- *When two exits are required, they are to be placed a distance apart equal to one-half the diagonal dimension of the space. The code does not specify where the dimensions are to be taken, but the most prudent measurement is to the centerline of the doorway.*
- *The diagonal measurement line can go outside the building footprint.*

- *There are exceptions to the exit layout for corridors and for sprinklered buildings.*

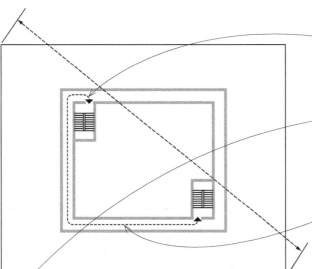

- *When a 1-hour fire-rated corridor is provided, the exit separation may be measured along a direct path of travel within the corridor. No minimum distance is required between stair enclosures, but*
- *"Scissor" stairs, which are interlocking stairs located in the same stair enclosure, can only be counted as one exit stairway per §1015.2.1. This provision addresses the required separation of exit enclosures, such as for stairs, in the core of office buildings.*
- *In such instances, the corridor connecting the tenant spaces to the exit enclosures allows placement of the exit enclosures closer together than half the diagonal distance of the floor without a corridor.*

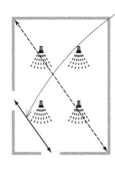
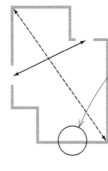
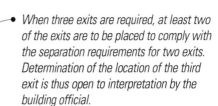
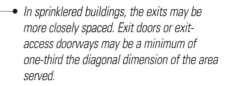

- *In sprinklered buildings, the exits may be more closely spaced. Exit doors or exit-access doorways may be a minimum of one-third the diagonal dimension of the area served.*

- *When three exits are required, at least two of the exits are to be placed to comply with the separation requirements for two exits. Determination of the location of the third exit is thus open to interpretation by the building official.*

Exit-Access Travel Distance

§1016.1 and Table 1016.1 provide the maximum length of exit-access travel distances, measured from the most remote point in the exit-access space to the entrance to an exit along a "natural and unobstructed path of egress travel."

Egress distances are measured in the exit-access space under consideration. They are not the total length of the means of egress but the distance an occupant must travel within the exit-access portion of the means of egress before entering the next higher level of protection in the means of egress, the exit. Thus these distances would apply on the upper floor of a high-rise building, but only until the occupant entered the exit stair enclosure and began moving through the exit portion of the means of egress.

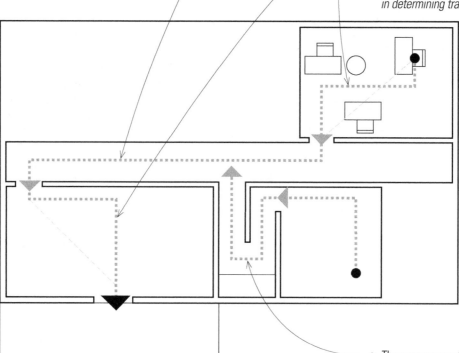

- *The path length is to be measured along the natural and unobstructed path of egress travel. The determination of the natural path must take fixed obstacles and obvious pathways into account. The potential location of furniture, especially large fixed furniture, such as library shelving, should be taken into account when measuring distances.*

- *On the other hand, in speculative office or commercial buildings, the designer may not have any idea of the ultimate use of the space and must use the 90° method. The designer should use common sense and care in determining travel distances.*

- *The measurement of exit-access travel distance must include the central path along unenclosed stairs and ramps and be measured from the most remote part of the building that makes use of the egress pathway.*

For most occupancies, the allowable exit-access travel distance is 200' (60 960) without a sprinkler system and from 250' to 300' (76 200 to 91 440) with a sprinkler system. The table below graphically represents the relative lengths of exit-access travel allowable for various occupancies.

Note that allowable exit access travel distances are increased in buildings with automatic sprinkler systems. This provision acknowledges the increased level of safety for occupants in sprinklered buildings. Note also that there are numerous footnotes to Table 1016.1 regarding occupancy-specific travel distances in such specialized building types as covered malls, atriums and assembly spaces.

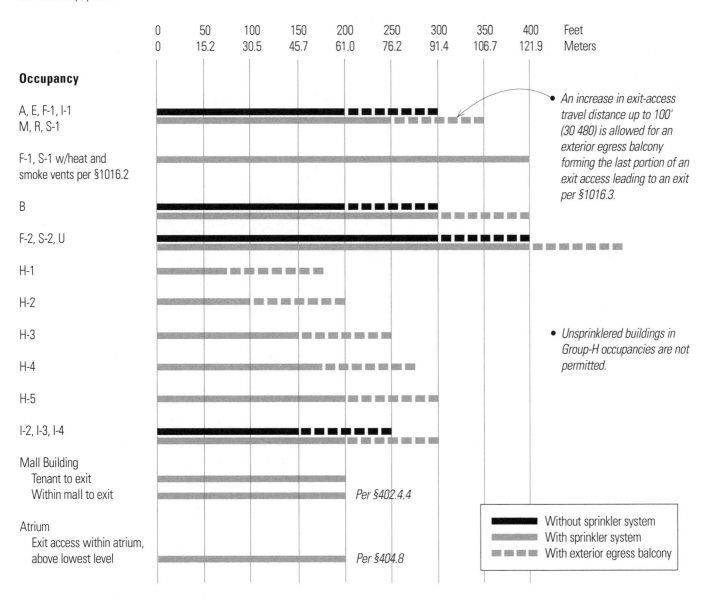

• An increase in exit-access travel distance up to 100' (30 480) is allowed for an exterior egress balcony forming the last portion of an exit access leading to an exit per §1016.3.

• Unsprinklered buildings in Group-H occupancies are not permitted.

EXIT-ACCESS DESIGN REQUIREMENTS

Corridors

§1017 provides for the design of corridors, which comprise the next level of occupant protection in the exit access above the aisle. **Corridor** is now a defined term in the Code as an enclosed exit access component that defines and provides a path of egress travel to an exit. It is important to note and remember that corridors are not exits; they are a component of the exit access.

The following table, based on Table 1017.1, provides information about the occupant load, fire-resistance rating and dimensions of corridors.

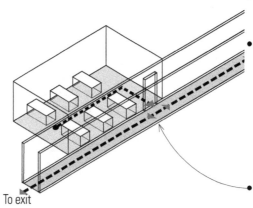

To exit

- *In exit-access systems with multiple components, it can be visualized that aisles lead to corridors that, in turn, lead to exits.*
- *While layouts are subject to interpretation by the building official, a corridor typically is a space longer than it is wide, separated from adjacent spaces by walls, and having two clear choices of egress leading to two exits. There may be variations in the details of this arrangement but the basic components will almost always be present in a corridor.*
- *Corridors usually have a fire-resistance rating of at least 1-hour, based on the occupancy group where they occur. Several occupancies allow the use of unrated corridors in sprinklered buildings.*

Occupancy	Corridor Required for Occupant Load	Fire-Resistance Rating Without Sprinklers	Fire-Resistance Rating With Sprinklers	Notes	Dead-End Distance [20' (6096) typical]	Minimum Corridor Width Per §1005.1 and §1017.2, 44' (1118) typical
A, B, F, M, S, U	>30	1	0	No rating required in open parking garages per §1017.1 , Ex. 3	50' (15 240) @ B & F when sprinklered	
B	>30 (49)*	1 (*if 1 exit per §1015.1)	0	§1017.1 , Ex. 4		
E	>30	1	0	No rating if one exterior door opens to outside		72" (1829) when occupant capacity >100
R	>10	Not permitted	½ hour	No rating at corridors inside dwelling unit or sleeping unit		36" (914) within dwelling unit
H-1. H-2. H-3	All	Not permitted	1			
H-4, H-5	>30	Not permitted	1			
I-2[1], I-4	All	Not permitted	0	1. See §407.3 for I-2		72" (1829) when serving I occupants not capable of self-preservation; 96" (2438) in I–2 when required for bed movement
I-1, I-3	All	Not permitted	1[2]	2. See §408.7 for I-3 sleeping areas	50' (15 240); see §1017.3, Exception 1 (only @ I–3, Conditions 2,3,4 per §308.4)	I occupants not capable of self-preservation
All					Unlimited when dead-end is less than 2½ times least corridor width	24" (610) typical for mechanical access; 36" (914) with < 50 occupants

Dead Ends

§1017.3 prescribes a maximum length for dead-end corridors. This is to avoid having occupants backtrack after they have gone some distance down a corridor before discovering that there is no exit or exit-access doorway at its end. Note that the allowable length of dead-end corridors may be modified in certain occupancies by provision of an automatic sprinkler system. See the immediately preceding table.

- *Dead ends in corridors are generally limited to 20' (6096) in length. See the table on the previous page for exceptions.*
- *Dead-end provisions apply where there are corridors that branch off the main egress path and may thus lead an occupant to proceed to the end of a side corridor which does not have an exit, before having to return to the main egress path.*
- *When only one exit is required, the provisions for dead-end corridors do not apply. Single exits have relatively low occupant loads and the premise of a single exit assumes a directional nature to the egress path where the way to the exit will be clear.*
- *Dead-end corridors are not limited in length when their length is less than 2½ times the least width of the dead end.*

width (w)

< 2½ w

Air Movement in Corridors

§1017.4 prohibits corridors from being used as part of the air-supply or return system. This is to allow the corridor to remain a separate atmosphere from the spaces it serves to the maximum extent possible. There are several exceptions to the corridor air-supply requirements:

1. There are allowances for the corridor to serve as a makeup-air source for adjacent spaces opening directly off the corridor, but only when the mechanical system is configured to have a positive pressure in the corridor to limit smoke migration into the corridor space.
2. Plenums are allowed in dwelling units.
3. Corridors may be used as return-air plenums within tenant spaces of 1,000 sf (93 m²) or less.

The plenum space created between the ceiling of the corridor and the floor or roof structure above may be used as a return air plenum under one or more of the following circumstances:

1. The corridor is not required to be of fire resistance-rated construction.
2. The corridor ceiling separates the plenum from the corridor with fire-resistance-rated construction.
3. The air-handling system serving the corridor shuts down upon activation of the air-handling smoke detector.
4. The air-handling system serving the corridor shuts down upon detection of sprinkler water flow.
5. The plenum space is part of an approved engineered smoke-control system.

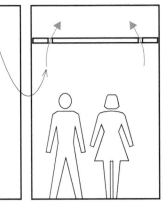
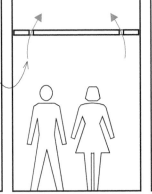

EXIT-ACCESS DESIGN REQUIREMENTS

Corridor Continuity

§1017.5 stipulates that when a fire-resistance-rated corridor is required, the fire rating is to be maintained throughout the egress path from the point that an occupant enters the corridor in the exit access until they leave the corridor at the exit.

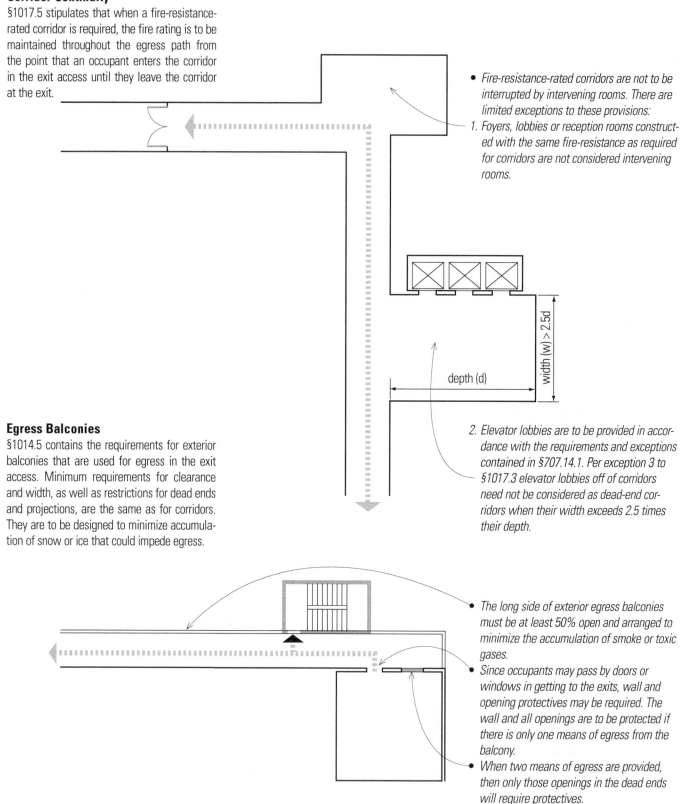

- *Fire-resistance-rated corridors are not to be interrupted by intervening rooms. There are limited exceptions to these provisions:*
  1. *Foyers, lobbies or reception rooms constructed with the same fire-resistance as required for corridors are not considered intervening rooms.*

width (w) > 2.5d

depth (d)

Egress Balconies

§1014.5 contains the requirements for exterior balconies that are used for egress in the exit access. Minimum requirements for clearance and width, as well as restrictions for dead ends and projections, are the same as for corridors. They are to be designed to minimize accumulation of snow or ice that could impede egress.

  2. *Elevator lobbies are to be provided in accordance with the requirements and exceptions contained in §707.14.1. Per exception 3 to §1017.3 elevator lobbies off of corridors need not be considered as dead-end corridors when their width exceeds 2.5 times their depth.*

- *The long side of exterior egress balconies must be at least 50% open and arranged to minimize the accumulation of smoke or toxic gases.*
- *Since occupants may pass by doors or windows in getting to the exits, wall and opening protectives may be required. The wall and all openings are to be protected if there is only one means of egress from the balcony.*
- *When two means of egress are provided, then only those openings in the dead ends will require protectives.*

The requirements of §1018 through §1023 apply to the level of fire-resistance, the dimensions, and the occupant capacity of exits. The exit is the intermediate portion of a means of egress, located between the exit access and the exit discharge. Unlike the exit access, where building functions share the occupied spaces with egress uses, exits are generally single-purpose spaces that function primarily as a means of egress. Once a certain level of occupant protection is achieved in the exit, it is not to be reduced until arrival at the exit discharge.

Buildings with One Exit per Table 1019.2

Occupancy	Maximum Height of Building above Grade Plane	Maximum Occupants (or Dwelling Units) per Floor /Travel Distance
A, B, E, F, M, U	1 story	49 / 75' (22 860)
H-2, H-3	1 story	3 / 25' (7620)
H-4, H-5, I, R	1 story	10 / 75' (22 860)
S	1 story	29 / 100' (30 480)
B, F, M, S	2 stories	30 / 75' (22 860)
R-2	2 stories	(4 dwelling units) / 50' (15 240)

Exit Design Requirements

§1018 through §1023 set out the general design requirements for exits.

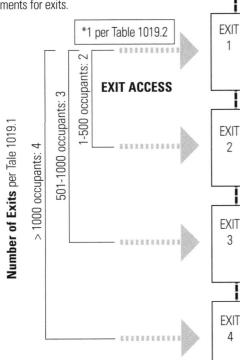

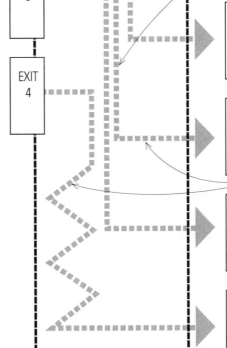

- Exits provide an additional level of protection above that for egress paths in the exit access.
- Exits are to be continuous from the point of entry from the exit access to the exit discharge.
- Exits may be either horizontal, as in passageways, or vertical, as in stairways, or a combination of both.
- Because exits are typically fully enclosed, fire-resistance-rated, single-purpose spaces, travel distance in exits is not limited.

- The minimum number of exits is related to occupant load and occupancy type. See the table on page 170 for the minimum number of exits.
- There are some limited uses where only one exit is required, as noted in Table 1019.2. These are situations in one-story and two-story buildings where there are relatively few occupants and travel distances are short. These situations are also related to occupancy type. This is tabulated in the table above right.

Exit Components

Various elements of the Exit are described in the following sections:

- §1018.2. Exterior exit doors that lead directly to an exit discharge or public way
- §1020. Vertical exit enclosures (stairs and ramps)
- §1021. Exit passageways
- §1022. Horizontal exits
- §1023. Exterior exit ramps and stairways

Vertical Exit Enclosures

§1020.1 requires that all interior exit stairways be enclosed. Vertical exit enclosures connecting four stories or more shall have a 2-hour fire-resistance rating. Vertical exit enclosures connecting less than four stories are to have a 1-hour rating. Enclosures at exit ramps have the same requirements as for stairways. The number of stories connected includes basements, but mezzanines are not counted as additional stories.

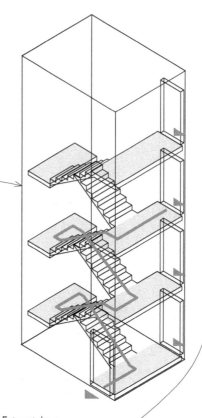

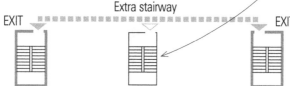

Extra stairway

EXIT · EXIT

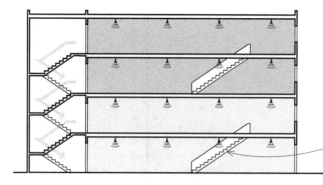

Some limited exceptions allow exit stairs to not be enclosed. These exceptions are very specific and limited:

1. Stairways serving less than 10 occupants and only one level above or below the level of exit discharge may be unenclosed in all but Group H and I occupancies.

2. Vertical exits need not be enclosed in Group A-5 buildings, such as outdoor stadiums, where the entire means of egress is essentially open.

3. Stairways inside of single-dwelling units or sleeping units in R-1 occupancies need not be enclosed.

4. Stairways that are not a part of the required means-of-egress system need not be enclosed. Such a stairway would be in addition to the means of egress required by the code. The stair opening must also comply with §707.2.

5. Stairways in open parking structures, which serve only the parking structure, need not be enclosed.

6. One exit stairway in Group I-3 occupancies, such as correctional facilities, may have a glazed partial enclosure per the detailed criteria contained in §408.3.6. This a very specialized condition not often encountered.

7. Stage-exit stairways, per §410.5.3, need not be enclosed.

8. In other than Group H and I occupancies, half of the required egress stairways may be unenclosed when the stairways serve only one adjacent floor and each of the two interconnected floors is provided with two means of egress. The two interconnected floors cannot open onto other floors. Thus there must be at least one intermediate floor between a pair of interconnected floors. This is typically used in office or commercial buildings where one egress stairway may be unenclosed to allow the stairs to be a part of the day-to day circulation within the space. In practice this is essentially the same as Exception 4. This exception may also occur in a two-story building with two exit stairs, one of which is enclosed and the other is open.

9. In other than Group H and I occupancies interior stairs serving only the first and second stories of a building with an automatic sprinkler system are not required to be enclosed as long as two means of egress are provided from both floors served by the unenclosed stairs. The unenclosed stairs shall not be open to other stories.

Openings and Penetrations

§1020.1.1 and §1021.4 limit openings and penetrations in exit enclosures and exit passageways to those necessary for exit access to the enclosure from normally occupied spaces and for egress from the enclosure. This recognizes the primary use of exit components as pathways for egress.

Opening protectives for both exit passageways and exit enclosures shall be as required by §715. Thus 1-hour fire-rated walls will require 1-hour-rated doors per Table 715.4. Where higher fire ratings for exit enclosures are required, the rating of exit-passageway construction and opening protectives should match as well.

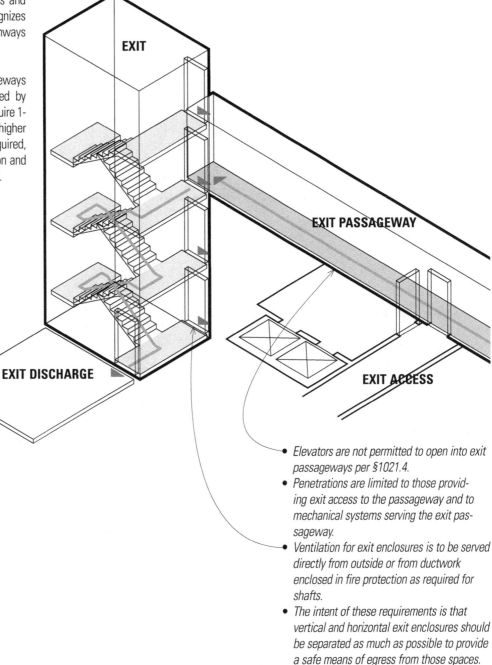

EXIT

EXIT PASSAGEWAY

EXIT DISCHARGE

EXIT ACCESS

- *Elevators are not permitted to open into exit passageways per §1021.4.*
- *Penetrations are limited to those providing exit access to the passageway and to mechanical systems serving the exit passageway.*
- *Ventilation for exit enclosures is to be served directly from outside or from ductwork enclosed in fire protection as required for shafts.*
- *The intent of these requirements is that vertical and horizontal exit enclosures should be separated as much as possible to provide a safe means of egress from those spaces.*
- *In malls areas such as mechanical and electrical rooms may open onto the exit passageway per §402.4.6 if they are provided with 1-hour fire-resistance-rated doors.*

EXIT DESIGN REQUIREMENTS

Vertical-Enclosure Exterior Walls

§1020.1.4 explains the requirements that apply to the exterior walls of vertical exit enclosures.

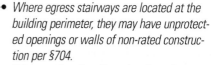

- Where egress stairways are located at the building perimeter, they may have unprotected openings or walls of non-rated construction per §704.
- Where stairwell walls project from the building at one side of the enclosure and are flush on the other, the walls flush with the enclosure need not be protected if the angle between the walls is equal to or greater than 180° (3.14 rad).
- Where this angle is less than 180° (3.14 rad), the building exterior wall must be of 1-hour rated construction with 3/4-hour opening protectives within 10' (3048) horizontally of the enclosure and vertically from the ground to a point 10' (3048) above the topmost landing of the stair or the roof, whichever is lower.

Discharge Identification Barrier

§1020.1.5 prohibits stairways in exit enclosures from extending below the level of exit discharge unless an approved barrier is provided to prevent occupants from proceeding down beyond the level of exit discharge.

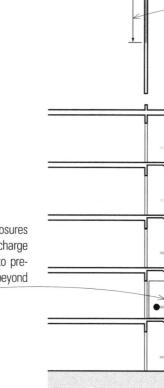

- Directional exit signs are to be provided per §1011, which governs exit signs.

Stairway Signage

§1020.1.6 requires that exit signage be provided inside of stairway enclosures serving more than three stories.

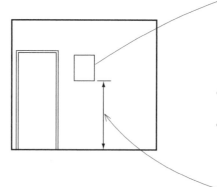

STAIR No. 1
Floors B through 4

3

No Roof Access
Exit at Floor 1

- These signs must identify the floor level at each floor landing.
- The exit signage must state whether the stair goes to the roof for fire access, as well as direct occupants to the level of and direction to the exit discharge.
- Exit signs are to be 5' (1524) above the floor and readily visible when the landing door is open and closed.

Smoke-proof Enclosures

§1020.1.7 requires smoke-proof enclosures or pressurized stairways for each of the required exits in high-rise buildings (§403), where floor surfaces are more than 75' (22 860) above the lowest level of fire-department vehicle access, and in underground buildings (§405), where floor surfaces are more than 30' (9144) below the level of exit discharge.

The stairway enclosures are to comply with the requirements of §909.20 for fire-protection systems in smoke-proof enclosures.

Exit Passageways

§1021.1 contains the requirements for exit passageways, which are similar to corridors in the exit access, but have more restrictions placed upon their use. They are not to be used for any other purpose than as a means of egress. They are similar to vertical stairway enclosures but are used for the horizontal portions of exits.

> 75' (22 860) above the lowest level of fire-department vehicle access

> 30' (9144) for underground

- The stairways shall exit into a public way or an exit passageway, yard or open space having direct access to a public way.
- The exit passageway is to have no other opening, which implies it is not used for any other purpose except for egress, and shall be of 2-hour fire-resistance-rated construction.
- Exceptions allow other openings into the exit passageway when they too are pressurized and protected as for the smoke-proof enclosure or the pressurized stairway.
- There are no limits on travel distance in vertical exit and exit-passageway enclosures so they could be many stories high or very long in horizontal distance. The intent is that occupants have the protection necessary to give them time in an emergency to safely traverse the egress travel distance.

- Width requirements for exit passageways are similar to other parts of the egress system, with a 44" (1118) minimum width for typical exit passageways and 36" (914) for those serving 50 or fewer occupants.
- Doors may not project more than 7" (178) into the passageway when fully open, nor reduce the required width by more than one-half in their fully open position.
- Exit passageways are to be of at least 1-hour fire-rated construction, or more if connected to an exit enclosure with a higher fire rating.

EXIT DESIGN REQUIREMENTS

Horizontal Exits

§1022 provides for horizontal exits, a unique concept for exiting because they provide an intermediate area of refuge rather than a complete means of egress to grade. The concept is based on the assumption that, by providing a high degree of fire-resistive separation between two segments of a building, occupants can pass through the separation, which is essentially a fire-compartment wall, and be safe from the danger that caused the need to egress.

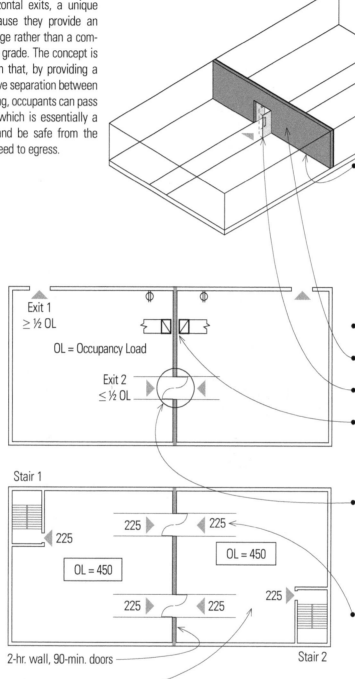

- *A horizontal exit is a wall that completely divides a floor of a building into two or more separate exit-access areas in order to afford safety from fire and smoke in the exit-access area of incident origin. The horizontal exit essentially creates separate buildings on each floor level to allow exiting to occur from one protected area into another without entering exit enclosures or exit passageways.*
- *The essential characteristics of a horizontal exit are as follows:*
- *The separation must be of at least 2-hour construction.*
- *Opening protectives are based on the fire-rating of the wall.*
- *There must be no interconnection of the ductwork or utilities between sides of the horizontal exit that would permit fire or smoke to enter the area of refuge.*

- *Horizontal exits may not serve as the only exit from a portion of a building. Where two or more exits are required, not more than one-half of the exits may be in the form of a horizontal exit. There are two exceptions which apply to I-2 or I-3 occupancies:*

- *In I-2 occupancies horizontal exits may comprise up to two-thirds of the required exits from a building or floor area. This recognizes that the occupants of I-2 occupancies may not be capable of unassisted self preservation and that they would likely be as well served in an emergency by getting into an area protected by a horizontal exit than they would being moved out of a building.*
- *In I-3 occupancies all exits are allowed to be horizontal exits. This recognized that occupants of these buildings are typically under restraint and cannot be allowed to freely exit a building in an emergency.*

- *For many buildings, using the concept of horizontal exit enables the omission of one stairway or egress pathway. If a building exceeds the allowable area and a fire wall is used to form a horizontal exit, only one stairway need be provided for each segment, and the horizontal exit can be used as the second exit for each segment.*

§1022.4 specifies that the refuge area to which a horizontal exit leads shall be of sufficient size to accommodate 100% of the occupant load of the exit access from which refuge is sought, plus 100% of the normal occupant load of the exit access serving as the refuge area. The capacity of such refuge floor area shall be determined by allowing 3 sf (0.28 m²) of net clear floor area per occupant, not including aisles, hallways and corridors. The area of stairs, elevators and other shafts shall not be counted.

In Group I-3 occupancies, the capacity of the refuge area shall be determined by allowing 6 sf (0.6 m²) of net clear floor area per occupant. In Group I-2 occupancies, the capacity of the refuge area shall be determined by allowing 15 sf (1.4 m²) of net clear floor area per ambulatory occupant and 30 sf (2.8 m²) of net clear floor area per nonambulatory occupant.

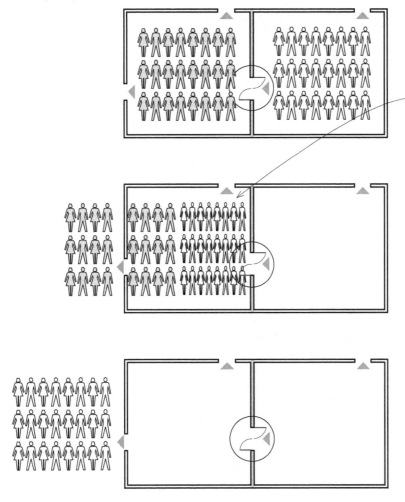

- The theory underlying the horizontal exit involves moving occupants from an area impacted by an emergency to another safe area of refuge. A building designed with a horizontal exit has two adjacent areas that may serve as both exit access and horizontal exit. To accommodate the anticipated number of occupants, the area to be considered as a horizontal exit must accommodate its own occupant load plus 100% of the load of the area being exited. This additional load is calculated at 3 sf (0.28 m²) per occupant of net clear area for other than hospitals. This recognizes that in an emergency people can be in tighter quarters and still egress safely. Where there may be nonambulatory patients the floor area requirements are 30 sf (2.8 m²) per occupant to allow more space for beds and wheelchairs. The application of the design criteria takes into account the direction of exit flow. The downstream area receiving the occupants from upstream must be able to accommodate both its normal occupant load and that of the occupants from the adjacent space. Only in cases where a horizontal exit may function in both directions must both areas have more exit capacity than is normally required for the area itself. A two-way exit is allowed when the exit-access design requirements are independently satisfied for each exit-access area.

- The design of the exit-access capacity for the building segment serving as the refuge area shall be based on the normal occupant load served and need not consider the increased occupant load imposed by persons entering such refuge area through horizontal exits.

Exterior Exit Ramps and Stairways

§1023 contains the requirements for exterior exit ramps and stairways that serve as a component of a required means of egress. Exterior exit ramps and stairways may be used in a means of egress for occupancies less than 6 stories or 75' (22 860) in height. Note that exterior exit ramps and stairways may not be used as an element of a required means of egress for Group I-2 occupancies.

- *Exterior exit stairways serving as part of a required means of egress must be open on one side, with an aggregate open area of at least 35 sf (3.3 m²) at each floor level and at each intermediate landing.*
- *Each opening must be a minimum of 42" (1067) high above the adjacent walking surface.*
- *The open areas must face yards, courts, or public ways.*

- To second stairway

Exterior exit ramps and stairways are to be separated from the building interior per §1020.1, with openings limited to those necessary for exit access. The exceptions to this protection requirement are:

1. In other than Group R-1 or R-2 occupancies, in buildings no more than two stories above grade, and where the exit discharge is at the first story, no protection is required. The path of travel is only one story high and is considered short enough to not need protection.

2. When there is an exterior egress balcony that connects two remote exterior ramps or stairways with a 50% open balcony and openings at least 7' (2134) above the balcony, protection is not required. This recognizes the inherent safety of having two means of egress from the balcony as well as the protection afforded by the openings, which minimize the chance of heat or toxic gases being trapped in the egress path.

3. Exterior ramps or stairs need not be protected where interior stairways may also be unprotected per §1020.1.

Open-Ended Corridors

The fourth exception to §1023.6 is used where there are open-ended corridors that are by their nature not separated from exit stairways. Additional separation from the interior to the stairways is not required if several conditions (not exceptions) are met.

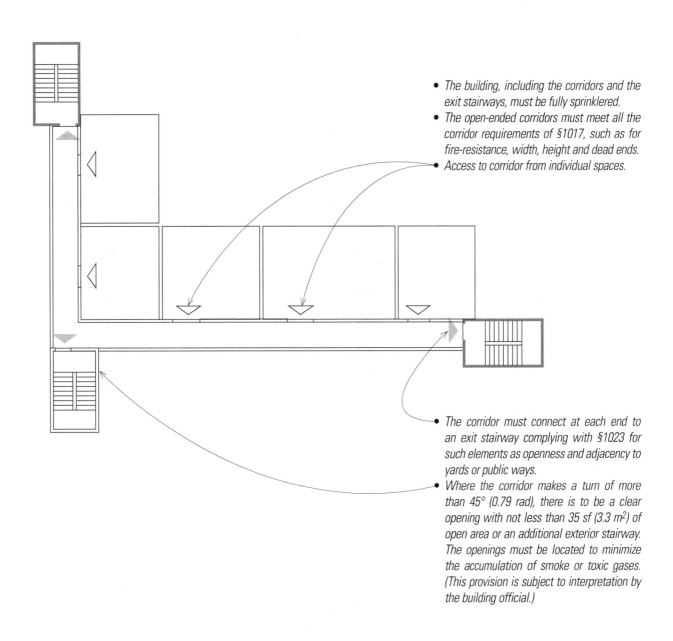

- *The building, including the corridors and the exit stairways, must be fully sprinklered.*
- *The open-ended corridors must meet all the corridor requirements of §1017, such as for fire-resistance, width, height and dead ends.*
- *Access to corridor from individual spaces.*

- *The corridor must connect at each end to an exit stairway complying with §1023 for such elements as openness and adjacency to yards or public ways.*
- *Where the corridor makes a turn of more than 45° (0.79 rad), there is to be a clear opening with not less than 35 sf (3.3 m²) of open area or an additional exterior stairway. The openings must be located to minimize the accumulation of smoke or toxic gases. (This provision is subject to interpretation by the building official.)*

EXIT DISCHARGE

§1024 contains the provisions for the design of the exit discharge. The exit discharge is the third and final portion of the means of egress. It begins where building occupants leave the exit portion of the means of egress and ends when they reach the public way. There is some flexibility on the part of the building official to determine when occupants have gone far enough from the structure in question to be safe, and in these cases, the exit discharge may lead through an egress court. Egress courts are defined as a court or yard that provides access to a public way.

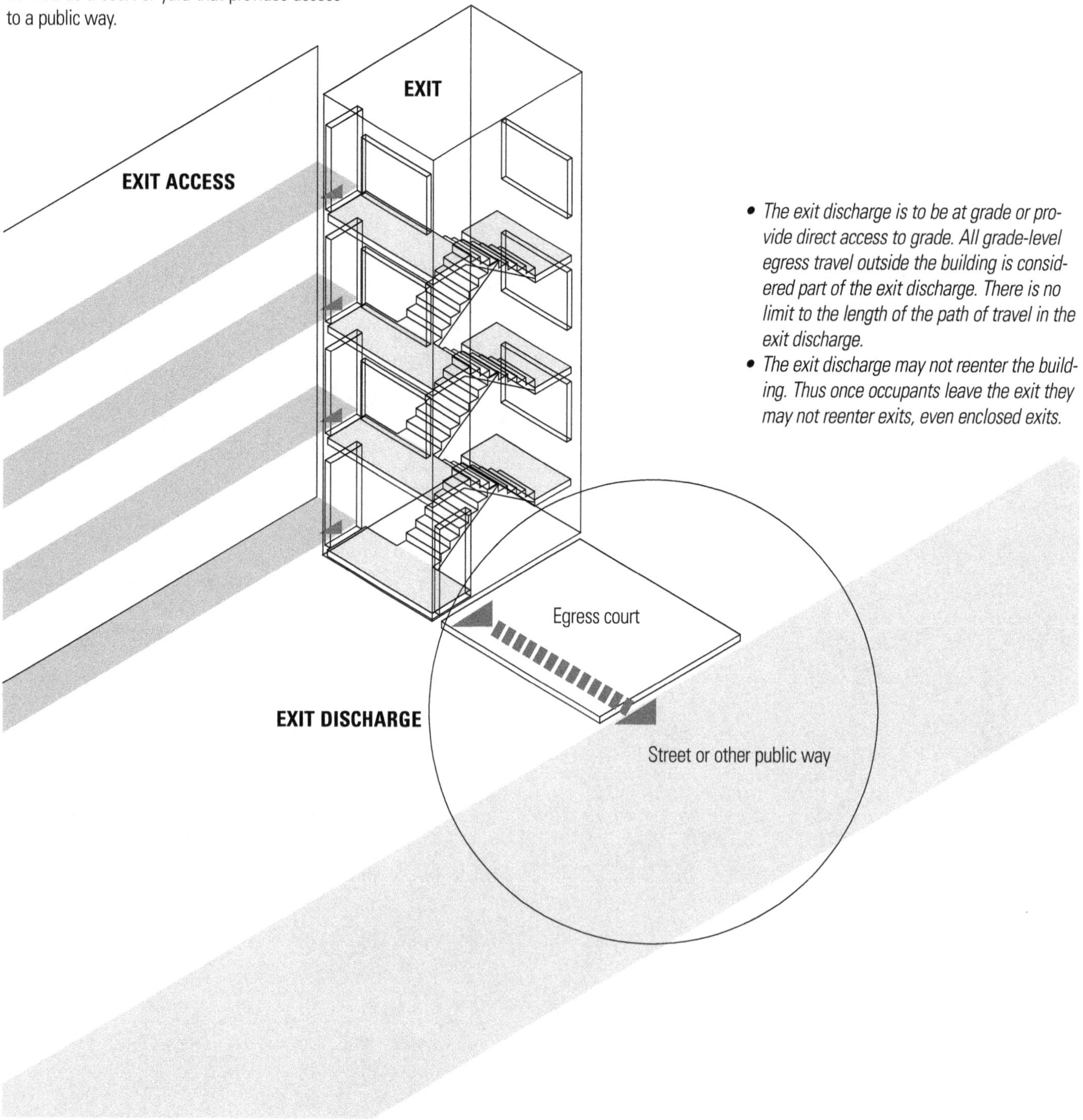

EXIT

EXIT ACCESS

- *The exit discharge is to be at grade or provide direct access to grade. All grade-level egress travel outside the building is considered part of the exit discharge. There is no limit to the length of the path of travel in the exit discharge.*
- *The exit discharge may not reenter the building. Thus once occupants leave the exit they may not reenter exits, even enclosed exits.*

Egress court

EXIT DISCHARGE

Street or other public way

Exit Discharge through Intervening Spaces

The exit discharge path must connect the exit to the public way without intervening spaces except under specific conditions:

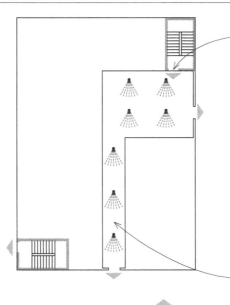

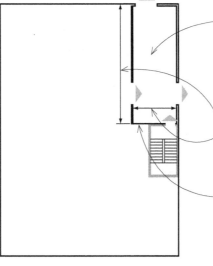

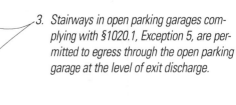

1. Up to 50% of the number and capacity of the required exit enclosures may exit through areas on the level of exit discharge when **all** of the following conditions are met:

1.1 Such exit enclosures egress to a free and unobstructed way to the exterior of the building.

1.2 The entire level of exit discharge is separated from the area below by fire-resistive construction complying with the required fire rating for the exit enclosure. (It is not clear why this applies only to levels below, nor what is to be done when no levels below exist.)

1.3 The egress path from the exit enclosure is protected throughout by a sprinkler system.

2. A maximum of 50% of the number and capacity of the exit enclosures may exit through a vestibule when **all** the following conditions are met:

2.1 The vestibule is separated from the areas below by fire-rated construction to match that of the exit enclosure.

2.2 The depth of the vestibule from the exterior of the building is not more than 10' (3048) and its width is not more than 30' (9144).

2.3 The vestibule is separated on the level of exit discharge by construction equivalent to wire glass in a metal frame. This is roughly equivalent to $3/4$-hour of fire-resistance-rated construction per §715.5.3.

2.4 The area is used only for means of egress and exits directly to the outside.

3. Stairways in open parking garages complying with §1020.1, Exception 5, are permitted to egress through the open parking garage at the level of exit discharge.

EXIT-DISCHARGE DESIGN REQUIREMENTS

Exit-Discharge Capacity

§1024.2 requires that the capacity of an exit discharge be not less than the required discharge capacity of the exits being served.

Exit-Discharge Location

§1024.3 requires that exit-discharge components be separated from adjacent property lines by at least 10' (3048) of space, and from other buildings on the same lot unless the adjacent elements are fire-rated per §704, based on fire-separation distance.

Exit-Discharge Components

Per §1024.4 exit-discharge components are presumed to be outside the building envelope and are required to be sufficiently open to prevent the accumulation of smoke or toxic gases.

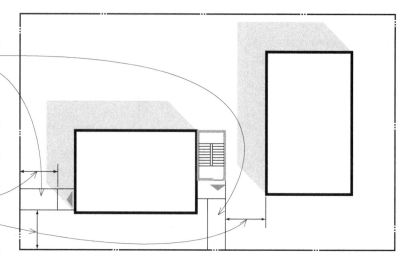

Egress Courts

§1024.5 requires egress courts to comply with corridor and exit-passageway requirements for clear width and door encroachment.

• When the egress court is wider than the required egress width, any reductions in width should be gradual and at an angle of less than 30° (0.52 rad). This is to prevent having inside corners where occupants could be trapped by the press of large groups of people exiting the building in an emergency. This is another example of the water-flow analogy for exiting.

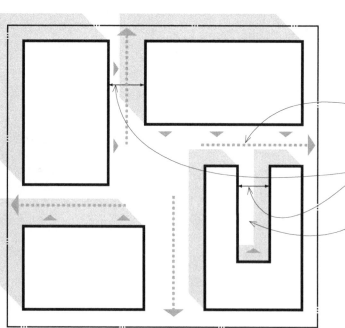

• Egress courts are open spaces that provide access to a public way from one or more exits.

• Egress courts less than 10' (3048) wide are to be of 1-hour-rated construction for 10' (3048) above the floor of the court.

• Openings in such egress courts are to have $3/4$-hour-rated opening protectives that are fixed or self-closing, except in Group R-3 occupancies or where the court serves an occupant load of less than ten.

§1025 contains egress requirements specific to assembly occupancies which supplement the general requirements located elsewhere in Chapter 10. Very large assembly occupancies are required to have smoke-protected seating. The number of such large assembly occupancies that are designed and constructed every year are so small as not to warrant treatment in this book. However, there are several basic principals that apply to assembly occupancies that should be understood.

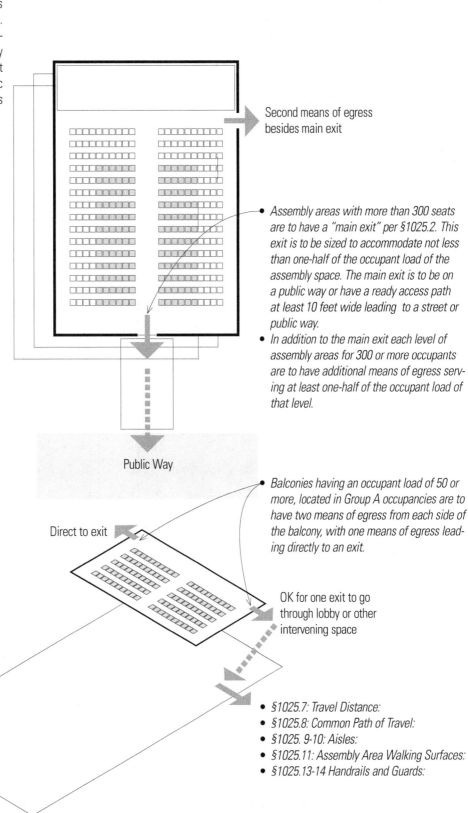

Second means of egress besides main exit

- *Assembly areas with more than 300 seats are to have a "main exit" per §1025.2. This exit is to be sized to accommodate not less than one-half of the occupant load of the assembly space. The main exit is to be on a public way or have a ready access path at least 10 feet wide leading to a street or public way.*
- *In addition to the main exit each level of assembly areas for 300 or more occupants are to have additional means of egress serving at least one-half of the occupant load of that level.*

Public Way

- *Balconies having an occupant load of 50 or more, located in Group A occupancies are to have two means of egress from each side of the balcony, with one means of egress leading directly to an exit.*

Direct to exit

OK for one exit to go through lobby or other intervening space

- *§1025.7: Travel Distance:*
- *§1025.8: Common Path of Travel:*
- *§1025. 9-10: Aisles:*
- *§1025.11: Assembly Area Walking Surfaces:*
- *§1025.13-14 Handrails and Guards:*

EMERGENCY ESCAPE

I-1 and R occupancies are required by §1026 to have egress openings for emergency escape and rescue. These are in addition to normal paths of egress leading out of rooms in such occupancies. These are to provide a way out of sleeping rooms for the occupants and a way into those rooms for rescue personnel in emergencies.

Basements and all sleeping rooms in the above noted occupancies below the fourth story are to have at least one emergency escape and rescue opening. These are to open directly onto a public street, public alley, yard or court. There are exceptions to this requirement applicable to certain occupancies, but we recommend provision of such openings in all residential occupancies where it is practical.

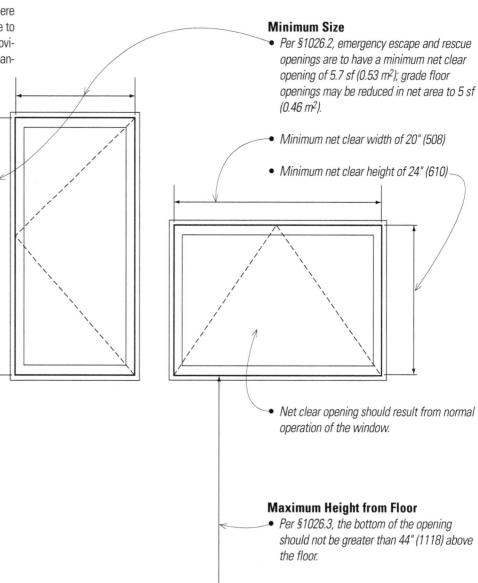

Minimum Size

- *Per §1026.2, emergency escape and rescue openings are to have a minimum net clear opening of 5.7 sf (0.53 m²); grade floor openings may be reduced in net area to 5 sf (0.46 m²).*

- *Minimum net clear width of 20" (508)*

- *Minimum net clear height of 24" (610)*

- *Net clear opening should result from normal operation of the window.*

Maximum Height from Floor

- *Per §1026.3, the bottom of the opening should not be greater than 44" (1118) above the floor.*

11

Accessibility

Chapter 11 regulates the design and construction of "facilities for accessibility to physically disabled persons." The scope definition for this section thus varies from other state and federal standards that discuss people with disabilities instead of the term used in the IBC. This subtle difference in definitions highlights the primary issue to be kept in mind about this chapter of the Code. The terms of federal law and of state code modifications often expand coverage of local codes far beyond the basic Code requirements. The designer must understand that many states have significantly different, locally developed accessibility requirements; in some cases, states have rewritten Chapter 11 completely. The basic Code provisions should therefore be reviewed carefully against federal criteria as well as state and local amendments. The likelihood of significant and extensive local variation from the basic IBC is higher for this chapter than for any other in the code. The designer must be certain to use the correct accessibility code for a specific project.

Because the defined scope for accessibility makes reference to §3409, Accessibility for Existing Buildings, that section will be examined as part of this chapter. Also, we believe that several portions of the "Supplementary Accessibility Requirements" of Code Appendix E apply in most conditions covered by the ADA or local amendments; we will therefore discuss Appendix E as part of this chapter. The designer should verify that status of the adoption of this Appendix by the local AHJ. Because of the interrelationship of scoping provisions for new and existing buildings, the authors assume that the existing building provisions of Chapter 34 will be enforced and that Appendix E will be adopted. We will therefore treat Chapter 11 as containing both the body of the Code and the Appendix.

Making buildings accessible to persons with disabilities is an increasingly important design requirement. It has become an even greater issue since the passage of the Americans with Disabilities Act (ADA) in 1990.

The model codes have incorporated requirements for accessibility that are intended to be coordinated with the requirements of the ADA. However, the designer must remember that plan review by the building official is only for compliance with the provisions of the building code. The model codes are typically not considered to be equivalent substitutions for the ADA and compliance with the Code is no guarantee of compliance with the ADA. Therefore, no designer should rely solely on the Code to determine access-compliance requirements. Every project should also be carefully reviewed against the provisions of the ADA to assure compliance with federal law. Remember that any approval by the building official has no bearing on the applicability of the ADA. The building official does not review for ADA compliance and has neither authority nor responsibility to enforce this federal law.

The code basis for access clearances and reach ranges is primarily for people who use wheelchairs. It is important to remember that the definition of disability also includes sensory and cognitive impairments, not just mobility impairments. Designers must also accommodate people with visual impairments and people with hearing impairments. Provision for access by mobility-impaired people will accommodate most disabilities, but the designer must also be certain to accommodate other disabled groups, such as people of short stature, in a coordinated fashion. A design solution for one group of disabled people should not adversely impact another group with different disabilities.

Accessibility to buildings is monitored closely by a large number of advocacy groups. They review the provision of, or lack of access to, buildings on an ongoing basis. A challenge to the accessibility of a building is among the most likely post-occupancy code or legal reviews that can happen after the completion of a project. Any decisions that lead to a lack of access are subject to scrutiny over the life of the project. Our advice is: If the applicability of accessibility requirements is in doubt, the designer and building owner should opt to provide the access.

The basic accessibility requirements in the Code have been developed from the predecessor model codes and scoping criteria. The Code requires that buildings and facilities be designed and constructed to be accessible in accordance with the Code and the detailed dimensional requirements contained in the referenced standard, ICC A117.1, referenced hereafter simply as A117.1.

The provisions of Chapter 11 focus on scoping, telling the designer and building official where provisions are to apply and in what quantities. For example, scoping sets forth such items as the number of required accessible rooms in hotel accommodations. The dimensions and technical requirements for such accessible hotel rooms are contained in A117.1. The designer **must** therefore have a current reference copy of A117.1 to be able to comply with the access requirements of the Code.

Designers should read and familiarize themselves thoroughly with the detailed requirements of A117.1. We will only touch on certain areas of that standard here. A thorough and detailed examination of A117.1 is the subject for another book. A117.1 has a strong resemblance to the Americans with Disabilities Act Access Guidelines (ADAAG), but they are not identical. Again, we reiterate that the designer must consult both the locally modified code and the ADA to be certain of compliance. There is currently a movement to coordinate A117.1 and ADAAG, but it is likely that there will always be subtle differences between A117.1, ADAAG, and local access codes. The designer must therefore make comparisons between the specific standards applicable for each project.

The designer must also be completely familiar with the Americans with Disabilities Act (ADA), particularly the ADA Accessibility Guidelines (ADAAG). ADAAG sets forth the fundamental federal design criteria for access. The ADA is a civil-rights law and applies under different circumstances than the Code. It also may be applied retroactively under different circumstances than the Code as well. It should be presumed from the beginning of a design project that both ADAAG and the Code, as amended by local code modifications, apply to the project. A detailed comparison of the applicability and potential conflicts between federal guidelines, local modifications and the Code is beyond the scope of this book, but designers must not only review the application of the Code to projects but also ADAAG and local modifications as well.

Designers should pay particular attention to the requirements for door clearances and landings contained in A117.1. The requirements for clear and level landings and for clearances at the latch side of doors depend on the direction of approach to the door by persons with mobility impairments. The provision of space in the floor plan for these clearances is essential during the initial design phases of a project.

The designer must consider the approach configuration and the side of approach to determine which criteria to apply. If these clearances are not provided at the outset of planning it will be very difficult to provide such clearances later without major revisions to the floor plans. It is essential that such clearance criteria for doors and for Type A and B dwelling units designed and constructed to be in accordance with A117.1 and the Federal Fair Housing Act be understood early and incorporated into the plans.

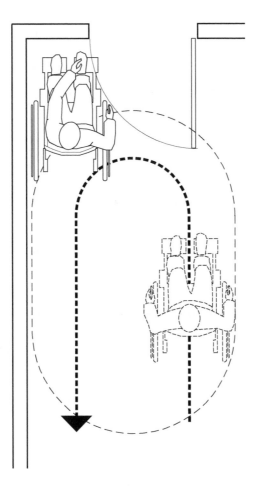

- *Study the diagrams in A117.1 to understand the criteria for provision of these clearances. The clearances are to allow wheelchairs to clear door swings, to allow maneuvering around obstacles, and to allow wheelchairs to pivot and turn.*
- *It is acceptable to demonstrate alternative means of providing such maneuvering room, but the designer must understand wheelchair movements and design criteria in the same way one understands how a car moves through a parking lot or a truck approaches a loading dock. The same understanding should be gained of reach ranges for people seated in wheelchairs to allow the designer to correctly place operating switches or equipment.*

Accessible Route

The intent of an accessible route is to allow persons with disabilities to enter a building, get into and out of spaces where desired functions occur, and then exit the building. It is also necessary that support functions such as toilets, telephones and drinking fountains be accessible. The goal is the integration of people with disabilities into the full function of a facility with a minimum of atypical treatment. Such unequal treatment could be considered to include provision of separate routes or auxiliary aids such as wheelchair lifts.

The concept of universal design is useful to keep in mind. The goal of universal design is to make facilities accessible to the widest possible range of people, regardless of mobility, physical ability, size, age or cognitive skills.

The designer should bear in mind all types of disabilities when designing accessible routes. Projections may affect people with visual impairments and, of course, changes in grade, such as ramps or steps, impact people in wheelchairs or with limited mobility.

The requirements for areas of refuge, discussed in Chapter 10, recognize that getting people with disabilities into a building under normal conditions may not allow them to exit under emergency circumstances. See the provisions for areas of refuge in §1007.6.

- The basis of accessible design is the idea of an accessible route. The definition of *accessible route* in §1102.1 uses the words *continuous* and *unobstructed*. Any usable path for people with disabilities must not cut them off from the spaces or elements of the building that they have a right to use. This route is often referred to in other code sections or access documents as the path of travel.

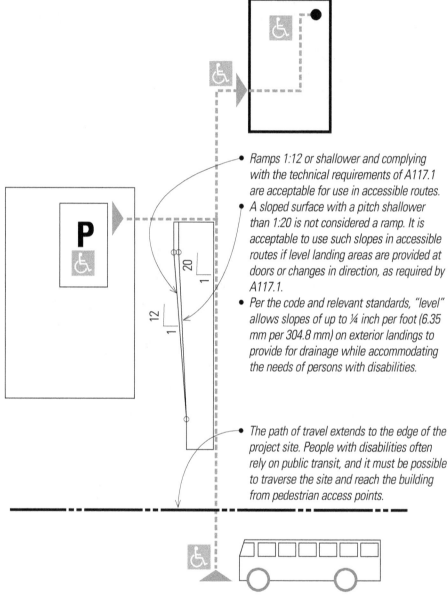

- Ramps 1:12 or shallower and complying with the technical requirements of A117.1 are acceptable for use in accessible routes.
- A sloped surface with a pitch shallower than 1:20 is not considered a ramp. It is acceptable to use such slopes in accessible routes if level landing areas are provided at doors or changes in direction, as required by A117.1.
- Per the code and relevant standards, "level" allows slopes of up to ¼ inch per foot (6.35 mm per 304.8 mm) on exterior landings to provide for drainage while accommodating the needs of persons with disabilities.

- The path of travel extends to the edge of the project site. People with disabilities often rely on public transit, and it must be possible to traverse the site and reach the building from pedestrian access points.

Dwelling Unit Types

The requirements for provision of both Type A and B dwelling units apply to a broad category of residential units. These requirements have significantly changed the space requirements for single-level multifamily housing and multiple-level buildings where elevators are provided. Designers of Group-R occupancies should study the scoping requirements carefully. Note also that while these code revisions are intended to conform to Federal Fair Housing requirements, those requirements should be reviewed as well, so that designers may assure themselves of compliance with both federal and local requirements.

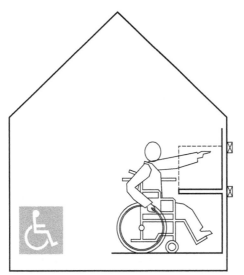

- *Definitions and requirements for Type A and B dwelling units have been incorporated into the Code to bring the code requirements in line with the federal Fair Housing Act Guidelines (FFHA). The designer has the option under given conditions of providing fully accessible units per A117.1 (Type A) or adaptable units (Type B) in residential projects. The scoping requirements for these dwelling units are contained in §1107. The plan dimension requirements are essentially the same for both types of units, as clearances must be provided for future accessibility in adaptable units.*

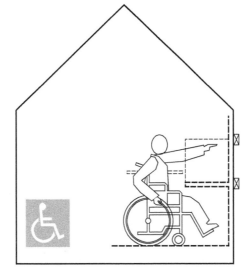

- *Certain accessibility features, such as grab bars, lowered counters and clear space under cabinets, may be provided later in adaptable units. Construction features, such as convertible cabinetry and wall blocking for grab bars, are necessary so that the conversion to an accessible unit may be made readily, without major remodeling.*

Technically Infeasible

The definition of technically infeasible should be studied carefully and used with great caution. This definition appears in Chapter 34 pertaining to existing structures and applies only to the alteration of existing buildings where provision of access would require removal of essential load-bearing members of a structure, or where some other site or physical constraint would prevent full compliance with code requirements for access. Application of this provision may lead to potential conflicts with the ADA or local access code modifications. Such regulations or legislation may have different criteria for determining technical infeasibility than this code section. Application of this provision should only be done after careful consultation with the client and the local AHJ.

SCOPING REQUIREMENTS

§1103.1 states that the requirements for accessibility apply both to buildings and to the sites where they are located. The accessible route is typically considered to extend to the boundaries of the site. Every path of travel may not need to be accessible, but a readily located accessible route must be provided inside and outside of the building and on the site.

Exceptions

§1003.2 specifies the exceptions to access requirements, and these exceptions are distributed among sections in other chapters of the Code. The exceptions apply in general to low occupancy spaces and those where provision of access would be disproportionate to the utility of the space. However, it is also worth noting once again that the exceptions noted in the Code may not correspond to federal or local laws and regulations, and must be carefully reviewed.

Specific exceptions to accessibility in the IBC include the following items:

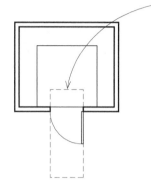

- *Walk-in coolers and freezers are not required to be accessible where they are intended for employee use only. Local access requirements should be verified with the local AHJ. Note that ramped walkways are available in such coolers and should be investigated by the designer as a means of providing access in the interest of universal design.*

When dealing with accessibility requirements, especially for new construction, the designer's first assumption should be that all areas and all functions used by the public should be accessible. It is also a prudent assumption that all employee areas should be accessible to people with disabilities except for specific service areas not typically occupied by either the public or employees.

- *Because of the potential for local or federal variations use the exceptions with great caution!*

- *Individual workstations are not required to be accessible but must be on an accessible route. Thus aisles leading to workstations must be of accessible width.*
- *Detached residences, such as single-family homes and duplexes, along with their sites and facilities, are not required to be accessible.*
- *Utility buildings (Group-U occupancies) not open to the general public or required to have accessible parking are exempt from accessibility requirements.*
- *Construction sites need not be accessible.*
- *Raised areas, areas with limited access, equipment spaces and single-occupant structures, such as guard stations, catwalks, crawl spaces, elevator pits and tollbooths, are not required to be accessible.*
- *R-1 occupancies with less than 5 sleeping rooms that are also the residence of the proprietor are not required to be accessible.*
- *Where a day-care facility is part of a dwelling unit only the portion of the structure used for day care is required to be accessible. The converse, however, is also that where day-care facilities are provided in dwelling units, the day-care portion must be accessible.*

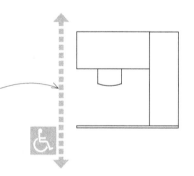

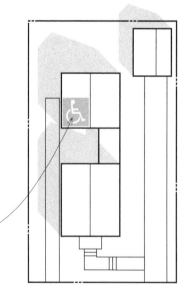

§1104 requires accessible routes to be provided at site arrival points, from public transportation stops, accessible parking and loading zones, public streets and sidewalks within the site. The only exception to this requirement is for a large site where elements are spread out along a vehicular access route that is not in itself accessible. This assumes that all people on the site are moving from one location to another by vehicle and that there is not a separate inaccessible pedestrian route for people without disabilities. The wording seems to require that an accessible route must be provided where a pedestrian path is included as part of the site path-of-travel system. This is consistent with the idea that persons with disabilities receive equal treatment as pedestrians. Note that this exception does not apply to buildings where there are Type B units; in such case an accessible route from site arrival points would need to be provided. This provision is included in the code to comply with the federal Fair Housing Act. Keep in mind that accessibility is to be provided for persons with many different types of disabilities, not just persons with mobility impairments.

Mezzanines and similar multilevel spaces must be connected by an accessible route unless they are under 3,000 sf (278.7 m²) in area. This minimum size exemption does not apply when these spaces house the offices of healthcare providers (B or I occupancies), transit facilities, or separate tenant spaces in multiple-tenant Group-M occupancies. There are also limited exceptions for other occupancies, see the provisions in §1107 that apply to specific healthcare related I and R occupancies. In general every effort should be made to provide at least one accessible route to all usable spaces within a building.

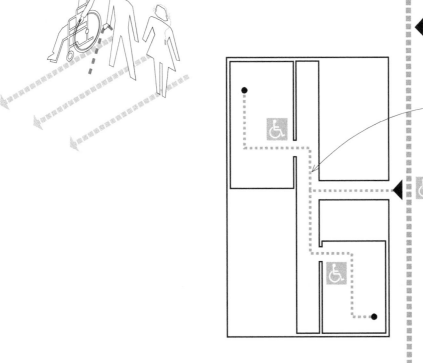

Within a building, an accessible path of travel is to be provided from accessible entrances to each portion of the building. If only one route is provided, it cannot pass through kitchens or service spaces. The accessible route must not be seen as a second-class way of moving about in the building. The accessible route should correspond to or be near the general circulation path. Where general circulation paths are interior and thus protected from the elements, the accessible route must be on the interior as well.

ACCESSIBLE ENTRANCES

§1105 requires that at least one main entrance, and at least 60% of the total of all entrances, to a building must be accessible. Where there are separate tenant space entries the same criteria apply to each tenant space. The only exceptions are entrances not required to be accessible and service entrances or loading docks that are not the only entry to a building or tenant space. Where service entrances are the only entrance, they are to be accessible.

When entrances to the building serve accessible adjunct facilities, such as accessible parking areas, passenger loading zones, transit facilities or public streets, then at least one of each of the entries serving those functions must be accessible. The design of the accessible route and entrance systems cannot require people with disabilities to traverse long distances to get from one accessible facility to another.

This building entrance also may not need to be accessible, because a total of at least 60% of the other entrances provide access to an accessible route. If there are only two entrances then both must be accessible to meet the 60% criteria.

Entrances into tenant spaces such as this one need not be accessible, because accessible entrances are already provided from within the building.

The requirements of §1106 highlight the nature of the how Chapter 11 operates. The requirements for parking are stated in terms of quantities, not as criteria for how to lay out accessible stalls. The stall dimensions are contained in the reference standard, A117.1. The scoping requirements for parking are per Table 1106.1, which stipulates the number of accessible parking stalls to be provided based on the total number of parking spaces provided.

- The accessible stalls are to be included in the total parking count, but they must be set aside for use by people with disabilities. The basic criterion is that at least 2% of parking spaces be accessible. Parking requirements at hospitals are higher, requiring that 10% of patient and visitor parking spaces serving hospital outpatient facilities be accessible. It may be difficult to determine how spaces in a large parking lot may be allocated in practice between inpatient and outpatient visitors and patients. This is another area where if there is any doubt or ambiguity the prudent course is to provide the highest standard of access that may potentially be applicable. For rehabilitation and outpatient facilities the requirements are increased again, being 20% of the total. This recognizes that the population of users of such facilities is much more likely to have disabilities that would require accessible parking.

- For every 6 accessible spaces, or fraction thereof, one space is to be a van-accessible space. This is a space with a larger access aisle than the typical accessible parking space. For small parking areas, where only one space is provided, it must be van accessible. The van space counts toward the total of required accessible spaces. All accessible parking counts in the grand total of parking spaces, both accessible and nonaccessible.

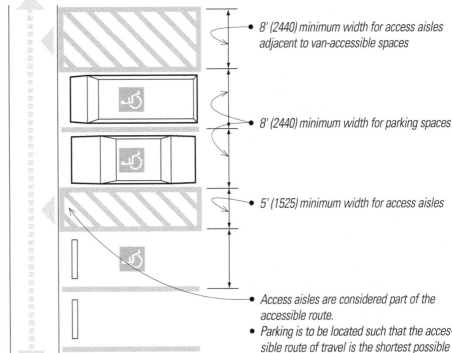

- 8' (2440) minimum width for access aisles adjacent to van-accessible spaces

- 8' (2440) minimum width for parking spaces

- 5' (1525) minimum width for access aisles

- Access aisles are considered part of the accessible route.

- Parking is to be located such that the accessible route of travel is the shortest possible path from the parking area to the nearest accessible building entrance. Note that this requirement can be construed to place accessible parking immediately adjacent to an entrance no matter how the circulation in the parking area is configured. The definition of "possible" can be the subject for debate and subject to non-uniform interpretation by building officials and disabled access advocates. For buildings with multiple accessible entrances and adjacent parking, accessible parking is to be dispersed in such a way as to have accessible parking nearby each accessible entrance.

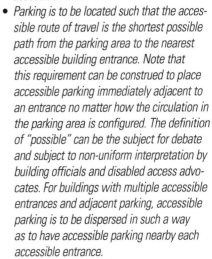

- Example: Parking for 400 vehicles; 8 accessible spaces minimum; 2 of the 8 must be van-accessible since there are more than 6 accessible spaces required. When the triggering number is exceeded by any fractional amount the requirement rounds to the next whole number of spaces.

- Accessible passenger loading zones are to be provided at medical facilities where people stay longer than 24 hours, whether as long-term residents or as medical patients. Loading zones are also required when valet parking services are provided.

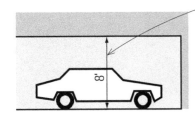

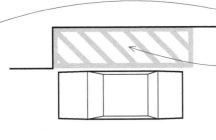

- Accessible passenger loading zones should have a minimum overhead clearance of 8' (2440).

- 8' (2440) minimum width for adjacent access aisle

The intent of the dwelling unit and sleeping unit requirements is that hotels, motels, multifamily dwellings and Group I institutional uses be accessible to the maximum extent that is practical, and in quantities that will serve the anticipated demand by people with disabilities. The determination of the quantities required is not always empirically based, but the designer must use the scoping required by the Code to design the facility. Common-use and recreational facilities open to all residents or to the public are to be accessible as well.

Accessible Dwelling Units and Sleeping Units

§1107.2 divides accessible dwelling units and sleeping units into accessible units, Type A or B units. We will use the word "units" here to include both dwelling and sleeping units unless indicated otherwise. Sleeping units are defined in Chapter 2 as rooms or spaces in which people sleep. This could include dormitory, hotel or motel guest rooms or suites, sleeping rooms in dwelling units or nursing home rooms.

Accessible Spaces

The requirements of §1107.3 arise from the requirements for accessible routes. Rooms and spaces that serve accessible units and that are available to the general public, or for the use of the residents in general, are to be accessible. This includes such areas as toilet and bathing rooms, kitchen, living and dining areas and exterior spaces including patios, terraces and balconies. There is an exception related to recreational facilities that we will discuss later in commentary about §1109.14.

Accessible Route

§1107.4 requires that in Group R-1 and R-2 occupancies at least one accessible route connect the primary building or site entrance with the front entry to each accessible dwelling unit. An accessible route should also connect the accessible units to areas that serve the units, such as laundry rooms or community rooms. The exceptions to this provision apply on steep sites where buildings with dwelling units are connected by nonaccessible vehicular routes, similar to the provisions of §1104. In such cases, accessible parking at each accessible facility is acceptable.

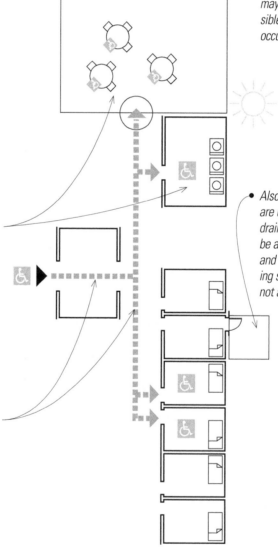

- *Accessible units are described in A117.1 and are meant to be fully accessible with all accessible provisions installed.*
- *Type A units are described in A117.1 and are meant to be laid out to be fully accessible. Not all access provisions may be implemented initially, but all access provisions can be readily implemented.*
- *Type B units are also described in the standard, but are defined to comply with the technical requirements of the Federal Fair Housing Act Guidelines administered by the Department of Housing and Urban Development (HUD). Type B dwelling units therefore provide a minimum level of accessibility and are more easily thought of as "adaptable" units where minor modifications may be made to make them more accessible, if necessary, to accommodate disabled occupants.*

- *Also, decks and patios in Type B units that are usually dropped at exterior doors for drainage and weather protection need not be accessible if the step is less than 4" (102) and the exterior area has an impervious paving surface. Note that this exception does not apply to Type A dwelling units.*

Group I Occupancies

Group I occupancies have detailed special requirements in §1107.5 that recognize the institutional nature of this occupancy group. Typically, the requirements for the number of accessible spaces are higher than for other occupancies as these facilities are expected to have a higher proportion of people with disabilities than other occupancies.

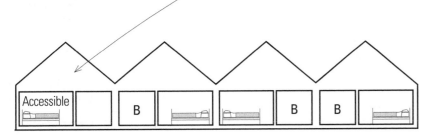

- Group I-1 occupancies are to have accessible units and Type B units.
- At least 4%, but not less than one of the dwelling units and sleeping units are to be accessible units.
- Where there are four or more dwelling or sleeping units intended to be occupied as a residence then every unit is to be a Type B unit. §1102 has the following definition for "INTENDED TO BE OCCUPIED AS A RESIDENCE": "This refers to a dwelling unit or sleeping unit that can or will be used all or part of the time as the occupant's place of abode."
- Thus, for most such facilities, which will typically have four or more units, all units are to be Type B and of the unit totals 4% are to be accessible units.
- Note that the number of Type B units may be reduced in accordance with §1107.7, which is further discussed later in this chapter.

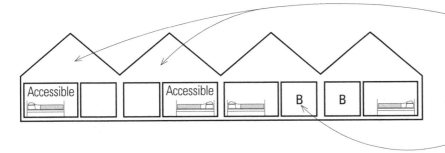

- Group I-2 nursing home requirements are described in §1107.5,2. These facilities, which have a high proportion of persons with disabilities, have high requirements for both accessible units and Type B units.
- At least 50%, but not less than one of the dwelling units and sleeping units, are to be accessible units.
- Where there are four or more units the requirements and criteria for Type B units are the same for I-2 nursing home occupancies as they are for I-1 occupancies as noted above.
- Thus, for most such facilities, which will typically have four or more units, all units are to be Type B and of the unit totals 50% are to be accessible units.
- Note that the number of Type B units may be reduced in accordance with §1107.7, which is further discussed later in this chapter.

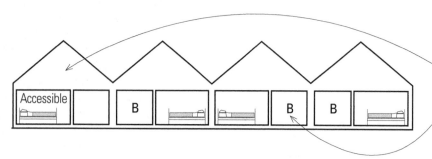

- Group I-2 requirements are described in §1107.5.3 which is misleadingly labeled as "I-2 hospitals." The provisions of this section apply not only to general care hospitals, but also to psychiatric facilities, detoxification facilities and residential care/assisted living facilities. Thus the residential requirements for I-2 which may seem out of place for a hospital do apply for other I-2 occupancies.
- At least 10%, but not less than one of the dwelling units and sleeping units are to be accessible units.
- Where there are four or more units the requirements and criteria for Type B units are the same for these I-2 occupancies as they are for I-1 occupancies as noted above.
- Thus, for most such facilities, which will typically have four or more units, all units are to be Type B and of the unit totals 10% are to be accessible units.
- Note that the number of Type B units may be reduced in accordance with §1107.7 which is further discussed later in this chapter.

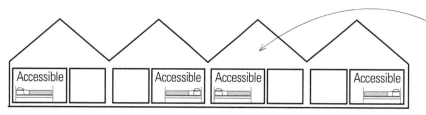

- In Group I-2 rehabilitation facilities per §1107.5.4, which may be a hospital or a separate facility or a part of either which specializes in treatment of conditions that affect mobility, 100% of the dwelling units and sleeping units are to be accessible units.

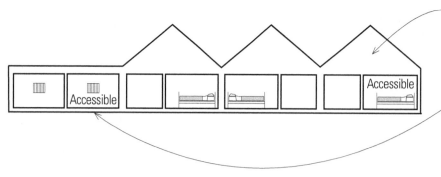

- Group I-3 occupancies have requirements that 2% of the dwelling units, but not less than one, of sleeping units be accessible units. There are no requirements for Type B units in I-3 occupancies.
- Where there are specialized cells, such as holding cells or detention cells, at least one of each type of cell is to be an accessible unit. These requirements are in addition to those for the normal population of the institution. Medical care facilities shall have accessible units, which are to be counted separately from typical cells and special function cells.

Group R Occupancies

The triggers for various scoping requirements in Group R occupancies are the number of sleeping spaces and the number of units. A percentage of each type of multi-unit residential facility must be accessible, with the number and the diversity of the types of accommodation increasing with the size and type of the residential facility. These provisions do not apply to individual single family residences, but do apply to single-family residences where there are more than 4 such units per building.

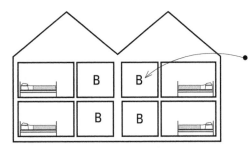

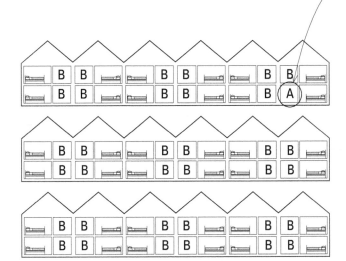

- *§1107.6.1.2 requires every dwelling unit to be a Type B dwelling unit in residential occupancies, such as Group R-2 and R-3 occupancies, where there are more than four dwelling units.*

- *As buildings get larger Type A dwelling units are required. For example, in Group R-2 occupancies containing more than 20 dwelling units, at least 2%, but not less than one, of the dwelling units must be a Type A dwelling unit. Because they have a higher degree of accessibility, Type A units may be substituted for Type B units, but not vice-versa.*

- *Per §1107.6.1.1 R-1 units are to have certain accessible units. As the number of units increases the number accessible units increases. Note that additional accessible features such as roll-in showers are required as the number of units increases. Thus Table 1007.6.1.1 requires 2 accessible units for up to 50 total units with no requirement for roll-in showers. However, when there are 51 or more units the 4 accessible units are required in which a roll-in shower is to be provided in at least one of the accessible units.*

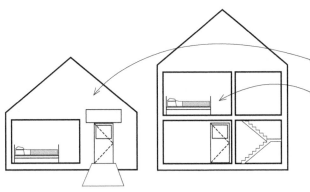

- *To apply the numerous exceptions to these requirements, the designer must further understand the distinction for Type B dwelling unit requirements, based on whether the units in questions are ground floor units or multistory units.*

- *A ground-floor dwelling unit has a primary entrance and habitable space at grade.*

- *A multistory unit refers to a single dwelling unit with multiple floors where there is habitable space or a bathroom space (which is not normally considered habitable space) on more than one floor level. Note that the definition of multistory does not refer to a single-level dwelling unit located above grade in a multistory building.*

DWELLING UNITS AND SLEEPING UNITS

General Exceptions per Section

The exceptions for accessible dwelling units depend on specific circumstances and do not apply equally to Type A and B dwelling units. The exceptions are:

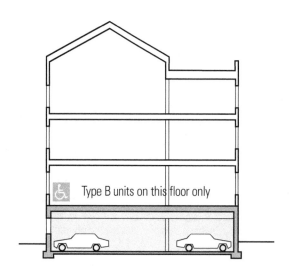

Type B units on this floor only

1. In walk-up type buildings, where no elevator is provided, neither Type A nor Type B dwelling units need be provided on floors other than the ground floor. The number of Type A dwelling units is to be provided based on the total of all units in the building, not just at the ground floor. It should be assumed that in buildings with more than 4 and less than 20 units, the ground-floor units should all be Type B units.

2. In podium type buildings without elevators, where the dwelling units occur only above the ground floor, only the dwelling units on the lowest floor need comply with the requirements of this section. Thus on the lowest level of a building with 4 to 20 units, all of the units on the lowest level of dwelling units need be Type B units, but only on that level. When the total dwelling unit count exceeds 20, then Type A dwelling units would need to be provided as required in this section, but again only on the lowest dwelling unit level.

3. A multilevel dwelling unit without an elevator is not required to comply with the requirements for Type B dwelling units. Thus townhouse-style multilevel dwelling units are not required to be Type B units. When elevator service is provided to one floor of the unit, however, that floor must meet Type B requirements and a toilet facility provided. It is not stated that this facility requires access, but it should be designed as such. The elevator may be inside the unit, or may serve corridors outside the unit, but if an elevator serves any level of the multilevel unit, that level must meet the requirements noted above.

• To meet the intent of accessibility provisions, we recommend that the ground floors of townhouses that are on an accessible route be made adaptable, and that an adaptable, or fully accessible, powder room be provided for use by guests on the ground floor. This allows for visitation with the unit's occupants by persons with disabilities.

4. *Where there are multiple buildings on a site that do not have elevators, and the site has varying grades, then the need to provide Type B dwelling units should be examined in the context of the entire site.*

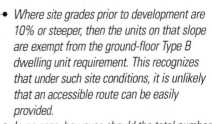

- *Where site grades prior to development are 10% or steeper, then the units on that slope are exempt from the ground-floor Type B dwelling unit requirement. This recognizes that under such site conditions, it is unlikely that an accessible route can be easily provided.*

- *In no case, however, should the total number of Type B dwelling units be less than 20% of the total number of ground-floor dwelling units on the entire site. So even where the site is steeply sloped throughout, accessible route provisions must be made to satisfy the 20% minimum requirement.*

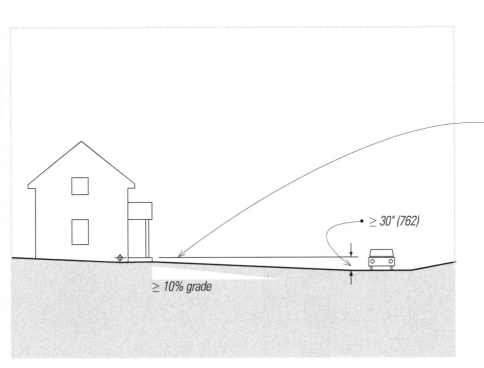

5. *Neither Type A nor Type B dwelling units need be provided where a site is susceptible to flooding and an accessible route cannot be provided such that the grade cannot be within 30" (762) of the floor within 50' (15 240) of the entry, or the grade exceeds 10% between the pedestrian or vehicular arrival point and the primary entrances of the units. This is very specific site criteria. This exception should be used with great caution as the conditions may be subject to design solution by site grading.*

- *The arcane and somewhat ambiguous nature of these and most other exceptions to accessibility provisions in this Code and ADAAG only reinforce our basic recommendation regarding provision of access in buildings. If it seems that access may be required by the Code, then it should be provided. Provision of access does the building and occupants no harm, while its absence can lead to lack of access and legal problems that are best avoided.*

SPECIAL OCCUPANCIES

§1108 establishes supplemental accessibility requirements for special occupancies. The wording "special occupancies" is somewhat misleading in that the occupancies discussed in this section are the same as those described throughout the Code, and their occurrence in buildings is not unusual or special in most cases. However, the occupancies described in this section do not necessarily correspond to the conventional occupancy groups discussed in other parts of the Code.

Assembly Areas

In assembly areas with fixed seating, the following provisions are required:

- *§1108.2.1 requires that services be accessible either by providing an accessible route or providing equal services on the accessible route as in nonaccessible areas.*
- *§1108.2.2 requires wheelchair spaces to be provided based on the number of seats in the assembly occupancy per Table 1108.2.2.1.*
- *§1108.2.3 requires the wheelchair seating to be dispersed. The purpose of the dispersal is to provide a variety of sightlines and a variety of seating prices for people with disabilities to choose from.*

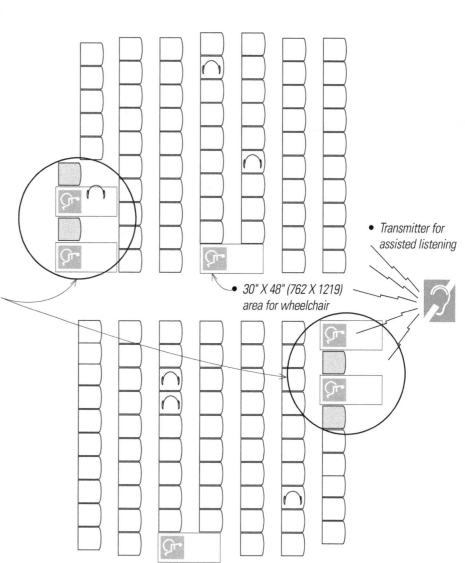

- *Transmitter for assisted listening*
- *30" X 48" (762 X 1219) area for wheelchair*

- *In assembly areas with fixed seats, where audible communications are integral to the use of the space; assistive listening devices are to be provided. The number of receivers is based upon the seating capacity in the assembly area. These requirements apply in small venues only when audio amplification is provided. However, in assembly areas with 50 or more occupants, assistive listening devices are required whether or not other amplification is provided. Assistive listening devices are required for all spaces where audio-amplification systems are installed. The best guideline is that if amplification is provided for some, it should be provided for all. Do not assume that assisted listening and wheelchair access always overlap. They often do not.*

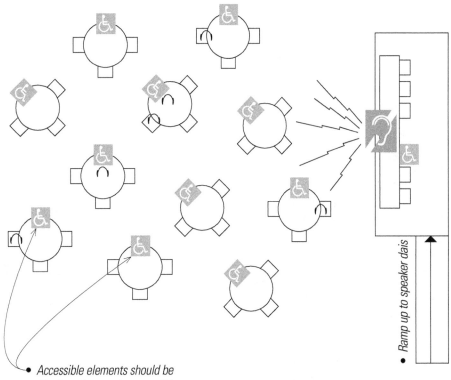

Ramp up to speaker dais

- *Accessible elements should be distributed as much as possible to avoid segregation of people with disabilities.*

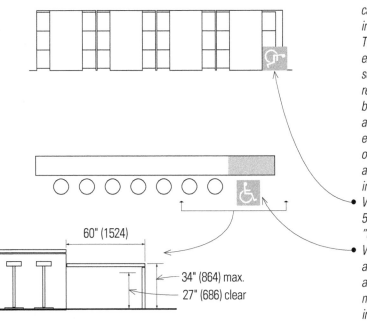

60" (1524)

34" (864) max.
27" (686) clear

- *Dining areas are to be accessible for the total floor area allotted for table and seating, except for mezzanine areas that contain less than 25% of the total area and if the same services are offered in both areas.*
- *The above exception should be used with caution due to the possible ambiguous interpretation of the term same services. The intent of the exception is that the same experiences of environmental character, service, and food be available to all patrons, regardless of disability. The designer must be certain that not only the physical character of the space but the operation of the establishment will not result in segregation of people with disabilities. The basis of access, especially under the ADA, is providing equal access as a civil right.*
- *Where seating or tables are built-in, at least 5% of the total but not less than one such "dining surface," is to be accessible.*
- *Where only counters are provided for dining and the counter exceeds 34" (864) in height, a 60" (1524) section of the counter is to be made accessible. Note that this is scoping information only. Clearance requirements for aisles between tables and chairs and at counters are contained in the reference standard A117.1.*

OTHER FEATURES AND FACILITIES

§1109 sets forth scoping provisions for access to various building parts and functions. These provisions apply to all areas except Type A and B dwelling units, which are to comply with A117.1. Where facilities for people without disabilities are provided, then essentially equal facilities are to be provided for people with disabilities. It is also possible and permissible to make all facilities usable by almost all potential users whether disabled or not. This is the underlying principle of what is called "universal design." Where there are multiple facilities, then percentages apply to determine the total number of such facilities to be provided. But in almost every case, at least one such facility is to be accessible as a base requirement. Where there is a distinctive use for facilities, such as bathing versus toilet facilities, or express checkout lines in addition to normal lines, then accessible facilities should be provided for each different function. The criteria in this section usually do not determine if the facility in question needs to be accessible, but only what criteria apply if accessibility is required by other parts of the code.

Equal access to facilities

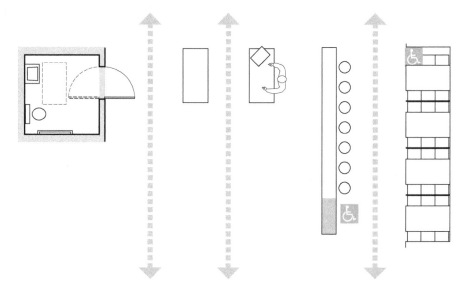

Toilet and Bathing Facilities

§1109.2 requires all toilet rooms and bathing facilities to be accessible. Where there are inaccessible floors that otherwise comply with the Code, the only available toilet and bathing facilities should not be located on that level. At least one of each type of fixture, element, control or dispenser in each toilet room is to be accessible. The facilities are to provide equal access to all of the functions provided in them.

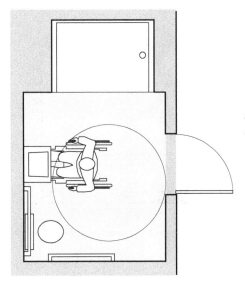

The exceptions for accessible toilet rooms and bathing facilities are very focused and very minimal:

1. *In private offices not meant for public use and meant for a single occupant, the toilet facilities need not be accessible, but provisions must be made for them to be adapted for use by a person with disabilities. Thus, the space layout must accommodate access in the initial floor plan.*

2. *These provisions do not apply to facilities that are not required to be accessible in §1107. This correlates with the exception to §1109.1, which refers the designer to A117.1 for information about Type A and B dwelling units.*

3. *Where multiple instances of single-user toilet facilities occur that are in excess of the plumbing requirements, then at least 5%, but not less than one in each cluster of such facilities, must be accessible. This is an instance of provision of equal access at each location of a group of facilities.*

4. *Children's facilities for use in day care or primary education need not be accessible if the fixtures are in excess of those required by the plumbing code. Thus children's-size fixtures that do not comply with accessible reach and height requirements may be provided, but only in addition to the required number of accessible fixtures, not in place of them.*

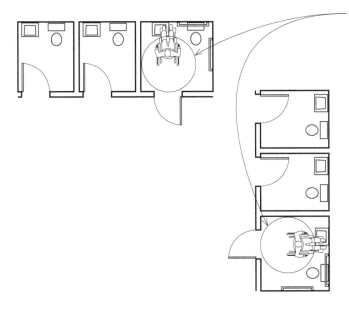

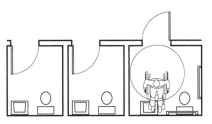

Unisex Toilet and Bathing Rooms

§1109.2.1 requires an accessible unisex toilet in assembly and mercantile occupancies when six or more water closets in total are required in the facility. This also applies to all recreational facilities where separate-sex facilities are provided, regardless of the count of water closets. This does not apply in recreational bathing facilities where there is only one bathing fixture in each separated-sex bathing room. These unisex facilities can be counted as part of the total number of fixtures to satisfy fixture-count requirements.

Sinks

§1109.3 requires 5% of sinks to be accessible, except for service sinks.

Kitchens

§1109.4 requires that, when provided in accessible space or rooms, kitchens, kitchenettes and wet bars must be accessible per the space criteria of A117.1.

Drinking Fountains

§1109.5 requires that 50% of the drinking fountains be accessible on floors where they are provided. Note that, given the wording of this section, this requirement will also apply for floors not otherwise considered or required to be accessible. Note also that each drinking fountain location has two fountains, one located at a low elevation for persons using wheelchairs and one higher for standing persons.

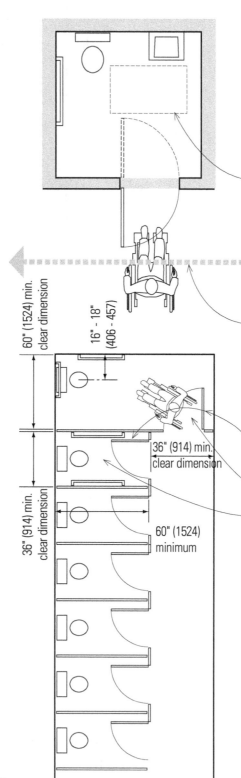

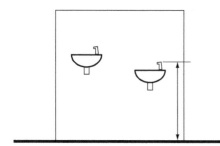

- Unisex facilities are to meet the criteria of A117.1 for space layout. Each facility is to contain only one water closet and one lavatory; however, in unisex bathing facilities, a shower or bathtub may be in the same room. Where lockers or similar storage units are provided in separate-sex bathing facilities, then accessible storage facilities are to be provided in the unisex bathing facilities as well.
- Unisex toilets are to have only one water closet; however, a facility with one water closet and a urinal is still allowed to be considered a unisex toilet room.
- In addition to the provisions of A117.1, a clear space of 30" by 48" (762 by 1219) is to be provided if the door swings into the room, clear of the door swing.

- Unisex facilities are to be on an accessible route. They are to be reasonably close to the separate-sex facilities, no more than one floor above or below them, and with the accessible route no more than 500' (152 400) in length. Practically speaking, in most instances, all of these facilities should be located adjacent to one another.
- Doors shall be 32" minimum clear width.

- When water closet compartments are provided, at least one should be accessible.
- When there are six or more toilet compartments in a toilet facility, then at least one compartment is to be an ambulatory-accessible stall per A117.1, in addition to the wheelchair-accessible compartment.

Elevators

IBC §1109.6 requires passenger elevators on an accessible route to be accessible per A117.1.

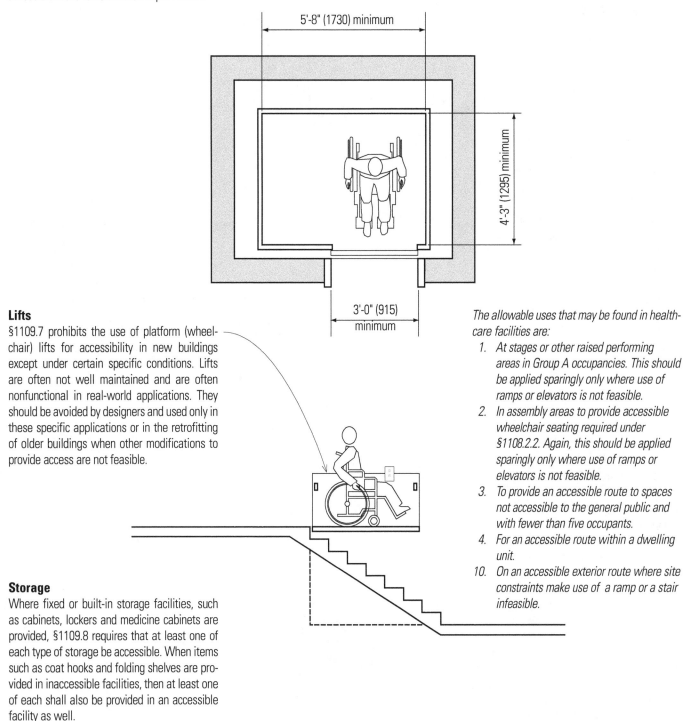

Lifts

§1109.7 prohibits the use of platform (wheelchair) lifts for accessibility in new buildings except under certain specific conditions. Lifts are often not well maintained and are often nonfunctional in real-world applications. They should be avoided by designers and used only in these specific applications or in the retrofitting of older buildings when other modifications to provide access are not feasible.

Storage

Where fixed or built-in storage facilities, such as cabinets, lockers and medicine cabinets are provided, §1109.8 requires that at least one of each type of storage be accessible. When items such as coat hooks and folding shelves are provided in inaccessible facilities, then at least one of each shall also be provided in an accessible facility as well.

The allowable uses that may be found in healthcare facilities are:

1. *At stages or other raised performing areas in Group A occupancies. This should be applied sparingly only where use of ramps or elevators is not feasible.*
2. *In assembly areas to provide accessible wheelchair seating required under §1108.2.2. Again, this should be applied sparingly only where use of ramps or elevators is not feasible.*
3. *To provide an accessible route to spaces not accessible to the general public and with fewer than five occupants.*
4. *For an accessible route within a dwelling unit.*
10. *On an accessible exterior route where site constraints make use of a ramp or a stair infeasible.*

Detectable Warnings

§1109.9 requires passenger transit platforms without guards to have detectable warnings at the edge to warn people with visual impairments of the falling hazard at that edge.

Assembly-Area Seating

§1109.10 is a cross-reference back to the provisions for seating and assistive listening contained in §1108.2.

Seating at Tables, Counters and Work Surfaces

§1109.11 is to be read in concert with §1108.2.8.1. The scoping requires 5% of seats at fixed or built-in tables or work surfaces to be accessible if they are on an accessible route. As is typical for such provisions, these accessible facilities are to be dispersed in the building or the space containing these features.

Customer-Service Facilities

§1109.12 provides for customer-service facilities for public use on accessible routes. These facilities include dressing rooms, locker rooms, check-out aisles in stores, point-of-sales stations, food service lines and waiting lines. They should be dispersed, and they should provide the same diversity of service facilities as those for nonaccessible facilities. These requirements would apply to healthcare related areas such as pharmacies, food service areas or accounting offices where payments are processed.

Controls, Operating Mechanisms and Hardware

Where controls such as light switches, thermostats, window hardware and convenience outlets are intended for operation by the occupant §1109.13 requires that they be accessible per the reach-range criteria contained in A117.1. These requirements apply to accessible spaces and along accessible routes. Only one window need be accessible in each space under these requirements. Also, accessible windows are not required in kitchens or bathrooms, recognizing the small size of these rooms and the potential access restrictions posed by fixtures and cabinets.

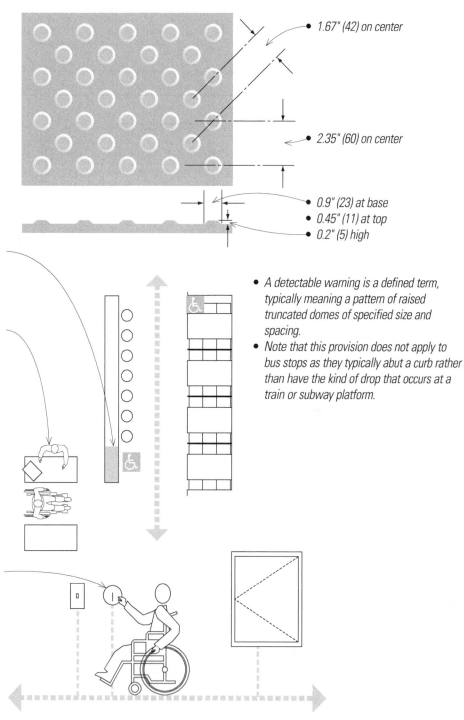

- 1.67" (42) on center
- 2.35" (60) on center
- 0.9" (23) at base
- 0.45" (11) at top
- 0.2" (5) high

- A detectable warning is a defined term, typically meaning a pattern of raised truncated domes of specified size and spacing.
- Note that this provision does not apply to bus stops as they typically abut a curb rather than have the kind of drop that occurs at a train or subway platform.

§1110 contains the requirements for the international symbol of accessibility used to identify required accessible elements.

Signs

§1110.1 requires the international symbol of accessibility to be located at accessible parking spaces per §1106.1, accessible areas of refuge per §1007.6, at accessible toilet locations, at accessible entries, accessible checkout aisles, and at accessible dressing and accessible locker rooms.

The ADAAG requires permanent room signage to be located in a prescribed location. Signs are also to have tactile raised lettering and Braille symbols. These requirements apply even though not stated in the Code. This is a good example of why designers must review ADA requirements along with code requirements. See signage requirements in the commentary on Appendix E, on page 219.

Directional Signage

Where not all elements are accessible, §1110.2 requires there be signage to direct people with disabilities to the nearest accessible element. These signs must have the international symbol of accessibility. Directional signage is required at inaccessible building entrances, inaccessible toilet facilities, inaccessible bathing facilities and elevators not serving an accessible route. Directions to the nearest accessible unisex toilet are to be provided where there are unisex facilities per §1109.2.1.

Other Signs

When special access provisions are made, then §1110.3 requires signage be provided to highlight those provisions. The specific requirements noted in the Code are:

1. When assistive listening devices are provided per §1108.2.6, signs to that effect are to be provided at ticket offices or similar locations.
2. Each door to an exit stairway is to have a sign in accordance with §1011.3. This section requires a tactile exit sign complying with A117.1.
3. At areas of refuge and areas for assisted rescue signage shall be provided in accordance with §1007.6.3 through §1007.6.5.
4. At areas for assisted rescue signage is to be provided per §1007.8.3.

§3409.1 applies to provision of access in existing buildings when certain actions occur, such as maintenance that requires a building permit, a change in use or occupancy, additions or alterations requiring provision of accessibility as noted in this section. The intent of access provisions for existing buildings is that as buildings are remodeled over time, they will eventually be made fully accessible as the whole building is eventually remodeled or additions made to it. §3409.3 clarifies that alterations do not impose a higher standard for accessibility than that which would be required for new construction. On the other hand, existing access is not to be reduced or compromised by alteration work.

Maintenance of Facilities

The title of §3409.2 is somewhat misleading in that it is intended to require that access be maintained, not that maintenance activities trigger access requirements. Only those maintenance activities that are more readily described as alterations will require a permit, and are not otherwise excepted by §3409.6, will trigger consideration of accessibility.

Change of Occupancy

When occupancy of an existing building is changed, §3409.4 requires that those portions of the building that are altered as a result of the change in occupancy be made accessible as required in §3409.6. In addition, there are accessibility requirements triggered by the change in occupancy that may require work not otherwise related to the change in occupancy. These accessible features are noted on the right.

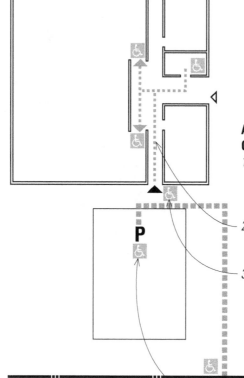

Accessible Features for a Change in Occupancy

1. At least one entrance must be made accessible. It should be the main entrance, but may be another if making the main entry accessible is not feasible.
2. An accessible route must be provided from an accessible entry to the primary function areas.
3. Signage complying with IBC §1110 must be provided. This will most likely be signage noted in IBC §1110.2, directing building users to accessible features.
4. Where parking is provided, accessible parking is to be provided. The reference to provision should be presumed to require alterations of existing parking lots to provide access, even if no other work is being done in the parking area. This would apply even when provision of additional parking is not required by the change in occupancy.
5. If loading zones are provided, either new or existing, then at least one is to be accessible.
6. An accessible route must connect accessible parking and loading zones to an accessible entrance.

- Where it is technically infeasible to meet the access standards for new construction, then other sections of §3409 may apply. This will be discussed in greater detail later.

Alterations

When a building is altered, the provisions of Chapter 11 and A117.1 apply to the area of alteration, unless this is technically infeasible. When compliance with §3409.6 is deemed to be technically infeasible, then the alteration is to comply to the maximum extent that is possible. As with other access provisions, the application of technical infeasibility and use of exceptions should be done with great caution and only in consultation with the building official. The definition of technical infeasibility is an invitation to negotiate with the building official. When the designer opts to use this definition to justify exclusion of an accessible element that would otherwise be required in new construction, the designer should seek the concurrence of the building official of this interpretation during the plan-review process.

The exceptions to the alteration access requirements are:

1. Access need not be provided when the altered area is not required to be on an accessible route, unless otherwise required by §3409.7 as discussed below.

2. A major exemption is that accessible means of egress are not required to be provided in alterations to existing buildings.

3. When individually owned Type A dwelling units in R-2 occupancies are altered they must comply with the Type B requirements of A117.1 and Chapter 11. This allows non-disabled persons to remodel accessible units to remove accessible features, but requires that these units remain adaptable so that the total stock of potentially accessible housing units is not decreased.

Alterations Affecting an Area Containing a Primary Function

Where an alteration affects the accessibility of the route to the area of primary function, or the alteration work involves the primary function area, §3409.7 requires that route be made accessible. This route must also provide access to toilet facilities and drinking fountains serving the area of primary function.

The intent of §3409.7 for alterations is to ensure that accessibility occurs to the maximum extent feasible. The extent of these requirements are directly related to their proximity to the area of alteration. Accessibility should be provided throughout the structure to the maximum extent feasible.

There is a cost cap of 20%, assignable to the cost of providing an accessible route to the area of primary function. This cost cap can be construed in two ways. One is that the cost of access is 20% of the total cost of the project, with a total cost of 100% for the project. However, the text states that the cost of providing an accessible route shall not exceed 20% of the cost of alterations. If the cost of alterations is x, then the total budget for the work would be x plus 20% of x, or 120% of the budget for the work. The interpretation of this section should be verified with the local authorities while the budget is being established. The exceptions also exclude alterations made solely to such components as electrical systems, mechanical systems, hardware, operating controls and fire-protection systems.

These exceptions are spelled out in the section pertaining to primary function areas, so the exemptions should properly be construed such that just mechanical work alone in an area of primary function should not trigger accessible route improvements. On the other hand, it seems reasonable to construe that the cost of mechanical work, if done in conjunction with other alteration work in a primary function area that does trigger access, would count toward the total cost of construction when determining the value assigned as 20% of the cost of alteration in determining what cost to assign to access improvements.

Since most owners do not initially consider the cost of providing access elements when setting their budgets, it is wise for the designer to use the 120% rule when establishing the initial construction budget.

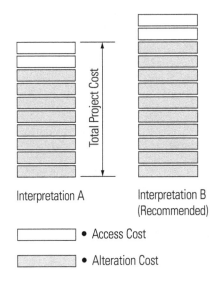

Interpretation A

Interpretation B (Recommended)

● Access Cost

● Alteration Cost

Scoping for Alterations

§3409.8 applies to scoping for alterations to existing buildings. These requirements recognize the interrelationships between building structure and vertical elements such as elevators, stairs, steps in paths of travel and ramps steeper than current code allowances. The scoping provisions apply first to the areas of alteration and then in limited ways to accessible routes leading to the areas that are altered.

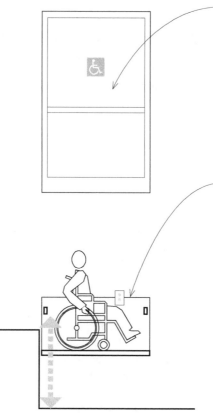

- *Entrances: Accessible entrances are to be provided per §1105, except when there is already an accessible entrance; then the altered entrance is not required to be accessible unless required by §3409.7. We recommend that altered entrances be made accessible whenever possible.*
- *Elevators: §3409.8.2 requires elevators serving altered portions of a building to be altered to comply with A117.1 and ASME A18.1.*
- *Platform Lifts: §3409.3 allows the use of platform (wheelchair) lifts in accessible routes in existing buildings. This stipulation recognizes that lifts are often a reasonable mitigation for technical infeasibility in existing buildings where steps or steep ramps cannot be altered due to structural complications.*
- *Stairs and Escalators in Existing Buildings: When new escalators or stairs are added in existing buildings where none existed previously, §3809.8.4 requires an accessible route be provided in accordance with the requirements of §1104.4 and 1104.5 for new buildings.*

- *Ramps, §3409.8.5: When existing ramps are steeper than allowed for accessible routes, or necessitated because of structural or space considerations that meet the criteria for technical infeasibility, then they may be used under limited circumstances per Table 3409.8.5.*

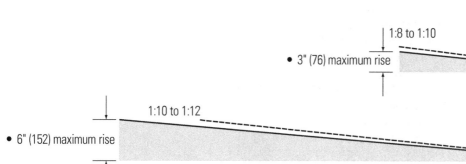

1:8 to 1:10

- 3" (76) maximum rise

1:10 to 1:12

- 6" (152) maximum rise

- *Ramps having slopes between 1:10 and 1:8 may be used for rises of up to 3" (76).*
- *Ramps with slopes between 1:10 and 1:12 may have a rise of 6" (152).*

- *The limitations recognize that short, steep ramps, though not acceptable or desirable in new construction, may still be usable for wheelchair users for very limited lengths and rises. In such circumstances, better access, though compromised, is still provided with a steep ramp than with a step.*

Existing unaltered areas | Altered or added areas

- *Dwelling or Sleeping Units: §3409.8.7 states that accessibility requirements apply only to the spaces being altered or added in Group I-1, 2 or 3 and Group R-1, 2 or 4 occupancies. Within altered or added areas, the quantity requirements for access per §1107 for accessible Type A units apply only to the quantity of spaces being altered or added.*

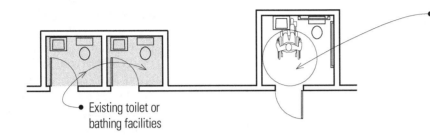

- Existing toilet or bathing facilities

- *Toilet Rooms: Where it is technically infeasible to provide access to existing toilet or bathing facilities, §3409.8.9 deems it acceptable to provide unisex facilities instead. Such facilities should be located on the same floor and in the same area as existing facilities.*

- Existing dressing, fitting or locker rooms

- *Dressing, Fitting and Locker Rooms: Where it is technically infeasible to provide access to existing dressing, fitting or locker rooms, §3409.8.10 requires one accessible room on the same level to be provided. If existing facilities are same-sex, then the accessible ones shall be same-sex as well. When the existing facilities are unisex, the accessible facilities may be unisex as well.*

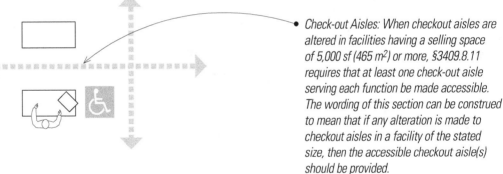

- *Check-out Aisles: When checkout aisles are altered in facilities having a selling space of 5,000 sf (465 m²) or more, §3409.8.11 requires that at least one check-out aisle serving each function be made accessible. The wording of this section can be construed to mean that if any alteration is made to checkout aisles in a facility of the stated size, then the accessible checkout aisle(s) should be provided.*

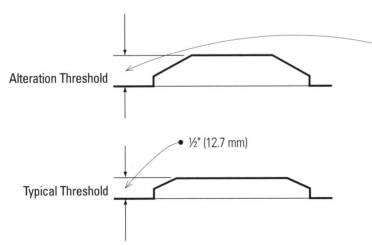

Alteration Threshold

½" (12.7 mm)

Typical Threshold

- *Thresholds: §3408.12 specifies that thresholds in existing buildings may be ¾" (19.1) high. Such thresholds must have beveled edges at each side to facilitate the passage of wheelchairs. The slopes of the bevels are not specified but should be similar to threshold dimensions set forth in A117.1 for new thresholds. Where feasible, thresholds fully compliant with new construction standards should be used.*

HISTORIC BUILDINGS

The provisions of §3409.9 apply to buildings that have been designated as historical by an authority having jurisdiction. The determination of what is technically infeasible to perform in an existing building is further complicated if the facility is also a qualifying historic building. In addition to the structural determinations made under the definition of technical infeasibility, the designer and the building official must confer regarding the impact of accessible elements on the historic fabric of the building. It is essential to understand that the fact that a building has historical significance does not justify lack of accessibility. The building official must verify any determinations regarding provision of less than full access under the doctrines of technical infeasibility or historical impact. Such determinations should be rendered in writing and kept on file by the jurisdiction, the designer and the building owner. If a building is determined to be a designated historical structure, then the following alternative requirements for accessibility may be applied.

- *Site Arrival Points: As for other renovations, §3409.9.1 requires that at least one accessible route of travel be provided from a site arrival point to an accessible entrance. This entrance need not be the main entrance if provision of access at that entry is technically or historically infeasible.*

- *Multilevel Buildings: §3409.9.2 requires that access be provided from the accessible entrance to public spaces on that level. Not expressed, but implied, is the fact that access is not required above or below the level of the accessible entrance.*

- *Entrances: §3409.9.3 requires that at least one main entrance be accessible. When this is not feasible for technical or historical reasons, another entrance may be used. If this is done, the accessible entry must remain unlocked while the building is in use with signage directing users to the accessible entrance. It may also be acceptable to provide a locked accessible entrance if a notification or remote monitoring system is provided.*

- *Toilet and Bathing Facilities: Where these facilities are provided, §3409.9.4 requires that at least one accessible toilet room complying with §1109.2.1 be provided.*

Appendix E

Provisions in the appendices of the IBC do not have any force or effect in local jurisdictions unless the local authority having jurisdiction specifically adopts them. We have included the supplementary accessibility provisions in this chapter as an example of how appendices work. The access section is included because the provisions in this appendix closely parallel ADA and local modifications made to access regulations. In our opinion, most of these requirements should be applied based on common sense or compliance with the ADA, whether the appendix is adopted locally or not. Note that the referenced standards in the appendix include the ADA Access Guidelines (ADAAG) and various federal publications related to access and access regulations in addition to A117.1.

Designers should be familiar with these provisions as it is likely that these requirements or similar ones will be encountered in the course of design work. Appendix requirements are typically ones that are not universally accepted or that are still under development. These provisions are placed in the appendix to give local authorities the option to adopt them or not without compromising the basic code.

Appendix E uses the same reference criteria as the basic Code, A117.1. The provisions of the appendix are designed to coordinate the basic code with supplemental requirements. If adopted, the section numbers are placed in the basic code in the numerical sequences noted in the appendix. The appendix requirements form an overlay with the basic code to add requirements for specific occupancies or functions beyond the requirements of the basic code.

Definitions

§E102 contains definitions of accessible elements.

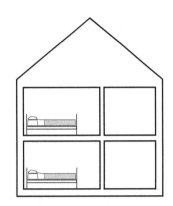

- *Transient Lodging: This is the primary additional term introduced in Appendix E. This function is typically a Group I or Group R occupancy. The major distinction is that the residents of these facilities are not long-term residents and thus are less familiar with their surroundings and may require supplementary services, clearances or devices to use the transient lodging functions. Longer-term residents may adapt their spaces or have personal or programmatic supplements to their activities to accommodate their disabilities. Note that the definition of Transient Lodging in §E102 excludes inpatient medical care facilities and long-term care facilities which are also I occupancies.*

Accessible Routes

§E103 adds requirements for access to raised speaker's platforms or lecterns in banquet areas. (Note that this is an ADAAG requirement.)

Special Occupancies

§E104 adds requirements for transient lodging facilities as defined.

Accessible Beds

§E104.2 adds requirements for accessible beds per A117.1, based on the total number of beds in the sleeping accommodation, once there are twenty-five or more beds. The space requirements apply to both sides of a bed, but the exception in §E102.4.2.1 allows the access spaces to overlap when two beds are side by side.

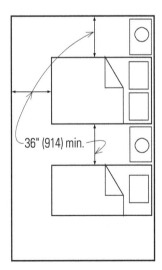

36" (914) min.

SUPPLEMENTARY ACCESSIBILITY REQUIREMENTS

Communication Features

§E104.3 provides for accessible communication features at sleeping accommodations in transient lodging per Table E104.3.1, based on the number of rooms. These features are to accommodate people with hearing disabilities and include visual notification of telephone rings, door knocks or bells. These functions are to be separate from the visual emergency alarm notifications. There is also to be a electrical receptacle near the phone to allow plugging in of a portable TTY for phone access. Telephones are also required to have volume controls.

It is not stated whether these rooms are to be combined with rooms made accessible for other disabilities, or separate. It is best to assume that the provision of mobility and auditory access is separate, and the requirements for numbers of rooms are additive. However, it is also reasonable to assume that the idea of dispersal and diversity of access will require certain rooms to provide both types of access in the same room. People with multiple disabilities should be accommodated. Also, rooms that are accessible for people with mobility impairments are usable by people without those disabilities. If adopted in the project's jurisdiction, then these requirements should be reviewed with the building official.

Other Features and Facilities

The facilities noted in §E105 are temporary in nature or equipment added after the construction of the building is complete. As such, they may not be under the control of the designer, but designers should be aware of these requirements in order to make physical layouts that can accommodate them.

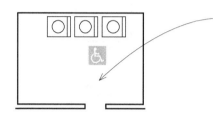

- *Laundry Equipment: When laundry equipment is provided in spaces that are required to be accessible, §E105.2 requires that at least one of each type of equipment should be accessible per A117.1.*

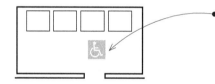

- *Depositories, Vending Machines and Other Equipment: §E105.3 requires that spaces containing vending machines be accessible per A117.1.*

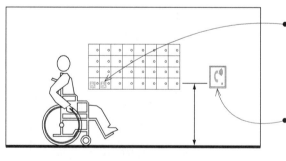

- *Mailboxes: §E105.4 requires that when mailboxes are provided at an interior location at least 5% of them be accessible per A117.1. Where mailboxes are provided at each unit then accessible mailboxes are to be provided at accessible units.*
- *Two-way communication systems: Where communication systems are provided, such as those located at apartment entrances to gain admittance, the system is to comply with A117.1.*

Telephones

The telephone access requirements contained in §E106, although in the appendix, should be applied in almost all cases. In addition, these detailed access requirements should be reviewed with the ADAAG and local code adoptions.

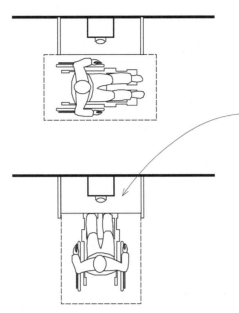

- *The access requirements illustrated are from A117.1. Note that while phones may be accessible from the front or the side as illustrated, at least one phone per floor is to provide for a forward approach, except at exterior phones available with dial-tone-first service. The general rule to apply is that where pay phones are provided, they should be accessible per Table E106.2.*
- *Each public phone is to have volume controls per A117.1.*

Phones per Floor or Level	Wheelchair-Accessible Phones
1	1 per floor or level
1 bank*	1 per floor or level
2 or more banks	1 per bank

*Bank consists of a row of two or more phones

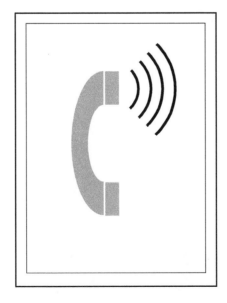

- *TTY's: §E106.4 requires that TTY's, telephone devices for hearing-impaired people using text or other nonverbal communication, are to be provided when there are more than four public telephones in a bank. The only exception applies to multiple banks of phones on one floor that are less than 200' (60 960) apart; then only one such bank need have a TTY. TTY locations are to be indicated by the international symbol of TTY and directional signs are to be provided to TTY locations. In addition to the TTY requirements, a shelf for portable TTY's along with an electrical receptacle is to be provided at each interior bank of three or more telephones.*

Signage

Where permanent signs for rooms or spaces are provided at doors, §E107 requires the signs to be tactile. Where pictograms are provided as permanent designations of permanent interior rooms, the pictograms are to have tactile text descriptors per A117.1. Directional signs for permanent interior locations except for building directories and personnel names and temporary signs should also be accessible per A117.1. This is another instance where these requirements should be followed regardless of local adoptions. The ADAAG should be reviewed along with A117.1 for signage requirements.

- *International TDD Symbol*

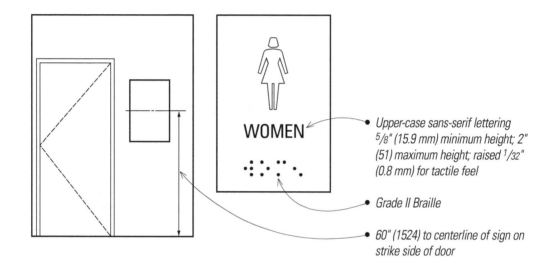

- *Upper-case sans-serif lettering ⁵⁄₈" (15.9 mm) minimum height; 2" (51) maximum height; raised ¹⁄₃₂" (0.8 mm) for tactile feel*

- *Grade II Braille*

- *60" (1524) to centerline of sign on strike side of door*

12

Interior Environment

The quality of interior environments is one of the primary factors in code development outside of the considerations of structural safety, fire-resistance and egress. Many of the environmental considerations in model codes came into being as a response to deplorable living conditions in buildings at the turn of the 20th century. Tenement housing and factory conditions often contributed to serious public-health problems for occupants. Tenants were often crowded together in tiny rooms with little light or ventilation and with inadequate sanitation or maintenance of facilities. Also, unregulated building materials, especially decorative materials, led to installation of highly flammable materials in egress paths, often making egress passages more dangerous than the spaces occupants were leaving in an emergency. Code provisions were written to address these quality-of-life issues for occupants. Recently, indoor air quality has become increasingly important, as buildings become more energy efficient and more airtight. The need to prevent accumulations of molds or toxic gases while at the same time achieving energy efficiency has been a challenge for recent code development.

The considerations for interior environment address several primary factors. Chapter 8 considers the flame-resistance and smoke-generation capacities of various building materials used in interior spaces; Chapter 12 addresses the remaining environmental considerations.

These include air quality, both for human health and also for building durability, which is governed by regulations for the ventilation of building and occupied spaces. Another consideration is the quality of the occupants' experience in a space, which is regulated by provisions for temperature controls, natural or artificial lighting levels, and air and sound isolation between spaces. Also covered are the sizes of the spaces to be occupied, access to unoccupied spaces, and ongoing sanitation and maintenance issues for building interiors, including the materials that can be readily maintained in toilet and bathing facilities. Chapter 12 is therefore something of a catchall for the Code with many provisions having only slight relationship to the chapter title.

INTERIOR ENVIRONMENT

This chapter has many disparate parts. One part governs quality-of-life issues for interiors and also includes measures to assist in mitigating the impact of climate on buildings. The standards set forth in this chapter also relate to how spaces work and feel for occupants, and how buildings can be maintained. This is in contrast to Chapter 8, which deals primarily with the fire safety of the occupants of interior spaces. The chapter title is somewhat misleading in that requirements of the chapter also pertain to spaces not normally occupied, although they are inside the building envelope.

Ventilation

§1203 deals with the ventilation of concealed attic and rafter spaces as well as with the ventilation of occupied spaces. The section is therefore divided among requirements for ventilation of building material areas and for ventilation of spaces used by building occupants.

Where building materials are enclosed on both the exterior and interior side of spaces, the concealed spaces, whether attics or rafter spaces, become susceptible to moisture intrusion.

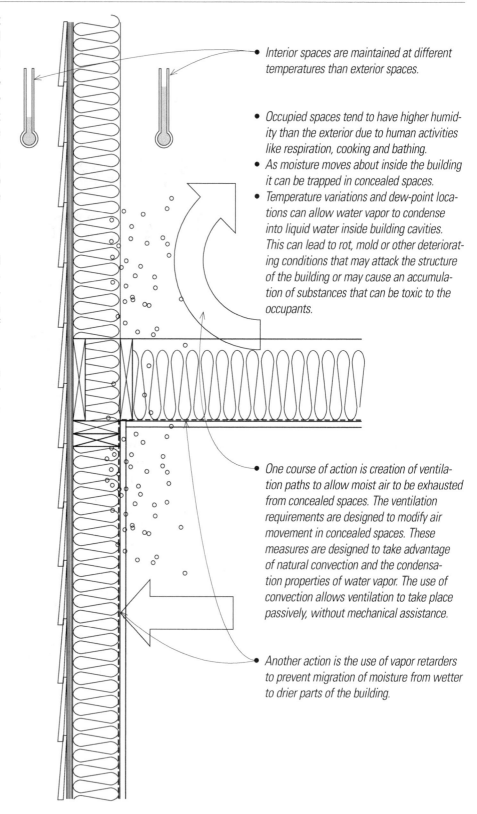

- *Interior spaces are maintained at different temperatures than exterior spaces.*

- *Occupied spaces tend to have higher humidity than the exterior due to human activities like respiration, cooking and bathing.*
- *As moisture moves about inside the building it can be trapped in concealed spaces.*
- *Temperature variations and dew-point locations can allow water vapor to condense into liquid water inside building cavities. This can lead to rot, mold or other deteriorating conditions that may attack the structure of the building or may cause an accumulation of substances that can be toxic to the occupants.*

- *One course of action is creation of ventilation paths to allow moist air to be exhausted from concealed spaces. The ventilation requirements are designed to modify air movement in concealed spaces. These measures are designed to take advantage of natural convection and the condensation properties of water vapor. The use of convection allows ventilation to take place passively, without mechanical assistance.*

- *Another action is the use of vapor retarders to prevent migration of moisture from wetter to drier parts of the building.*

Attic Spaces

§1203.2 requires that ventilation openings be provided at the bottom and top of sloped roof conditions to allow for convection in attic spaces and cathedral ceilings.

- With this provision the Code assumes that ventilation for insulation will be above the insulation. Designs should make this assumption as well. Vents are to be distributed to promote convection, with half high and half low in sloped conditions.

- A minimum of 1" (25.4 mm) of space is to be provided between insulation and roof sheathing.
- Blocking and bridging must be arranged to not interfere with the flow of ventilation.

Ventilation opening requirements are expressed in terms of fractions of the area to be ventilated. Note that ventilation area requirements refer to free area. Free area is the amount of space allowing actual airflow. Vent louvers and screened openings do not provide 100% free area. A good rule of thumb for louvers at vent openings is to assume 50% free area. Consult manufacturers' data for actual free areas of any vent coverings selected to be certain of their actual free area.

- Typically ventilation is to be 1/150 of the area to be ventilated, with 50% located at least 3' (914) above eave or cornice vents that provide the other half of the ventilation.
- The amount of ventilation area can be reduced in half, to 1/300, when a vapor barrier is installed on the warm side of the concealed space, and the vertical height separations for inlets and outlets are maintained.

- Vents are to be covered with screens to keep out vermin and birds as well as be protected from the entry of rain and snow.

Under-Floor Ventilation

§1203.3 requires that, in buildings that are not slab-on-grade structures, the space between the underside of floor joists and the earth below be ventilated.

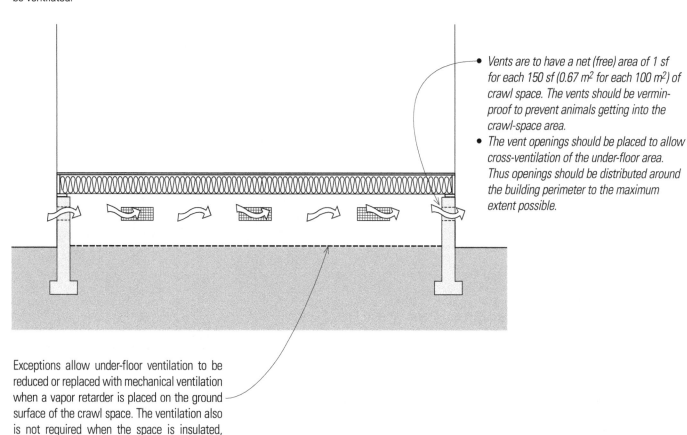

*   *Vents are to have a net (free) area of 1 sf for each 150 sf (0.67 m² for each 100 m²) of crawl space. The vents should be vermin-proof to prevent animals getting into the crawl-space area.*
*   *The vent openings should be placed to allow cross-ventilation of the under-floor area. Thus openings should be distributed around the building perimeter to the maximum extent possible.*

Exceptions allow under-floor ventilation to be reduced or replaced with mechanical ventilation when a vapor retarder is placed on the ground surface of the crawl space. The ventilation also is not required when the space is insulated, climate conditioned and provided with a vapor retarder. Any of these measures is considered equivalent protection from vapor intrusion and collection when combined with a given quantity of passive or mechanical ventilation.

Temperature Control

§1204 states that all habitable spaces are to be provided with space heating. The heating system must be capable of sustaining a temperature of 68°F (20°C) at a point 3' (914) above the floor of the space. This capability is to be available on the coldest anticipated design day of the year.

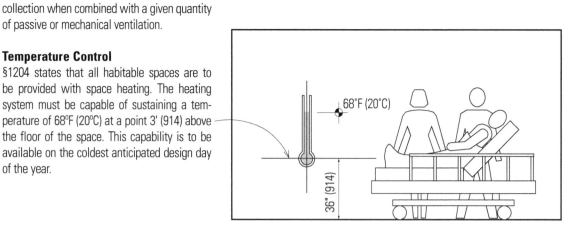

Natural Ventilation

§1203.1 requires natural ventilation when mechanical ventilation is not provided. Mechanical ventilation is to be per the International Mechanical Code. The intent of these requirements is that habitable spaces be provided with light and air, even when those interior spaces are windowless and adjoin another room.

②

• *8% of Room 2 floor area*

• *4% of Room 1 and Room 2 floor area*

①

• *Per §1203.4.1 rooms are to have an area openable to the outdoors of at least 4% of the floor area of the space being ventilated. In this illustration that would be 4% of the total area of both spaces shown.*

• *§1203.4.1.1 states that adjoining rooms without openings to the outdoors must have an opening of at least 8% of the floor area of the inner room, or 25 sf (2.3 m²), whichever is larger, into the adjacent room with the window.*

• *The openable area in the room facing the outdoors shall be based on the total area of all of the rooms relying on the window for ventilation. Thus in this example, area number 2 is considered part of room number 1.*

Where there are additional contaminants in the air in naturally ventilated spaces, then additional mechanical ventilation may be required by the code.

In all cases, the Code requires mechanical ventilation of rooms containing bathtubs, showers, spas and similar bathing fixtures. Note that natural ventilation is not considered an alternate to mechanical ventilation in bathrooms. This recognizes the need for positive ventilation of interior moisture to prevent moisture buildup in the structure. Many times, especially in cold climates, natural ventilation is not used and moisture can build up in buildings when mechanical ventilation is not provided.

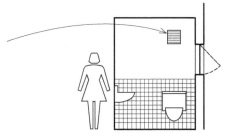

LIGHTING

§1205 requires all habitable spaces to have natural or artificial light. Note that while artificial light is to provide 10 foot-candles (107 lux) over the area of the room at a height of 30" (762), no specific light-quantity criteria is set for natural light. While highly unlikely, it is thus possible under a literal interpretation of the Code to have a building that could be occupied only during daylight as long as there were no internal rooms lacking natural light.

10% of this room's floor area

8% of both room's floor area

Net exterior glazed opening areas are to be 8% of the floor area of the room, or rooms, served. Thus it is two times the requirement for ventilation area.

The Code allows for interior rooms to borrow light from adjacent rooms. For an adjoining room to serve another, at least half of the common wall between them must be open. The opening must be at least 10% of the floor area of the interior room, or 25 sf (2.3 m²), whichever is larger.

10 foot-candle (107 lux) average illumination for artificial illumination

30" (762)

Light level measured at 30" (762 mm) above the floor.

Stairways in dwelling units must have specified light levels, measured at the nosing of every stair tread. The basic light level is to be 1 foot-candle (11 lux) minimum at the floor level. Stairs in other occupancies are to be lit per Chapter 10. See §1006 for emergency lighting requirements at means of egress. The requirements in Chapter 10 are essentially the same as for residential stairs as noted above.

Yards or Courts

Exterior openings are to open to the outdoors to yards and courts of sizes specified in §1206. Yard and court sizes are set to provide some time minimum dimensions for light wells and backyards of multistory buildings so that these areas provide real light and air to the spaces they serve.

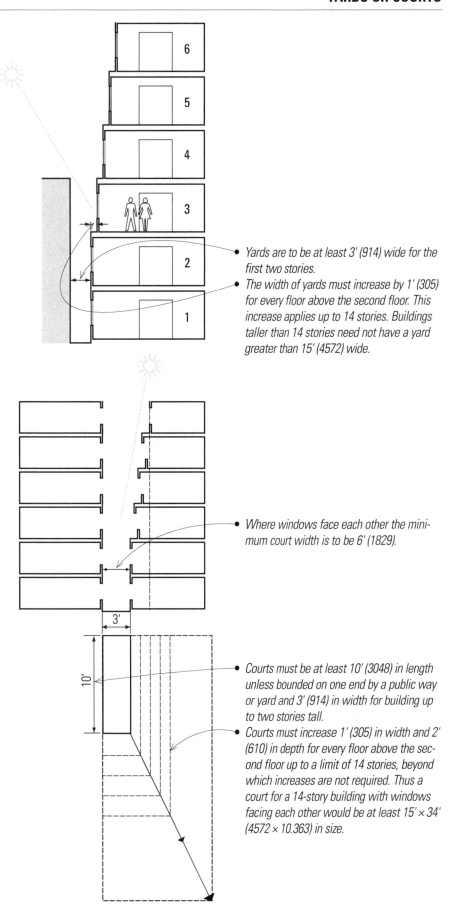

- *Yards are to be at least 3' (914) wide for the first two stories.*
- *The width of yards must increase by 1' (305) for every floor above the second floor. This increase applies up to 14 stories. Buildings taller than 14 stories need not have a yard greater than 15' (4572) wide.*

- *Where windows face each other the minimum court width is to be 6' (1829).*

- *Courts must be at least 10' (3048) in length unless bounded on one end by a public way or yard and 3' (914) in width for building up to two stories tall.*
- *Courts must increase 1' (305) in width and 2' (610) in depth for every floor above the second floor up to a limit of 14 stories, beyond which increases are not required. Thus a court for a 14-story building with windows facing each other would be at least 15' × 34' (4572 × 10.363) in size.*

Sound Transmission

Sound transmission can severely impact the quality of life in multi-tenant buildings, especially in residential occupancies. §1207 sets standards for sound-transmission reduction at walls and floor/ceiling assemblies between adjoining "dwelling units." These criteria also apply at the perimeter of dwelling units where they abut other units, public areas or service areas. Many types of I and R occupancies associated with healthcare facilities may be considered to contain "dwelling units" as defined in Chapter 2. The definition is based on the provision of independent living facilities for one or more persons where there are permanent provisions for "living, sleeping, eating, cooking and sanitation." The definition, by including the word "and" would seem to require that all of the listed elements must be present for a space to be considered a "dwelling unit."

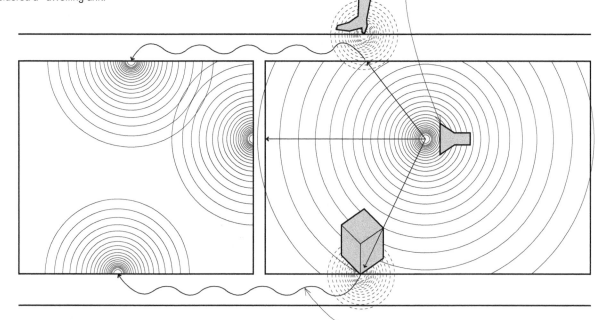

• *Airborne sounds are those such as music, speech or equipment sounds that travel through the air.*

• *Structure-borne sounds are those such as footfalls, dropped items or equipment vibrations that are conducted by the structure and resonate in other spaces than the space of origin.*

Criteria are established using ASTM test criteria. The code sets a higher standard for design criteria based on test data versus field-testing. These criteria recognize that test conditions are difficult to achieve in the field, and thus the expected minimum criteria are established such that field-measured conditions will meet the minimum necessary criteria. Criteria are established for both airborne sound (ASTM E 90) and for structure-borne sound (ASTM E 492).

Based on the abuses of tenement housing during the early part of the industrial revolution, model codes as well as §1208 of this Code set minimum room dimensions for each anticipated use of a building.

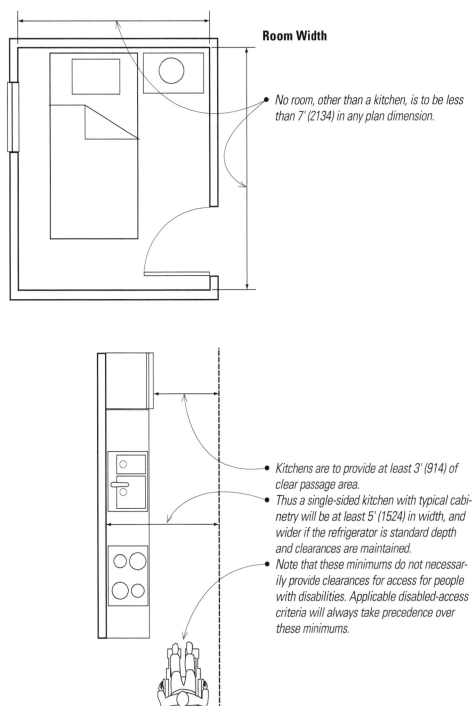

Room Width

- *No room, other than a kitchen, is to be less than 7' (2134) in any plan dimension.*

- *Kitchens are to provide at least 3' (914) of clear passage area.*
- *Thus a single-sided kitchen with typical cabinetry will be at least 5' (1524) in width, and wider if the refrigerator is standard depth and clearances are maintained.*
- *Note that these minimums do not necessarily provide clearances for access for people with disabilities. Applicable disabled-access criteria will always take precedence over these minimums.*

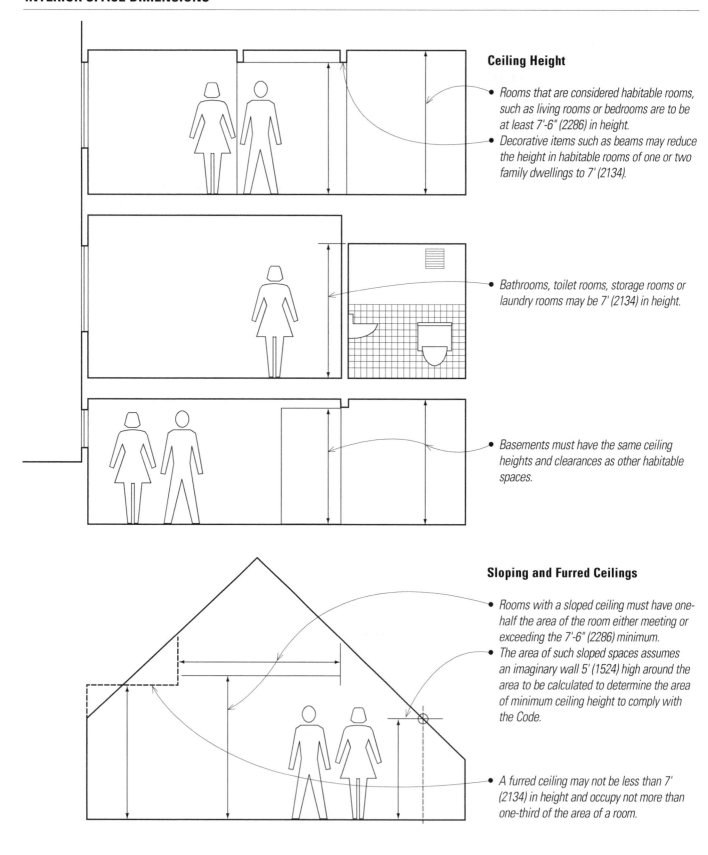

Ceiling Height

- *Rooms that are considered habitable rooms, such as living rooms or bedrooms are to be at least 7'-6" (2286) in height.*
- *Decorative items such as beams may reduce the height in habitable rooms of one or two family dwellings to 7' (2134).*

- *Bathrooms, toilet rooms, storage rooms or laundry rooms may be 7' (2134) in height.*

- *Basements must have the same ceiling heights and clearances as other habitable spaces.*

Sloping and Furred Ceilings

- *Rooms with a sloped ceiling must have one-half the area of the room either meeting or exceeding the 7'-6" (2286) minimum.*
- *The area of such sloped spaces assumes an imaginary wall 5' (1524) high around the area to be calculated to determine the area of minimum ceiling height to comply with the Code.*

- *A furred ceiling may not be less than 7' (2134) in height and occupy not more than one-third of the area of a room.*

Room Area

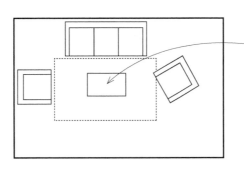

- Each dwelling unit must have one room of at least 120 sf (13.9 m²) in net area.
- Other rooms, except kitchens, are to have a minimum net area of 70 sf (6.5 m²).

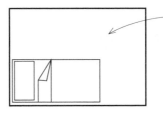

- Note that the definition of habitable spaces excludes bathrooms and toilet rooms, but includes spaces for cooking, living, sleeping or eating. Thus, bedrooms must be at least 10' by 7' (3048 by 2134) to meet the space criteria of this section. It is open to interpretation whether this is intended to apply in buildings that do not contain dwelling units.

Efficiency Dwelling Units

§1208.4 covers low- and moderate-income housing and senior housing, which have special needs that can often be accommodated by small units meeting minimum space and layout criteria. These are often called studio, efficiency or single-room-occupancy units. These criteria could be used for assisted living units as well. The title of this section and the text includes the term "dwelling unit" and is based upon the definition of dwelling unit contained in Chapter 2. This definition is based on permanent inclusion of all of the following elements in the unit: "living, sleeping, eating, cooking and sanitation."

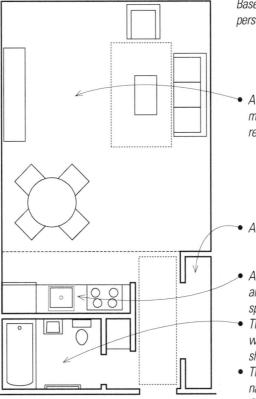

Based on the criteria of §1208.4, a single-person unit would contain:

- A living/dining sleeping room of 220 sf (20.4 m²). An additional 100 sf (9.3 m²) of area is required for each occupant in excess of two.

- A separate closet (of undefined size).

- A cooking area with a sink, cooking appliance and refrigerator with 30" (762) of clear space in front.
- There is also to be a separate bathroom with a water closet, lavatory, and bathtub or shower.
- The unit is to meet all of the ventilation and natural-light requirements contained in the Code.
- Note that access criteria must also be examined regarding doorway and path-of-travel clearances.

Access to Unoccupied Spaces

Per §1209 access is to be provided to mechanical appliances installed in crawl spaces, in attics or on roofs. Equipment must be provided with maintenance access both as a good design practice and per the International Mechanical Code.

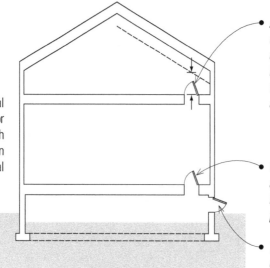

- Attic spaces that have more than 30" (762) of clear height inside the attic are to have an access opening of at least 22" by 30" (559 by 762). The opening is to be located such that there is at least 30" (762) of headroom above the opening.

- Crawl spaces under a building must have an access opening of at least 18" by 24" (457 by 610). This is of sufficient size to allow a person to get into the crawl space.

- The access opening may be either inside or outside of the building.

Surrounding Materials

These provisions in §1210 apply to toilet and bathing facilities in uses other than dwelling units. They are to address the maintenance, cleanliness and health issues associated with public toilet and bathing facilities. Note that while these provisions do not apply to dwelling units except as indicated, their application should be carefully considered by the designer for all types of building as good design practice.

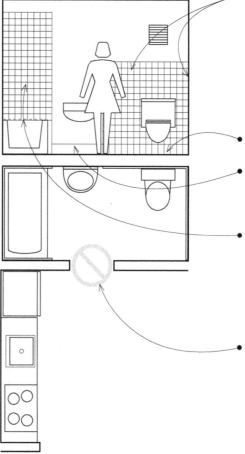

- Walls within 2' (610) of urinals and water closets must have a smooth, hard, nonabsorbent surface at least 4' (1219) high. The wall materials, apart from the structure, must also be of materials not adversely affected by moisture. Items such as grab bars and dispensers must be sealed to protect the underlying structure from moisture.

- Floors are to be smooth, hard and nonabsorbent.

- The surface treatment must extend up the wall at least 6" (152).

- Shower compartments and walls above tubs must be finished with a hard, nonabsorbent finish to a height of 70" (1778) above the drain inlet elevation. Built-in tubs with showers are to have sealant joints between the tub and the adjacent walls.

- Toilet rooms are not to open directly into a room used for preparation of food for the public. This obviously would include kitchens, but would also include serving areas or food-preparation stations outside of the kitchen in a food-service venue. A vestibule or hallway needs to be provided in such circumstances.

13
Energy Efficiency

The requirements for energy-efficient design are contained in a separate companion code, the International Energy Conservation Code (IECC). This code volume must be adopted by local jurisdictions to allow its enforcement to accompany the International Building Code. We will use this code as a brief example of how the family of International Codes interacts. Similar correlations occur with several other I-codes such as the International Mechanical Code and the International Plumbing Code.

Our discussion will focus on the building-design implications of the requirements contained in the Energy Conservation Code. The energy conservation requirements for mechanical and plumbing installations are typically contained in the mechanical and plumbing codes rather than in the building code. These are outside the scope of this book.

Note that many states and local jurisdictions have adopted their own energy codes. These local codes may have quite different standards than this model code. As is the case for other possible local modifications, the designer should always verify the status of local code adoptions.

ENERGY-EFFICIENT DESIGN

The calculations for energy use are inherently site specific. They depend heavily upon geography, climate and local environmental conditions. Obviously a building that performs well in a hot, humid climate near sea level, such as in Florida, will have quite different environmental-control and energy-use patterns than a building in the hot, dry high desert climate of central Arizona. In order to clearly discuss the concepts of the energy conservation calculation methods and criteria in the code, one must know where the building is, what its function is and what type of construction it uses.

For discussion purposes this chapter makes use of an assumed building model. This simplified model building is used to demonstrate the different analytical methods that may be used for energy analysis under this code. We will take the reader through the various methods using the same model to demonstrate the differences in construction materials, and insulation and glazing requirements that occur based upon different calculation methods. We will use the same model building for both residential and commercial analyses to highlight the variation in criteria between building types that is inherent in the Code's assumptions about how differing building types consume energy. These models illustrate principals that could apply equally to any building, whether in institutional occupancies or in multi-family residential occupancies.

• *The model building is assumed to be in Kansas City, Missouri. This location is near the geographic center of the 48 contiguous United States and represents most, if not all, conceptual conditions that designers will encounter in any locale. This area was also selected because it has weather with wide variations through all four seasons. It also falls into the middle range of design criteria such as degree-days of heating and cooling, climate zones and susceptibility to termite infestation.*

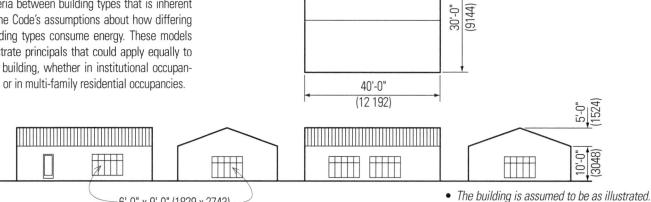

6'-0" x 9'-0" (1829 x 2743) windows

• *The building is assumed to be as illustrated. It is assumed to be a very simple 1200 sf (111.5 m²), single-story, pitched-roof building of wood-frame construction without a basement. The building has openings on all four walls but no interior partitions to simplify illustration of concepts and to allow use of the building for both residential and commercial analysis. The results of the calculations are intended only to show the use of the calculation methods and to allow comparison with each other to show the variations in energy conservation measures inherent in the different compliance methods allowable by the code. These are illustrative only and by no means meant to be applicable to real conditions.*

Before we go into the detailed energy analyses, it is worthwhile to discuss the general concepts that underlie the International Energy Conservation Code. We will refer to the International Energy Conservation Code in this chapter as the "IECC."

§101.3 Intent

This code regulates building envelope design and construction to promote the effective use of energy. The methods in the Code are meant to be flexible and to provide alternate methods of determining compliance using a range of prescriptive or performance criteria. It is to be read and used in concert with the other codes in the "I Code" family and this code is not meant to abridge requirements from other codes.

§101.4 Applicability

§101.4.1 & 3 Existing Buildings, Alterations and Additions

These requirements were developed using new buildings as the basis for design. They are also meant to apply to alterations, additions or repairs of systems. It is not the intent that the entirety of existing structures be made to comply with the Code when additions, alterations or repairs are undertaken. However, when work is done, the new work is to comply with the Code.

§101.4.2 Historic Buildings

Historic buildings listed in a recognized register such as a State Register of Historic Places are exempt from the IECC.

§101.4.4 and 5 Change in Occupancy and Mixed Use Buildings

When a change in occupancy occurs that will result in increased demand for fossil fuel or electricity then the building must comply with the IECC. In mixed-use buildings the appropriate criteria for residential (Chapter 4) or commercial (Chapter 5) occupancies are to be applied to each portion of the building.

§101.5 Compliance

Residential buildings are to comply with the provisions contained in Chapter 4 of the IECC. Commercial buildings are to comply with the provisions contained in Chapter 5. Compliance is broken into two broad categories: one is prescriptive and the other is performance based.

The code official is permitted to designate acceptable computer programs or other calculation methods that are deemed to verify compliance with the IECC. Building areas with very low energy use or without conditioned space need not comply if they are adequately separated from the remainder of the building under consideration.

§102.1.3

The solar heat gain coefficient (SHGC) for fenestration (which includes windows, doors and skylights) is to be the U-factor determined in accordance with National Fenestration Rating Council procedures. For fenestration without ratings the designer may use default U-factors determined from Tables 102.1.3(1). We will use these factors in our calculations for our design model.

Residential

The methods allow for use of either performance or prescriptive criteria. All residential projects must comply with all of the mandatory provisions contained in Sections 401, 402.4, 402.5, 402.6 and 403. Where one and two family dwellings are regulated by the International Residential Code verify with the AHJ to determine if energy conservation measures from the IRC apply to your project.

- §402 (Prescriptive): Design of systems by application of simplified prescriptive requirements (an application of simple cookbook approaches designed to minimize building energy consumption while reducing design complexity).
- §404 (Performance): Design by simulating performance (a calculation modeling method with a performance basis, usually performed by a computer program).

Commercial

Commercial methods are similar to those for residential buildings, but the criteria and requirements have different bases, coming from different technical resources. Mandatory provisions must be satisfied in a similar manner as for residential buildings. The mandatory provisions are contained in §s 502.4, 503.5, 503.2, 504, 505.2, 505.3, 505.4, 505.6 and 505.7.

- §502 (Prescriptive): Design of systems by application of simplified prescriptive requirements (an application of simple cookbook approaches designed to minimize building energy consumption while reducing design complexity). Note that the commercial standards have many more alternatives for HVAC systems recognizing the level of potential complexity of such systems in commercial buildings.
- §506 (Performance): Design by simulating performance (a calculation modeling method with a performance basis, usually performed by a computer program).

Chapter 2: Definitions

The definitions noted here include some contained in the IECC as well as additional definitions used in reference documents and in energy compliance calculations.

Building Thermal Envelope

The basement walls, exterior walls, floor, roof, and any other building element that enclose conditioned space. This boundary also includes the boundary between conditioned space and any exempt or unconditioned space.

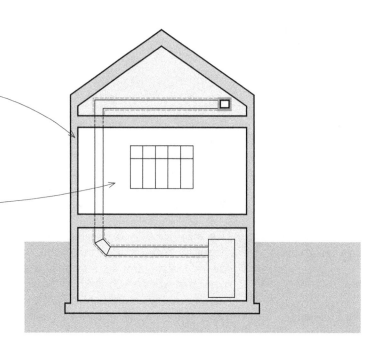

Conditioned Space

A heated or cooled space, which may contain uninsulated ductwork or which opens into another adjacent conditioned space.

Degree-Days Cooling, Degree-Days Heating

Units for measuring cooling or heating energy consumption based upon deviations in daily mean temperature from an assumed comfortable baseline temperature of 65°F (18°C).

- For any single day, there are as many degree-days as the difference in degrees of the mean temperature and 65°F (18°C).
- The mean temperature for cooling degree-days will be higher than 65°F (18°C), and for heating degree-days the mean temperature will be lower. Thus a day with a mean temperature of 45°F (7°C) would have 20 heating degree-days (65–45).
- The annual degree days are the sum of heating degree-days and cooling degree-days respectively. These two sets of numbers are not added together as there is no relevance to the sums of the two types of energy use.

73°F (23°C) Mean temperature above 65°F baseline

8°F = 8 cooling degree-days (73 - 65 = 8)

65°F (18°C) Baseline

7°F = 7 heating degree-days (65 - 58 = 7)

58°F (14°C) Mean temperature below 65°F baseline

Energy Analysis

A method for estimating the annual energy use of the proposed design and standard reference design based on estimates of energy use.

Residential Building

For this code, includes R-3 buildings, as well as R-2 and R-4 buildings three stories or less in height above grade. Where one- and two-family dwellings are regulated by the International Residential Code verify with the AHJ to determine if energy conservation measures from the IRC apply.

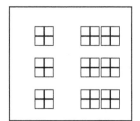

Standard Reference Design
A version of the proposed design that complies with the minimum requirements of the code. It is used to set the energy budget for maximum energy use for evaluation of compliance of designs using alternative building performance criteria.

Solar Heat Gain Coefficient (SHGC)
The ratio of the solar heat gain entering the space through the fenestration assembly to the incident solar radiation. Solar heat gain includes directly transmitted solar heat and absorbed solar radiation which is then reradiated, conducted or convected into the space.

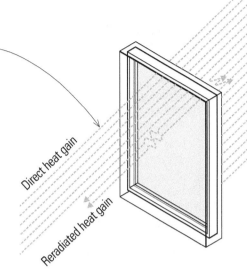

Direct heat gain

Reradiated heat gain

Thermal Conductance (C)
The time rate of heat flow through a unit area of a given material, expressed in terms of unit temperature difference between two surfaces, under steady conditions.

Thermal Resistance (R)
The reciprocal of thermal conductance, equal to the sum of the weighted R-values for each layer in the component, such as air films, thermal insulation, framing and glazing.

- R = (h per sf per °F/Btu) or [(m² per K/W].

(THERMAL TRANSMITTANCE).
(Btu/h • ft² • °F) [W/(m² • K)].

1 unit

1 unit

C = 1/R

Thermal Transmittance (U)
The coefficient of heat transmission (air to air) through a building component or assembly, equal to the time rate of heat flow per unit area and unit temperature difference between the warm side and cold side air films. Similar to thermal conductance and essentially equal to the inverse of the thermal resistance, R.

- U= (Btu per hour per sf per °F) or [W/m² per °K]

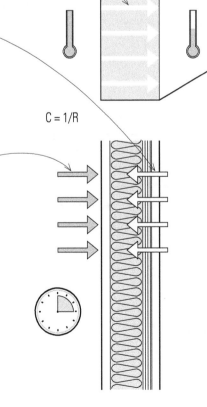

Chapter 3: Design Conditions

The criteria in Chapter 3 of the International Energy Conservation Code are adopted by the local Authorities Having Jurisdiction (AHJ) to establish the design criteria for use with Chapters 4 and 5.

For our design example we will fill in the data in Table 301.1 using information for Kansas City, Missouri. Kansas City is on the border between Zones 4 and 5, based on Figure 301.1. For our calculations we will assume our project is in Zone 4.

We will assume a demonstration building located in Kansas City, Missouri, to perform analyses for performance-based and prescriptive energy analyses for residential buildings and for prescriptive energy design compliance for commercial buildings. These should give you a feeling for the processes used for each type of design analysis. We will not cover the use of renewable energy resources for residential buildings, not because we do not support their incorporation into projects (which we very much support) but because including analysis of the wide range of potential renewable resource options is beyond the scope of this book. Once you become adept at the concepts used in energy analysis you can apply them to renewable-resource use analyses.

It is almost certain that the actual energy calculations you will use in practice will be performed using computer models. We are showing the input assumptions that go into determining base criteria for use in the modeling programs. We will perform only limited numbers of manual calculations to give a feeling for the use of the criteria and how their application varies between energy-use design methodologies. We do not presume to give the reader all the information to manually perform energy-use calculations as they in all likelihood will be done using a computer model. Our goal is not to duplicate the calculation process used by the computer models but to give you an understanding of the concepts that underlie use of the models.

Exterior Design Conditions

Condition	Value (for Kansas City, MO)
Winter, design dry-bulb (°F) per ASHRAE Handbook of Fundamentals. 97½% value	8°F at 97½% value
Summer, design dry-bulb (°F) per ASHRAE Handbook of Fundamentals, 2½% value	97°F at 2½% value
Degree-Days, heating (Per NOAA data)	5,000
Degree-Days, cooling (Per NOAA data)	1,250
Climate Zone, per Code Figure 302.1(26)	Zone 11B

For SI: °C = [(°F–32)]/1.8

Interior Design Conditions

Per §302 the interior design temperatures used for heating and cooling load calculations shall be a maximum of 72°F (22°C) for heating and minimum of 75°F (24°C) for cooling.

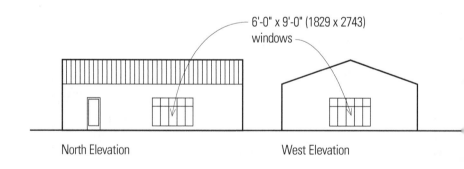

6'-0" x 9'-0" (1829 x 2743) windows

North Elevation   West Elevation

As illustrated, the building is very simple, with no interior partitions and simple glazing patterns on four faces aligned with the points of the compass. The basic building statistics for our assumed structure are as follows:

Gross Floor Area:	1,200 sf	(111.5 m²)
Gross Wall Area:	1,550 sf	(144 m²)
Gross Roof Area:	1,265 sf	(117.5 m²)
Glazing Area:	300 sf	(27.9 m²)
Door Area:	21 sf	(2 m²)
Total Fenestration:	321 sf	(29.8 m²)
Volume:	15,000 c.f.	(42.5 m³)

19.35% of gross wall area
25% of building gross floor area
1.75% of building gross floor area
1.35% of gross wall area
26.75% of building gross floor area

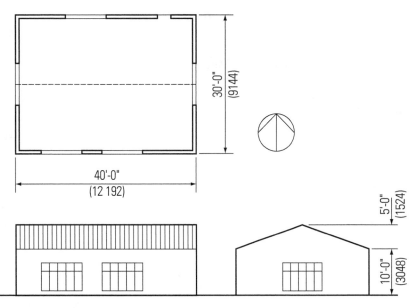

We will modify the base case glazing, shading and insulation criteria as necessary to demonstrate the trade-offs that can be used in complying with energy conservation requirements.

South Elevation   East Elevation

Chapter 4: Residential Energy Efficiency

Chapter 4 analyses total energy use for residential buildings, including all of its systems, such as heating, cooling, water heating and electrical energy use. The allowable energy use is based upon the annual energy use of a building thermal envelope based on the climate zone in which it is located per Table 402.1.1. The designer is free to manipulate variables such as insulation values of walls, heat-transfer coefficients of glazing and shading to meet the total energy-use budget established for the thermal envelope criteria.

Prescriptive Approach

§402.1.1 sets the basic criteria that the building thermal envelope is to meet the requirements of Table 402.1.1 based on the climate zones specified in Chapter 3.

The relationship of R-values and U-factors is accommodated by Table 402.1.3, which sets U-factors for substitution for R-values listed in Table 402.1.1.

Based on the criteria contained in Table 402.1.1, the requirements for our location in Kansas City, Missouri are as follows for Climate Zone 4: The equivalent U-factors from Table 402.1.3 are shown in italics. These factors are to be used in trade-offs for total UA analysis as discussed below.

	T 402.1.1	**(T 402.1.3)**
Fenestration U-Factor:	0.40	*(U- Factor : 0.40)*
Ceiling R-Value :	38	*(U-Factor: 0.030)*
Wood Frame Wall R-Value*:	13*	*(U-Factor: 0.082)*
Mass Wall R-Value:	5**	*(U-Factor: 0.141)*
Floor R-Value:	19	*(U-Factor: 0.047)*
Basement Wall R-Value:	10 if continuous or 13 for framing cavity insulation	*(U-Factor: 0.059)*
Slab R-Value and depth:	10@2 feet (610) (add R-5 for integrally heated slabs)	
Crawl Space R-Value:	10 if continuous or 13 for framing cavity insulation	*(U-Factor: 0.065)*

* See the discussion below regarding §402.2.4 and Table 402.2.4 for walls with steel-stud framing.

** Per §402.3.3 Mass Walls are considered walls of concrete block, concrete, insulated concrete form (ICF), masonry cavity, brick (other than brick veneer), earth (adobe, compressed earth block, rammed earth) and solid timber/logs.

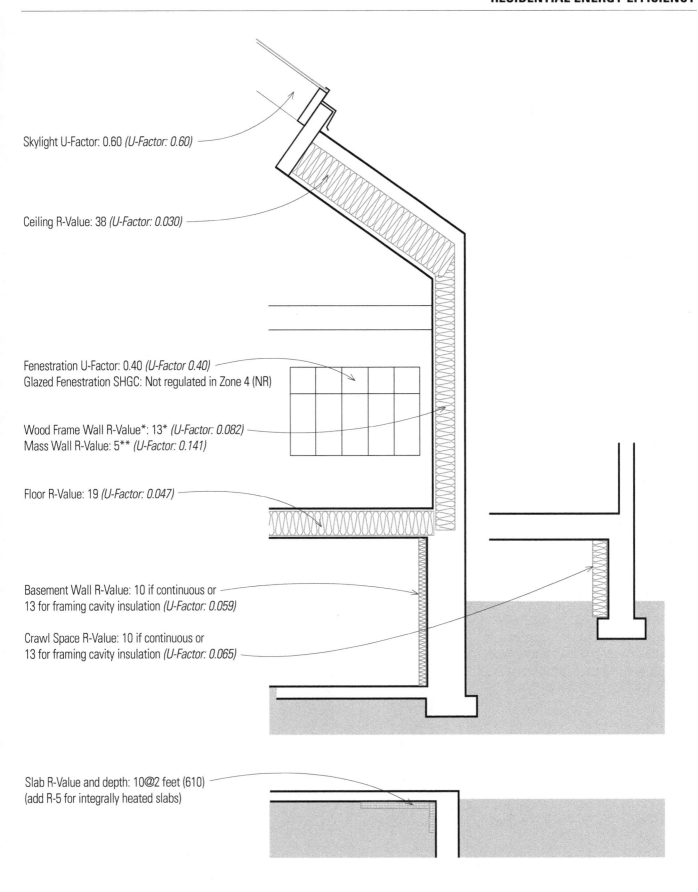

Skylight U-Factor: 0.60 *(U-Factor: 0.60)*

Ceiling R-Value: 38 *(U-Factor: 0.030)*

Fenestration U-Factor: 0.40 *(U-Factor 0.40)*
Glazed Fenestration SHGC: Not regulated in Zone 4 (NR)

Wood Frame Wall R-Value*: 13* *(U-Factor: 0.082)*
Mass Wall R-Value: 5** *(U-Factor: 0.141)*

Floor R-Value: 19 *(U-Factor: 0.047)*

Basement Wall R-Value: 10 if continuous or
13 for framing cavity insulation *(U-Factor: 0.059)*

Crawl Space R-Value: 10 if continuous or
13 for framing cavity insulation *(U-Factor: 0.065)*

Slab R-Value and depth: 10@2 feet (610)
(add R-5 for integrally heated slabs)

§s 402.1.3 & 4 UA Alternative analysis

These two sections allow use of U-factors to determine energy performance versus the R-factors used for assemblies in Table 402.1.1. The U-factor alternative under 402.1.3 allows use of innovative wall types that may have different thermal characteristics than the basic wood frame walls listed in Table 402.1.1. §402.1.4 allows trade-offs across all of the envelope components. If the total building thermal enveloped UA (sum of U-factor times assembly area) is less than or equal to the total UA resulting from using the U-factors from Table 402.1.3 using equal assembly areas the building will be considered in compliance with the requirements of Table 402.1.1. The UA calculation must be done in accordance with the ASHRAE Handbook of Fundamentals and include thermal bridging effects where framing materials transfer heat. The Solar Heat Gain Coefficient requirements of Table 402.1.1 are to be met in addition to meeting the UA requirements.

This concept is critical to understand for both prescriptive or performance alternatives. It allows use of larger window areas if offset by higher insulation values for wall assemblies. The total U-factor for the subject building must be less than or equal to the U-factor calculated using Table 402.1.1. The Department of Energy REScheck program (available online at http://www.energycodes.gov/rescheck/) uses the UA concept to analyze building compliances.

Using the REScheck program and the specified prescriptive values from Table 402.1.1 our demonstration building does not comply with the allowable overall code U value of 244. We recommend downloading the program and learning the basics of how it works and how it can be used very quickly to analyze basic system trade-offs to achieve energy compliance.

Component	Assembly	Gross Area	T 402.1.1	REScheck	UA
		(sf)	Cavity Insulation R Value	U-Factor	
Ceiling	Wood studs at 16" o.c.	1550	38	0.03	36
Wall		1200	13	0.082	103
Glazing	Double glazed, Wood frame	300	---	0.40	120
Allowable UA					244
Calculated UA					259
Pass/Fail					Fails

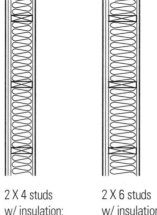

2 X 4 studs
w/ insulation:
R-13

2 X 6 studs
w/ insulation:
R-19

If we reduce the window area to 255 square feet or decrease the fenestration U value to 0.35 with 300 sf of glazing then the building will comply. Simplified calculation programs such as REScheck allow trade offs between systems to achieve compliance. The designer has the option of reducing window size, increasing window performance to comply. Other options would be to increase wall or ceiling insulation while keeping the windows the same. For our example using R-19 wall insulation would allow the windows to remain 300 sf and the building to be in compliance.

§402.2 Prescriptive Insulation Requirements

§402.2 lists specific prescriptive insulation requirements and alternative details to the prescriptive requirements in §402.1.1 that are listed in Table 404.1.1.

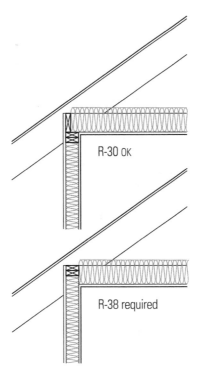

R-30 OK

R-38 required

- *Per §404.2.1 Where R-38 insulation is required R-30 insulation may be used if the uncompressed R-30 insulation extends over the wall top plate at the eaves. Similarly R-38 insulation installed in the same way meets the requirements for R-49 insulation.*

- *Per §402.2.2 allows use of less than R-30 insulation in areas of up to 500 square feet (46 m²) in ceilings without attics where the design of the roof/ceiling assembly does not allow sufficient space for installation of the necessary insulation.*

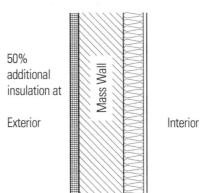

50% additional insulation at

Exterior

Mass Wall

Interior

- *Per §402.2.3 mass walls must have 50% of the insulation value required by Table 402.1.1 on their exterior to comply with the requirements of §402.1.1 or the walls must meet the insulation requirements for wood stud walls. Mass walls in Climate Zones 1, 2 and 3 are allowed reduced exterior insulation by exceptions to this section.*

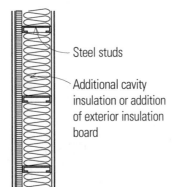

Steel studs

Additional cavity insulation or addition of exterior insulation board

- *§402.2.4 and Table 402.2.4 is to be used for walls with steel-stud framing. These have a factor that must be included that increase the thermal transmittance factors. This is to account for the thermal bridging effects of steel studs, which conduct thermal energy rather than resisting energy conductance as do wood studs. Thus steel-stud walls require additional thickness of cavity insulation or addition of exterior insulation board to reduce the thermal transmittance of the studs.*

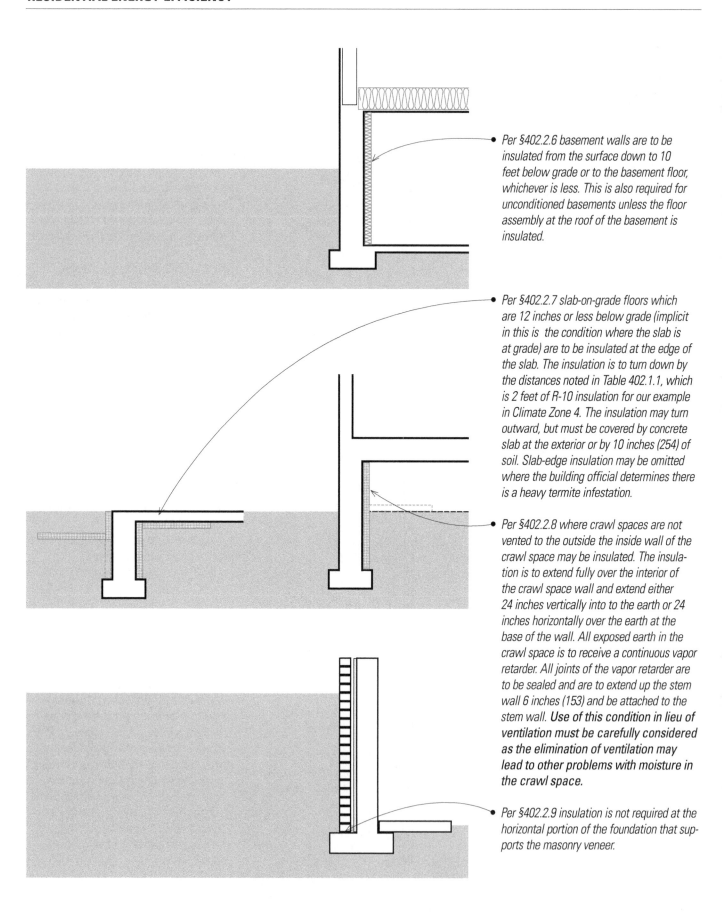

Per §402.2.6 basement walls are to be insulated from the surface down to 10 feet below grade or to the basement floor, whichever is less. This is also required for unconditioned basements unless the floor assembly at the roof of the basement is insulated.

Per §402.2.7 slab-on-grade floors which are 12 inches or less below grade (implicit in this is the condition where the slab is at grade) are to be insulated at the edge of the slab. The insulation is to turn down by the distances noted in Table 402.1.1, which is 2 feet of R-10 insulation for our example in Climate Zone 4. The insulation may turn outward, but must be covered by concrete slab at the exterior or by 10 inches (254) of soil. Slab-edge insulation may be omitted where the building official determines there is a heavy termite infestation.

Per §402.2.8 where crawl spaces are not vented to the outside the inside wall of the crawl space may be insulated. The insulation is to extend fully over the interior of the crawl space wall and extend either 24 inches vertically into to the earth or 24 inches horizontally over the earth at the base of the wall. All exposed earth in the crawl space is to receive a continuous vapor retarder. All joints of the vapor retarder are to be sealed and are to extend up the stem wall 6 inches (153) and be attached to the stem wall. **Use of this condition in lieu of ventilation must be carefully considered as the elimination of ventilation may lead to other problems with moisture in the crawl space.**

Per §402.2.9 insulation is not required at the horizontal portion of the foundation that supports the masonry veneer.

§402.3 Prescriptive Fenestration Requirements

An area-weighted average of the U-factor requirements may be used to satisfy the standards of §402.1.1. For example specialty windows may have a higher U-Factor which is offset by other windows of higher U-factor as long as the overall average is compliant. The same principal applies to Solar Heat Gain Coefficients (SHGC).

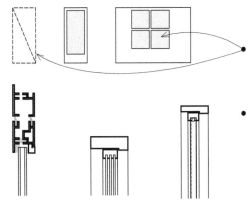

- One window up to 15 sf (1.4 m²) is exempt from the U-factor and SHGC requirements of §402.1.1. One opaque door assembly is also exempted.
- Replacement fenestration is to meet the requirements of §402.1.1 for U-factor and SHGC. This is in accord with the general principle in the code that alterations are to comply with current code requirements.

Aluminum window w/ thermal break　　Triple glazing　　Double glazing w/ low e-coating

§402.4 Mandatory Air Leakage Requirements

Buildings are to have measures installed to minimize air leakage into and out of the building envelope and around windows and doors. Window and door assemblies are to have maximum allowable air-infiltration rates. Sealing methods are to allow for differential expansion and contraction of construction materials.

§402.4.1 Air Leakage at Thermal Envelope

Items required by §402.4.1 to be caulked, gasketed, weatherstripped or otherwise sealed with an air barrier material, suitable film or solid material include:

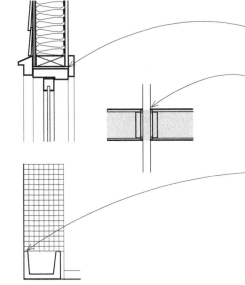

1. All joints, seams and penetrations.
2. Site-built windows, doors and skylights.
3. Openings between window and door assemblies and their respective jambs and framing.
4. Utility penetrations.
5. Dropped ceilings or chases adjacent to the thermal envelope.
6. Knee walls.
7. Walls and ceilings separating a garage from conditioned spaces.
8. Behind tubs and showers on exterior walls.
9. Common walls between dwelling units.
10. Other sources of infiltration.

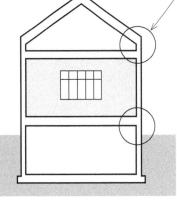

§402.4.2 Fenestration Air Leakage

§402.4.2 requires installation of measures to minimize air leakage at openings such as windows and doors.

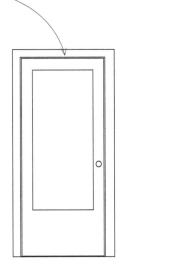

Windows and Skylights (sfm per square foot of window area)	Doors (cfm per square foot of door area)	
	Sliders	Swinging
0.3	0.3	0.5

For SI: 1 cfm/sf = 0.00508 m³/(s × m²)

§402.4.3 Recessed Lighting

Recessed fixtures that puncture the conditioned building envelope are to be sealed or gasketed to prevent air movement between conditioned spaces and unconditioned spaces. The seals may be a part of the fixture or the fixtures are to have a ½" (12.7 mm) gypsum board enclosure or other air-tight enclosure built over them. These enclosures are to maintain at least a ½" (12.7 mm) clearance from combustible materials and not less than a 3" (76) clearance to insulation materials.

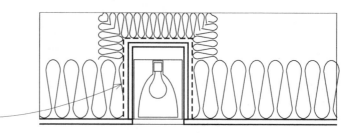

§402.5 Mandatory Moisture Control

Because energy compliance requires tighter buildings with less air infiltration and exfiltration it is essential for designers to think about moisture control to prevent buildups of condensation in building elements. Ventilation of spaces allows moisture to escape but also allows energy to escape. Tight buildings need careful attention to condensation control. Above-grade frame walls, floors and ceilings not ventilated to allow moisture to escape are to have an approved vapor retarder. The vapor retarder is to be installed on the warm-in-winter side of the thermal insulation.

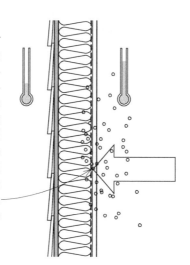

§402.6 Mandatory Maximum Fenestration U-Factor and SHGC

The area weighted average maximum fenestration U-factor permitted using trade-offs from §402.1.4 or §404 shall be 0.48 for vertical fenestration in our building in Climate Zone 4 and 0.75 in zone 4 for skylights. The area weighted average maximum fenestration SHGC does not apply for Climate Zone 4 so it is not regulated.

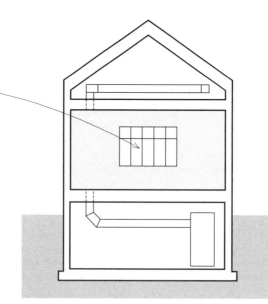

§403 Mandatory System Requirements
The code requires certain mandatory measures for heating and cooling systems.

§403.1 Controls
There is to be a thermostat for controlling each separate heating and cooling system.

§403.2 Duct Insulation
Supply and return ducts are to be insulated with a minimum of R-8 insulation. Ducts in floor trusses may have R-6 insulation. This requirement is waived for ductwork located completely inside a building thermal envelope. Thus it is acceptable to have un-insulated exposed ductwork inside a space since it is inside the building thermal envelope.

§403.3, 4 Mechanical System Piping Insulation
Piping such as hot water heating system pipes that carry fluids above 105°F (41°C) or cooling systems pipes that carry fluids below 55°F (13°C) are to be insulated with a minimum of R-2 insulation. Circulating hot water systems are also to have the same level of insulation. These systems are also to have a readily accessible switch to turn off the circulation system when the system is not in use.

§403.5 Mechanical Ventilation
Outdoor air intakes and exhausts are to have automatic or gravity dampers that close when the ventilation system is not operating.

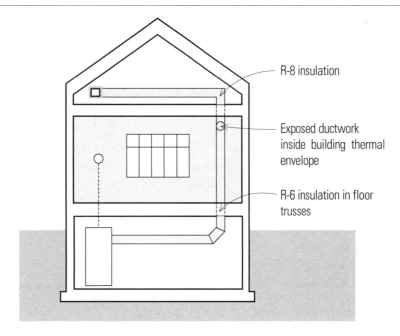

R-8 insulation

Exposed ductwork inside building thermal envelope

R-6 insulation in floor trusses

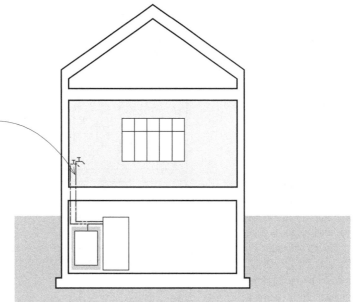

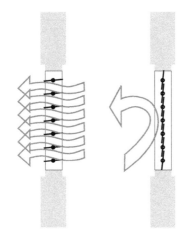

SIMULATED PERFORMANCE ALTERNATIVE

§404 Simulated Performance Alternative

The code allows use of performance modeling software to simulate energy performance of a proposed building. The analysis must include heating, cooling, and service water heating analyses. All the mandatory requirements of §s 402.4 (Air leakage), 402.5 (moisture control) 402.6 (maximum fenestration U-factor) and 403 (systems) must be met.

Compliance is based on demonstrating that the proposed design has an annual energy cost that is less than or equal to the annual energy cost of the standard reference design. The cost of energy shall be factored into the analysis. Time of use pricing may be factored in where there are time related offsets for energy use in alternative systems. It is acceptable to convert energy units from Btu to kWh as long as the unit conversions are taken into account.

The analysis must generate a report demonstrating compliance. The report shall be based on the specifications contained in Table 404.5 which allow comparison of performance analyses of designated building components for the standard reference design and the proposed design.

Software used for a full performance based analysis takes into account many more factors than the simpler "cook book" or basic system trade-off approaches used for the prescriptive methods. Performance analysis takes into account such things as: seasonal variations in heating and cooling loads over the course of a year; location and quantities of fenestration based on compass orientations of the glazing to determine actual heating or cooling losses; colors of materials; heat transfer rates for various materials to allow heat storage during the day and heat release at night; shading and tinting of glazing inside the glazing assembly; shading of the glazing by architectural elements or by vegetation. These powerful computer programs allow analysis of many interlocking considerations for building design to maximize design and utility flexibility while at the same time reducing energy consumption.

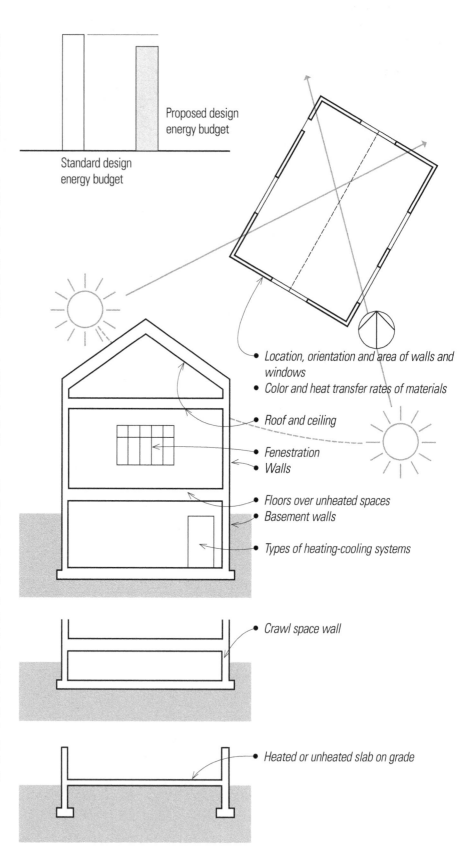

Proposed design energy budget

Standard design energy budget

- Location, orientation and area of walls and windows
- Color and heat transfer rates of materials

- Roof and ceiling

- Fenestration
- Walls

- Floors over unheated spaces
- Basement walls

- Types of heating-cooling systems

- Crawl space wall

- Heated or unheated slab on grade

IECC Chapter 5: Commercial Energy Efficiency

The IECC, itself a reference from the body of the International Building Code, adopts another code by reference. The IECC present two alternatives for energy compliance for commercial buildings. The first uses the requirements set forth in the Standard 90.1 - Energy Standard for Buildings Except for Low-Rise Residential Buildings developed by the American Society of Heating, Refrigeration and Air-Conditioning Engineers (ASHRAE) and the Illuminating Engineers Society of North America (IESNA). This will typically involve use of a computer-modeling program to analyze the overall performance of the building in a given climate. The other alternative is to use the provisions contained in Chapter 5 of the IECC.

This section of the code is similar to the residential section in that there are prescriptive standards and performance standards. Both alternative pathways assume that certain mandatory measures are met as well.

We will touch on the basic envelope requirements for commercial buildings, but we will not go in detail into mechanical or electrical requirements, as we anticipate that most calculations for these buildings will entail the use of prescriptive based design requirements only for very simple buildings and the use of computer modeling programs for most commercial buildings.

For comparison between residential and commercial requirements we will assume our case study building is a commercial structure. Below is a compariso n of the prescriptive residential requirements for Climate Zone 4 taken from Table 402.1.1 and the corresponding commercial prescriptive requirements for the same building elements, selected from Chapter 5.

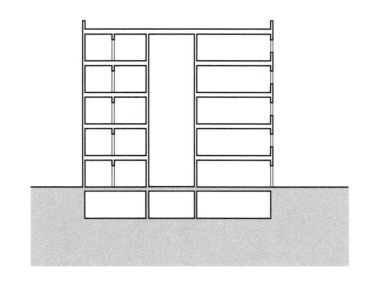

Table 402.1.1

T 402.1.1		Commercial Requirements
Fenestration U-Factor:	U = 0.40	40% maximum of wall area U = 0.40 (non-metal) 0.60 (metal)
Skylight U-Factor:	U = 0.60	U-glass = 0.60; U-plastic = 1.30 3% maximum of roof area
Glazed Fenestration SHGC:	Not Regulated in Zone 4	SHGC glass = 0.40, SHGC plastic = 0.62
Ceiling R-Value :	R-38	R-30 w/attic, R 15 if above roof deck
Wood Frame Wall R-Value:	R-13	R-13
Mass Wall R-Value:	R-5	
Floor R-Value:	R-19	
Basement Wall R-Value:	R-10 if continuous, 13 for framing cavity	NR
Slab R-Value and depth:	R-10@2 feet (+R-5 for integrally heated slabs)	R 7.5 at heated, NR at unheated
Crawl Space R-Value:	R-10 if continuous, R-13 for framing cavity	NR

COMMERCIAL ENERGY EFFICIENCY

The following mandatory requirements affect the design of the building envelope:

§502.4 Mandatory Air Leakage Requirements

§502.4.1: Window and door assemblies are to comply with the requirements of stated standards for door and window manufacturers. Site constructed windows and doors must meet the sealing requirements of §502.4.2.

§502.4.2: Commercial storefront and curtain wall glazing and openings are to comply with the air leakage requirements of ASTM E 283 when tested at 1.57 psf (75 pa).

§502.4.3: Openings and penetrations in the building envelope are to be sealed with caulking materials or closed with gasketing systems compatible with their construction materials and location. Sealing materials are to be designed to accommodate expansion and contraction of construction materials.

§504.2.4: Outdoor air intakes and exhaust openings such as stair and elevator shaft vents and other air intakes are to be equipped with not less than a Class I motorized, leakage-rated damper with a maximum leakage rate of 4 cfm per sf (6.8L/s – C m^2) at 1.0 inch water gauge (1250 pa). Gravity dampers are acceptable in buildings less than three stories in height above grade.

§502.4.5: Cargo doors and loading docks are to have weather seals to restrict infiltration when vehicles are parked in the doorways when the doors are open.

§502.4.6: To prevent mixing of conditioned and unconditioned air, enclosed vestibules separating conditioned space from the exterior are to be provided at entry doors. The vestibules are to be designed so the exterior and interior doors need not be open at the same time. Vestibules are not required in mild climate zones such as Zones 1 and 2.

§502.4.7: Similar to the requirements for residential construction recessed luminaires which are installed in the building envelope boundary are to be enclosed in gypsum board enclosures or be furnished with integral seals, gaskets or enclosures to prevent air movement between the conditioned space and the ceiling cavity.

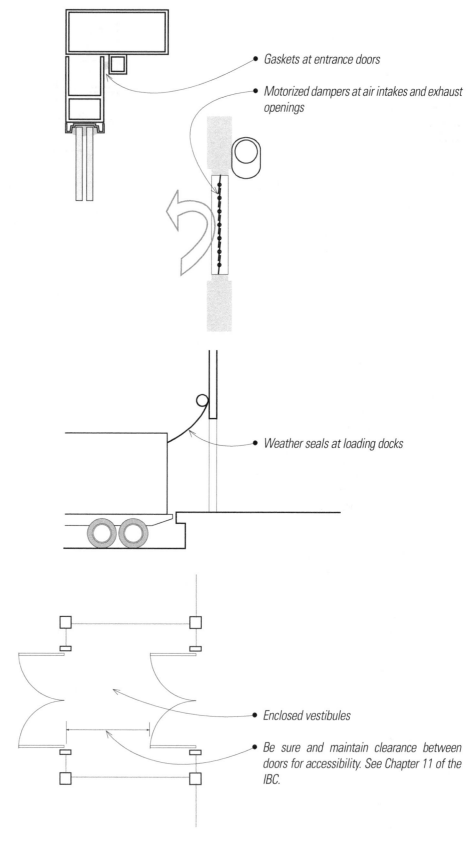

- Gaskets at entrance doors
- Motorized dampers at air intakes and exhaust openings
- Weather seals at loading docks
- Enclosed vestibules
- Be sure and maintain clearance between doors for accessibility. See Chapter 11 of the IBC.

§502.5 Mandatory Moisture Control

All framed walls, floors and ceiling that are not ventilated to allow moisture to escape are to have a vapor retarder having a permeance of 1 perm ($5.7 \times 10\text{-}11$ kg/Pa-s-m^2) or less. The vapor retarder is to be installed on the warm-in-winter side of the insulation. Buildings in Climate Zones 1, 2, and 3, construction where freezing will not damage the materials, and locations where approved means to avoid condensation in unventilated spaces are provided, are exempt from this requirement.

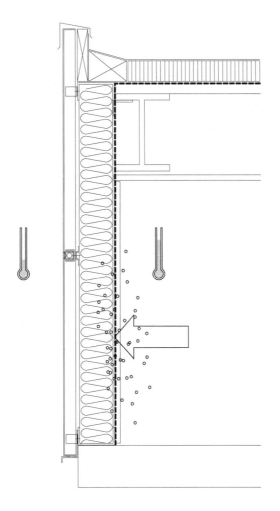

• *As we have noted elsewhere, the provision and location of ventilation and moisture barriers should be carefully addressed in all climate zones. The designer must accommodate the needs for both adequate ventilation and protection of spaces where moisture may gather with the needs to conserve energy. As buildings become tighter with less air exchange between the interior and exterior trapping of moisture becomes of greater concern and must be considered in detailing.*

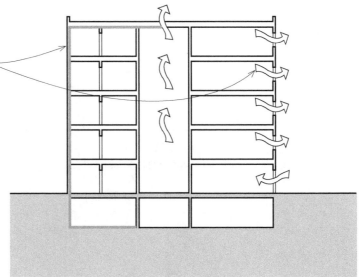

ENERGY CODE §505: Mandatory
Electrical Power and Lighting Systems

Lighting-control criteria for commercial buildings are based upon the assumption that most commercial facility use, such as in offices and retail establishments, occurs primarily in the daytime, or during defined periods of the 24-hour day. Lighting is to be switchable to be able to reduce the connected lighting load by approximately 50% while maintaining a reasonably uniform illumination pattern. Also, exterior lights not intended for 24-hour use are to have automatic switches or time-clock controls. Exterior lights of more than 100 watts are required to have a source efficiency of at least 60 lumens per watt.

Table 505.5.2 sets forth power-level criteria based upon building use. For an office building, for example, the overall building is allowed to use 1.0 watts per square foot to power interior lighting for the entire building. The tenant portion of the building may use up to 1.5 watts per square foot, assuming that light power levels will be lower elsewhere to allow the entire building to meet the lower overall building criteria.

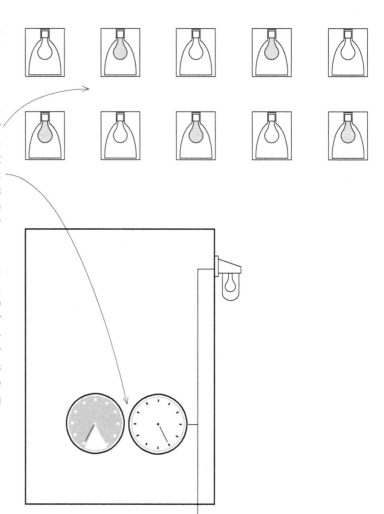

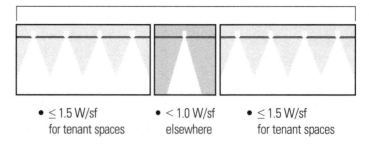

- 1.0 W/sf <u>average for entire building</u>

- ≤ 1.5 W/sf   - < 1.0 W/sf   - ≤ 1.5 W/sf
  for tenant spaces   elsewhere   for tenant spaces

(1 W/sf = 1W/0.0929 m²)

14

Exterior Walls

Chapter 14 of the International Building Code establishes requirements for exterior walls. It also sets standards for wall materials and wall coverings, as well as for wall components, such as windows, doors and trim. The provisions of this chapter are meant to apply primarily to weather protection and moisture control. The definition of exterior wall specifically excludes fire walls, which have additional fire-resistive requirements.

It is the intent of the Code that exterior-wall requirements for weather protection and moisture control not compromise the overriding structural bearing and fire-resistance requirements for walls. On the other hand, the Code does not intend for fire walls or structural bearing walls to not provide the required degree of weather resistance. As we will see as we proceed through the chapter, the intent is that weather resistance be built into exterior wall systems and that bearing walls need not have added envelope protection if they are designed as both load-bearing and weather-resistant systems.

EXTERIOR-WALL ENVELOPE

§1402 requires an "exterior wall envelope" to protect the structural members of a building, including framing and sheathing materials, as well as conditioned interior spaces from detrimental effects of the exterior environment.

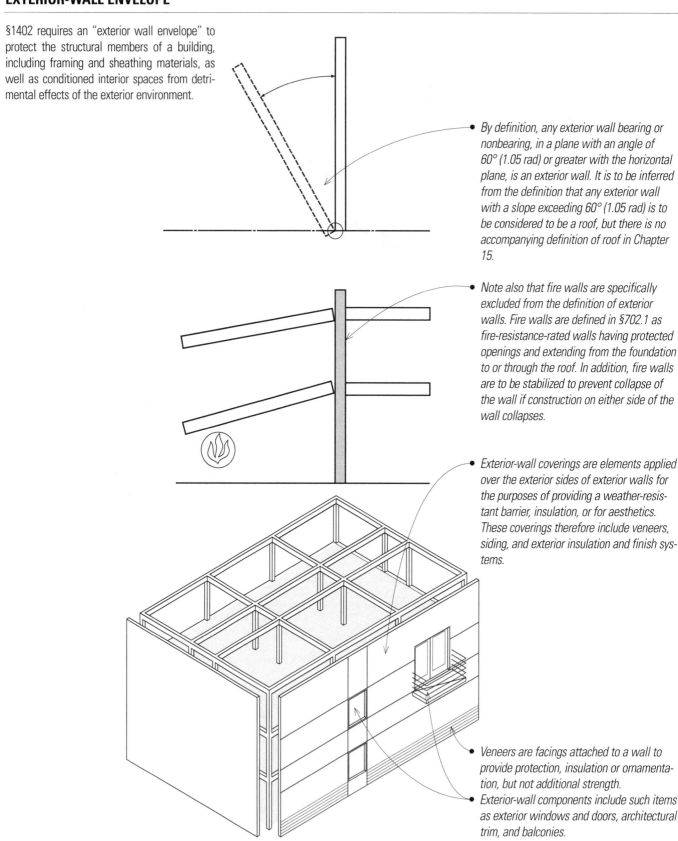

• By definition, any exterior wall bearing or nonbearing, in a plane with an angle of 60° (1.05 rad) or greater with the horizontal plane, is an exterior wall. It is to be inferred from the definition that any exterior wall with a slope exceeding 60° (1.05 rad) is to be considered to be a roof, but there is no accompanying definition of roof in Chapter 15.

• Note also that fire walls are specifically excluded from the definition of exterior walls. Fire walls are defined in §702.1 as fire-resistance-rated walls having protected openings and extending from the foundation to or through the roof. In addition, fire walls are to be stabilized to prevent collapse of the wall if construction on either side of the wall collapses.

• Exterior-wall coverings are elements applied over the exterior sides of exterior walls for the purposes of providing a weather-resistant barrier, insulation, or for aesthetics. These coverings therefore include veneers, siding, and exterior insulation and finish systems.

• Veneers are facings attached to a wall to provide protection, insulation or ornamentation, but not additional strength.
• Exterior-wall components include such items as exterior windows and doors, architectural trim, and balconies.

§1403 sets out the performance criteria for exterior walls, wall coverings and components.

Weather Protection

§1403.2 states that the basic performance criterion for exterior walls is providing an exterior-wall envelope that is weather-resistant. Exterior-wall systems must therefore be designed to prevent water penetration, the accumulation of water behind exterior veneers, and to provide drainage for any moisture that does enter the wall assemblies to the exterior of the veneer. The systems must also protect against condensation inside wall assemblies as required by the International Energy Conservation Code which is contained in a separate document from the IBC.

- *Exception 1: Weather-resistant systems are not required when exterior concrete walls or exterior masonry walls are designed for weather resistance per Chapters 19 and 21, respectively.*
- *Exception 2: Drainage requirements for wall assemblies may be waived when acceptable testing is performed on envelope assemblies including joints, penetrations and intersections per the test procedures of ASTM E 331. The basic criteria for this test require that a 4' by 8' (1219 by 2438) wall section be tested. The test section must contain a typical opening, a control joint, a wall/eave intersection, and one wall sill. The test must simulate pressure differentials of 6.24 psf (0.297 kN/m²) and last for at least 2 hours. These criteria are established to require that the tested wall assembly with all of its typical parts be exposed to simulations of wind-driven rain conditions that can be reasonably anticipated in actual installations. Such assemblies are usually tested by a manufacturer of wall systems to verify the weather-tight performance of their wall systems so that they can validate their performance claims and induce the use of their product.*

Other Requirements

Exterior walls are to be designed to resist structural loads per Chapter 16, and meet the applicable fire-resistance and opening-protection requirements of Chapter 7. In flood-hazard areas, designated per §1612.3, exterior walls extending below the design flood elevation are to be resistant to water damage. Any wood used in these areas is to be pressure-treated or naturally decay resistant.

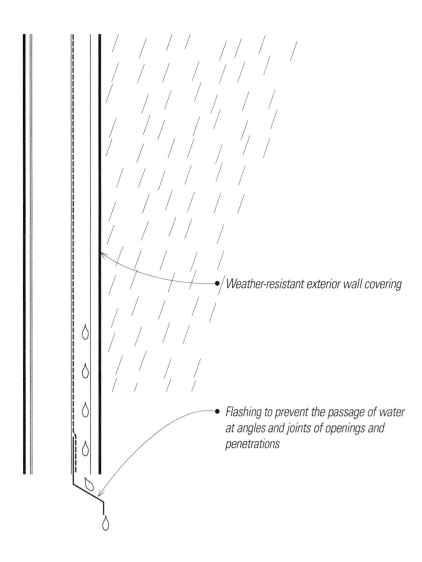

Weather-resistant exterior wall covering

Flashing to prevent the passage of water at angles and joints of openings and penetrations

WALL MATERIALS

§1403 and §1404 require that all materials used in exterior walls are to meet the detailed provisions for specific materials contained in Chapters 19 through 26.

- Concrete: Chapter 19
- Aluminum: Chapter 20 Aluminum siding must satisfy the requirements of AAMA 1402.
- Masonry and Glass-unit Masonry: Chapter 21
- Steel: Chapter 22
- Wood: Chapter 23
- Glass: Chapter 24
- Gypsum Board and Plaster: Chapter 25
- Plastics: Chapter 26
- Vinyl siding must conform to the requirements of ASTM D 3679.

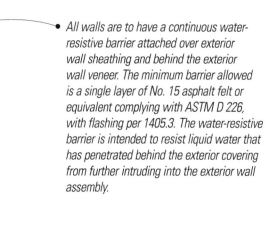

- *All walls are to have a continuous water-resistive barrier attached over exterior wall sheathing and behind the exterior wall veneer. The minimum barrier allowed is a single layer of No. 15 asphalt felt or equivalent complying with ASTM D 226, with flashing per 1405.3. The water-resistive barrier is intended to resist liquid water that has penetrated behind the exterior covering from further intruding into the exterior wall assembly.*

- *§1405.2 requires exterior wall coverings to provide weather protection for the building. The minimum thicknesses of weather coverings are specified in Table 1405.2. These thicknesses range from thin steel (0.0149" or 0.38 mm) to anchored stone masonry (2.625" or 66.67 mm).*
  The minimum thicknesses are based on conventional construction techniques and recorded performance of these materials in actual installations. These are minima, not maxima. Thicker assemblies or thicker individual components should be acceptable to the AHJ as long as the assemblies are designed to hold the weight of the covering materials.

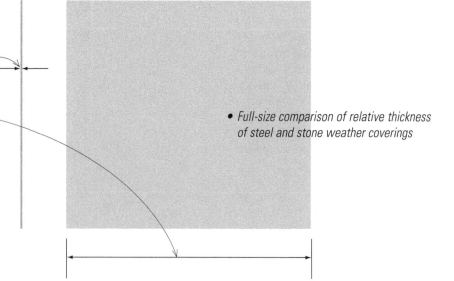

- *Full-size comparison of relative thickness of steel and stone weather coverings*

Flashing

§1405.3 requires the installation of flashing to prevent moisture from entering at locations such as those illustrated to the right. The intent of the list, that contains examples, not just the conditions to be satisfied, is that flashing be provided at any condition where it would prevent moisture from entering the structure.

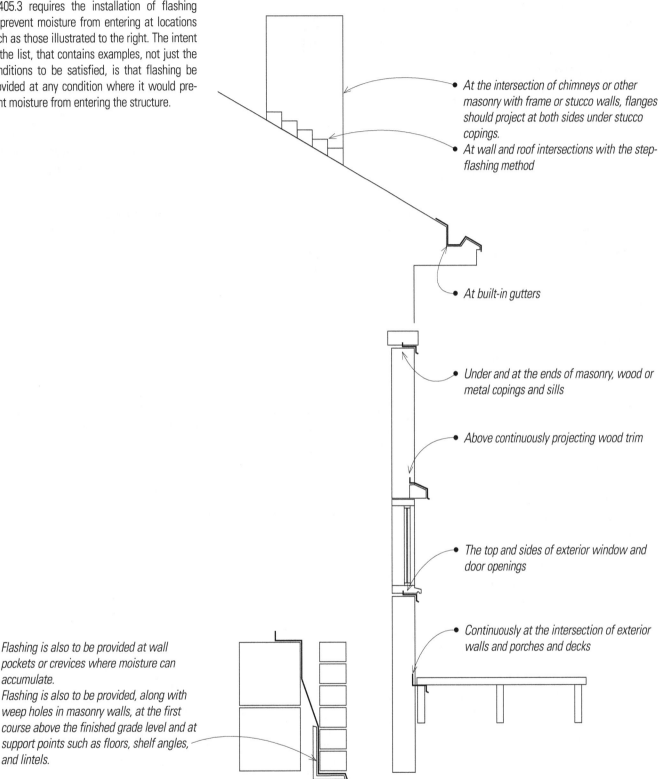

- At the intersection of chimneys or other masonry with frame or stucco walls, flanges should project at both sides under stucco copings.
- At wall and roof intersections with the step-flashing method

- At built-in gutters

- Under and at the ends of masonry, wood or metal copings and sills

- Above continuously projecting wood trim

- The top and sides of exterior window and door openings

- Continuously at the intersection of exterior walls and porches and decks

- Flashing is also to be provided at wall pockets or crevices where moisture can accumulate.
- Flashing is also to be provided, along with weep holes in masonry walls, at the first course above the finished grade level and at support points such as floors, shelf angles, and lintels.

INSTALLATION OF WALL COVERINGS

Wood Veneers

§1405.4 prescribes the requirements for installing wood veneers.

Notwithstanding the requirements of Table 1405.2 for the minimum thickness of weather coverings, wood veneer used in Type I, II, III and IV construction is to be of thickness called out in §1405.4.

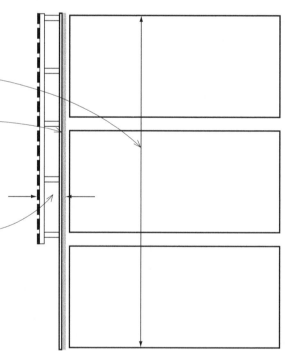

• *Wood veneer is to be 1" (25.4 mm) nominal thickness, versus ½" (12.7 mm) per the table.*

• *Hardboard siding is to be 0.438" (11.1 mm) versus ¼" (6.4 mm) per the table.*

• *Wood structural panels are to be ⁵⁄₈" (9.5 mm) versus 0.313" (8 mm) thick.*

The extent of wood veneer in other than Type V construction is also limited. The requirements are not exceptions, and all must be met for wood veneer to be used in these types of construction. The requirements are:

1. The veneer cannot exceed three stories in height; four stories is the limit if fire-retardant treated wood is used.
2. The wood is attached to or furred out from a noncombustible backing that is fire-resistance rated per the Code.
3. Where open or spaced wood veneers without concealed spaces are used, they do not project more than 24" (610) from the building wall. See the discussion later regarding combustible materials at exterior walls contained in §1406.

Anchored Masonry Veneer

§1405.5 prescribes specific requirements for the use of anchored veneers, whether of masonry, stone or terra-cotta. These provisions apply to materials anchored to backing materials using mechanical fasteners. This is distinct from adhered veneers that are attached to the backing by adhesives. These requirements are broken down by material into separate sections.

§1405.5 also requires anchored masonry veneers to comply with the American Concrete Institute standards contained in Sections 6.1 and 6.2 of ACI 530, but the text applies these requirements to several types of veneers: masonry, stone, slab-type and terra-cotta.

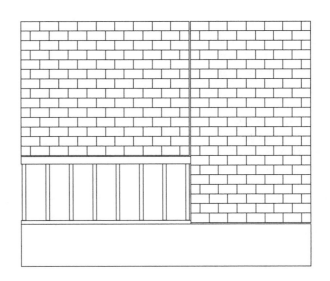

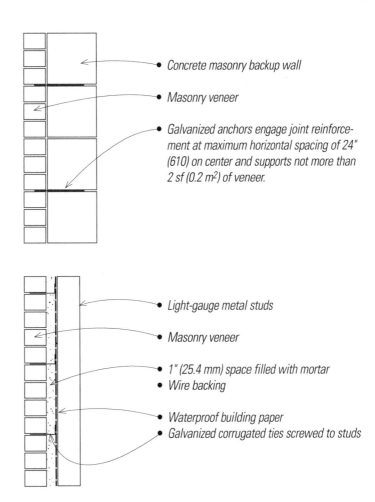

- Concrete masonry backup wall

- Masonry veneer

- Galvanized anchors engage joint reinforcement at maximum horizontal spacing of 24" (610) on center and supports not more than 2 sf (0.2 m²) of veneer.

- Light-gauge metal studs

- Masonry veneer

- 1" (25.4 mm) space filled with mortar
- Wire backing

- Waterproof building paper
- Galvanized corrugated ties screwed to studs

Stone Veneer

§1405.6 requires stone veneer units up to 10" (254) thick to be anchored to masonry, stone or wood construction by any one of the following methods. The methods are based upon the structural system backing the veneer:

1. With concrete or masonry backing, anchor ties of dimension and spacing as indicated on the illustration are to be provided. These ties are to be laid in mortar joints and are to be located in such a manner that no more than 2 sf (0.2 m²) of the wall is unsupported by the backing. In addition to the anchor ties in mortar joints attached to the backing with mechanical fasteners there is to be a minimum of 1" (25.4 mm) of continuous grout between the backing and the stone veneer.

2. With stud backing, wire mesh and building paper are to be applied and nailed as indicated on the illustration. Anchors similar to those required at concrete or masonry backing are to be installed in the mortar joints and anchored to the backing to support every 2 sf (0.2 m²) of veneer. Here too a minimum of 1" (25.4 mm) of grout is to fill the cavity between the backing and the stone veneer.

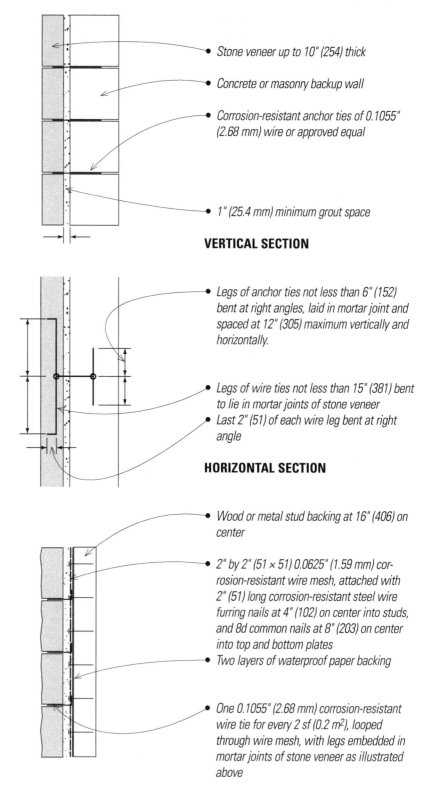

- Stone veneer up to 10" (254) thick

- Concrete or masonry backup wall

- Corrosion-resistant anchor ties of 0.1055" (2.68 mm) wire or approved equal

- 1" (25.4 mm) minimum grout space

VERTICAL SECTION

- Legs of anchor ties not less than 6" (152) bent at right angles, laid in mortar joint and spaced at 12" (305) maximum vertically and horizontally.

- Legs of wire ties not less than 15" (381) bent to lie in mortar joints of stone veneer
- Last 2" (51) of each wire leg bent at right angle

HORIZONTAL SECTION

- Wood or metal stud backing at 16" (406) on center

- 2" by 2" (51 × 51) 0.0625" (1.59 mm) corrosion-resistant wire mesh, attached with 2" (51) long corrosion-resistant steel wire furring nails at 4" (102) on center into studs, and 8d common nails at 8" (203) on center into top and bottom plates
- Two layers of waterproof paper backing

- One 0.1055" (2.68 mm) corrosion-resistant wire tie for every 2 sf (0.2 m²), looped through wire mesh, with legs embedded in mortar joints of stone veneer as illustrated above

VERTICAL SECTION

Slab-Type Veneer

§1405.7 covers marble, travertine, granite and other stone veneer units in slab form. These units are to be not more than 2" (51) in thickness and are to be anchored directly to concrete or masonry backings, or to wood or metal studs. Dowels are to be placed in the middle third of the slab thickness, a maximum of 24" (610) apart with a minimum of four per unit. The slab units are to be no larger than 20 sf (1.9 m²) in area. They are to have wire ties to the backing materials of size and spacing to resist a force in tension or compression of two times the weight of the attached veneer.

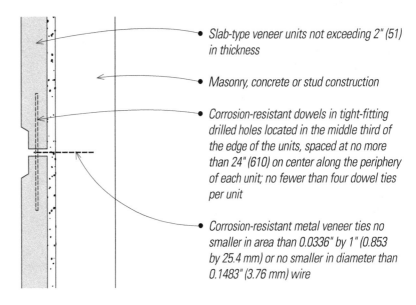

- Slab-type veneer units not exceeding 2" (51) in thickness

- Masonry, concrete or stud construction

- Corrosion-resistant dowels in tight-fitting drilled holes located in the middle third of the edge of the units, spaced at no more than 24" (610) on center along the periphery of each unit; no fewer than four dowel ties per unit

- Corrosion-resistant metal veneer ties no smaller in area than 0.0336" by 1" (0.853 by 25.4 mm) or no smaller in diameter than 0.1483" (3.76 mm) wire

Terra-cotta

§1405.8 specifies anchoring terra-cotta veneers in a manner similar to that for slab veneers. The terra-cotta must be at least 1.625" (41) in thickness with projecting dovetail webs on 8" (203) centers. Metal anchors are to be placed at 12" to 18" (305 to 457) spacing in the joints between units and tied to the backing materials by pencil rods passing through loops in the wire ties and embedded in at least 2" (51) of grout.

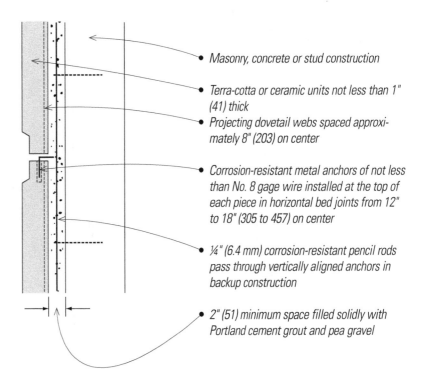

- Masonry, concrete or stud construction

- Terra-cotta or ceramic units not less than 1" (41) thick
- Projecting dovetail webs spaced approximately 8" (203) on center

- Corrosion-resistant metal anchors of not less than No. 8 gage wire installed at the top of each piece in horizontal bed joints from 12" to 18" (305 to 457) on center

- ¼" (6.4 mm) corrosion-resistant pencil rods pass through vertically aligned anchors in backup construction

- 2" (51) minimum space filled solidly with Portland cement grout and pea gravel

INSTALLATION OF WALL COVERINGS

Adhered Masonry Veneer

§1405.9 considers adhered masonry veneer units to be held to the backing by the adhesion of bonding materials to the substrate. The exterior adhesion requirements contained in the 2000 IBC have been removed from the body of the code. The reference in this section refers to the next section, which applies only to interior veneer.

§1405.9.1 applies only to the installation of interior masonry veneers, which must comply with the same installation requirements and weigh no more than 20 pounds per square foot (0.958 kg/m^2). Where supported by wood framing, the supporting members are limited to deflection to 1/600 of the span of the supporting members.

Metal Veneers

§1405.10 states that metal veneers must be of noncorrosive materials or must be coated with anticorrosive coatings such as porcelain enamel. Connections to the backing are to be made with corrosion-resistant fasteners. Wood supports for metal veneers are to be of pressure treated wood. Joints are to be caulked or sealed to prevent penetration of moisture. Masonry backing is not required for metal veneers except as may be necessary for fire-resistance per other code sections.

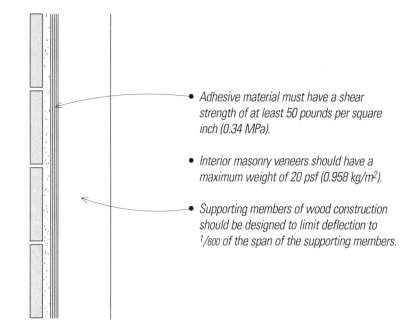

- Adhesive material must have a shear strength of at least 50 pounds per square inch (0.34 MPa).

- Interior masonry veneers should have a maximum weight of 20 psf (0.958 kg/m^2).

- Supporting members of wood construction should be designed to limit deflection to $^1/_{600}$ of the span of the supporting members.

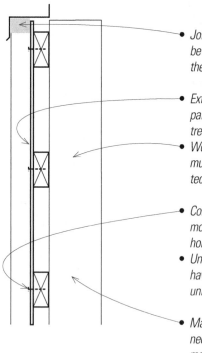

- Joints and edges exposed to weather must be caulked or otherwise protected against the penetration of moisture.

- Exterior metal veneer must be protected by painting, galvanizing or equivalent coating or treatment.
- Wood studs, furring strips or wood supports must be of pressure-treated wood or protected as required by §1403.2.

- Corrosion-resistant fastenings spaced not more than 24" (610) on center vertically and horizontally
- Units exceeding 4 sf (0.4 m^2) in area must have no fewer than four attachments per unit.

- Masonry backup not required except when necessary to meet fire-resistance requirements.

- Metal-veneer walls need to be grounded for lightning and electrical discharge protection per Chapter 27 of the IBC and per the ICC Electrical Code.

Glass Veneers

§1405.11 permits glass panels to be used as thin exterior structural glass veneer. The pieces of such glass, used as finish and not as glazing, may not exceed 10 sf (0.93 m²) in areas up to 15' (4572) above grade and may not be larger than 6 sf (0.56 m²) above that height. The glass is to be bonded to the substrate with a bond coat material that effectively seals the backing surface. Over this, mastic cement is to be applied to at least 50% of the glass surface to adhere it directly to the backing.

Where the glass veneer is installed at sidewalk level it is to be held above the adjacent paving by at least ¼" (6.4 mm) and the joint thus created is to be sealed and made watertight. When located more than 36" (914) above grade, glass units are to be supported with shelf angles. When installed at a height above 12' (3658), the glass units are to be fastened at each vertical and horizontal edge in addition to the required adhesive mastic and shelf angles. Edges of the glass panels are to be ground square and are to be flashed at exposed edges.

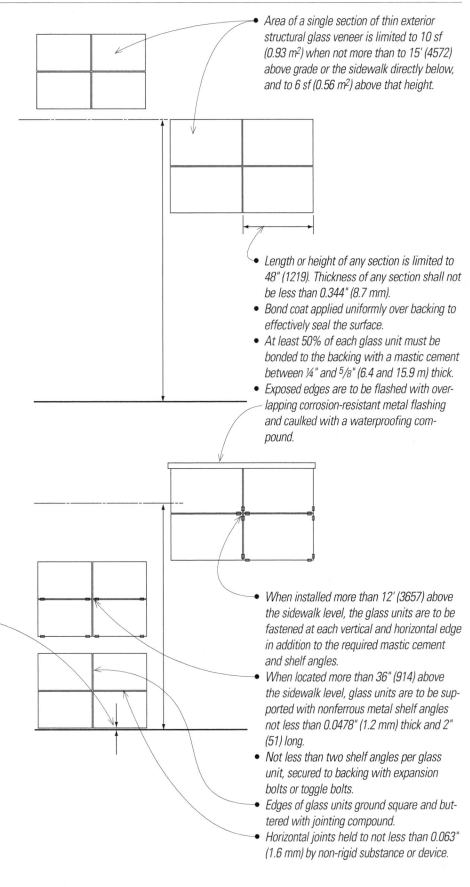

- Area of a single section of thin exterior structural glass veneer is limited to 10 sf (0.93 m²) when not more than to 15' (4572) above grade or the sidewalk directly below, and to 6 sf (0.56 m²) above that height.

- Length or height of any section is limited to 48" (1219). Thickness of any section shall not be less than 0.344" (8.7 mm).
- Bond coat applied uniformly over backing to effectively seal the surface.
- At least 50% of each glass unit must be bonded to the backing with a mastic cement between ¼" and ⅝" (6.4 and 15.9 m) thick.
- Exposed edges are to be flashed with overlapping corrosion-resistant metal flashing and caulked with a waterproofing compound.

- Glass veneer installed at sidewalk level is to be held by an approved metal molding at least ¼" (6.4 mm) above the highest point of the adjacent paving.
- The joint thus created is to be caulked and made watertight.

Vinyl Siding

Doors, windows and manufactured items such as vinyl siding are to be installed according to manufacturer's instructions. These instructions typically include fastening and flashing information for use by the designer, the AHJ and the installer. Fastenings are to be made with corrosion-resistant fasteners. Attachments should conform to applicable wind and seismic criteria for the area of installation per structural code requirements or manufacturers' installation instructions.

- When installed more than 12' (3657) above the sidewalk level, the glass units are to be fastened at each vertical and horizontal edge in addition to the required mastic cement and shelf angles.
- When located more than 36" (914) above the sidewalk level, glass units are to be supported with nonferrous metal shelf angles not less than 0.0478" (1.2 mm) thick and 2" (51) long.
- Not less than two shelf angles per glass unit, secured to backing with expansion bolts or toggle bolts.
- Edges of glass units ground square and buttered with jointing compound.
- Horizontal joints held to not less than 0.063" (1.6 mm) by non-rigid substance or device.

§1406 applies not only to building materials but also to appendages that project beyond the plane of the exterior wall. Such appendages include balconies, bay windows and oriel windows. The criteria for plastics in this application are contained in Chapter 26. The standards for ignition resistance of exterior materials are set forth in NFPA 268. Note, however, that the exceptions to §1406.2.1 cover such large areas that the criteria from NFPA 268 may not apply in many cases. The exceptions include: wood or wood-based products, combustible materials other than vinyl siding, and aluminum at least 0.019" (0.48 mm) thick and at exterior walls of Type V construction. Thus these requirements apply to classes of construction other than Type V. Note that wood is excluded since it is the material against which combustibility is measured.

Table 1406.2.1.2 shows the relationship of combustible veneers to fire-separation distance from a property line. When located closer than 5' (1524) to a property line, materials (those not covered by the exceptions) shall not exhibit sustained flaming per NFPA 269. As the distance to the property line increases, the radiant-heat energy flux decreases. Thus the heat impact on the materials decreases with distance and the types of material that will not exhibit sustained flaming increases.

Architectural Trim

Architectural trim and such items as balconies may be of wood where permitted by §1406.3 in buildings of Types I, II, III and IV construction if the buildings are less than three stories or 40' (12 192) in height. Where such buildings are less than 5' (1524) from the property line, no more than 10% of the exterior wall may be of combustible materials unless of fire-retardant treated wood. Where such trim items are located more than 40' (12 192) above grade they must be of noncombustible materials and attached with noncombustible fasteners.

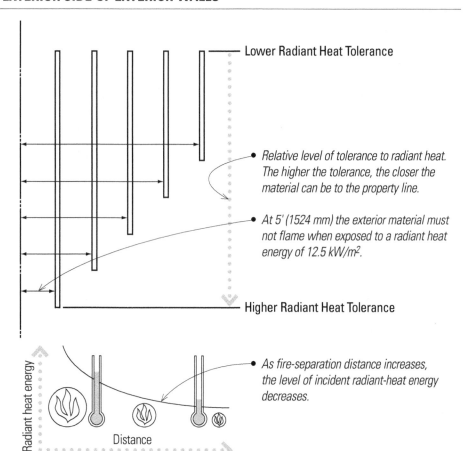

Lower Radiant Heat Tolerance

• *Relative level of tolerance to radiant heat. The higher the tolerance, the closer the material can be to the property line.*

• *At 5' (1524 mm) the exterior material must not flame when exposed to a radiant heat energy of 12.5 kW/m².*

Higher Radiant Heat Tolerance

• *As fire-separation distance increases, the level of incident radiant-heat energy decreases.*

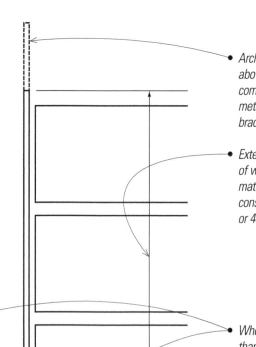

• *Architectural trim more than 40' (12 192) above grade must be constructed of non-combustible materials and be fastened with metal or other approved noncombustible brackets.*

• *Exterior-wall coverings may be constructed of wood or other equivalent combustible material in buildings of Types I, II, III and IV construction that do not exceed three stories or 40' (12 192) in height.*

• *Where the fire-separation distance is less than 5' (1524), combustible exterior-wall covering, except for fire-retardant-treated wood, shall not exceed 10% of exterior wall surface.*

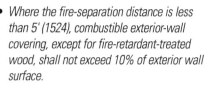

• Property line

Fire-Blocking

When combustible wall coverings are furred from the wall and form a solid surface, §1406.2.4 requires the space between the back of the covering and the wall to not exceed $1^5/8"$ (41), and the space must be fire-blocked per §717 so that no open space exceeds 100 sf (9.3 m²). Note that Exception 3 to §1405.4 allows open or spaced veneers to project up to 24" (610) from the backing wall. Fire-blocking may be omitted in certain circumstances, but good practice should dictate its inclusion in almost every condition where feasible.

- *$1^5/8"$ (41) maximum between back of exterior wall covering and wall.*

- *Fire-blocking per §717 with no open space exceeding 100 sf (9.3 m²).*

- *Fire-blocking is not required where the exterior wall covering is installed over noncombustible framing and the face of the exterior wall finish exposed to the concealed space is faced with aluminum, corrosion-resistant steel or other approved noncombustible material.*

- *Aggregate length of balconies not to exceed 50% of building perimeter on each floor.*

Balconies and Similar Projections

§1406.3 requires balconies to meet the requirements of Table 601 for floor construction or to be of Type IV construction. The aggregate length of balconies may not exceed 50% of the building perimeter on each floor. This applies to all construction types. The exceptions again have a significant impact on how this section is applied in practice. The exceptions are:

1. On buildings of Type I or II construction that are three stories or less in height fire-retardant-treated wood may be used for elements such as balconies or exterior stairs that are not used as required exits.
2. Untreated wood may be used for pickets and rails at guardrails that are under 42" (1067) in height.
3. Balconies and similar appendages on buildings of Type III, IV or V construction may be of Type V construction. Also fire sprinklers may be substituted for fire-resistance ratings when sprinklers are extended to these areas.

- *Fire-retardant-treated wood may be used for balconies, decks and porches on buildings of Types I and II construction that are three stories or less in height.*

- *On buildings of Types III, IV and V construction, balconies and similar appendages may be of Type V construction.*

METAL COMPOSITE MATERIALS

Metal composite materials (MCM) are a recently developed type of exterior finish material combing metal skin panels with plastic cores. These materials are regulated in two ways, both as exterior finishes under §1407 and as plastics under Chapter 26. They are required to meet all of the water-resistance and durability criteria set forth in this chapter for other finish materials. Their installation in rated construction is controlled by testing and approval data establishing flame spread and smoke generation. Per §14107.10 there are limitations on how these systems are used in Type I, II, III and IV construction.

The use of this material will be governed by tests and approvals for individual materials. These tests and approvals must be examined in light of the detailed requirements for generic types of exterior enclosure. As new materials are tested and approved, they must be compared to the code requirements to determine detailed responses to their installation. An example of a typical exterior MCM assembly is illustrated.

Successful installations of innovative materials such as MCM will be dependent upon close conformance with manufacturers' recommendations. This is true of other innovations as well. The code endeavors to incorporate new materials and new ways of assembling familiar materials such as the composite of aluminum and plastics to create MCMs. Innovations usually precede code development, so such innovative materials must be incorporated into buildings using manufacturers' data in concert with close consultation with the AHJ.

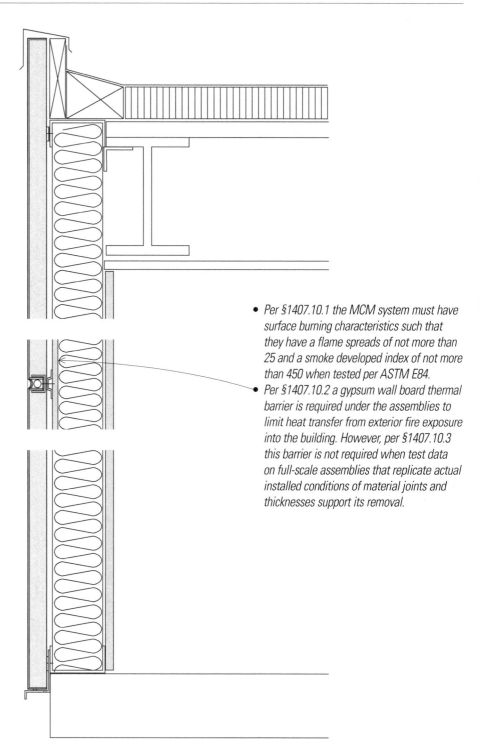

- *Per §1407.10.1 the MCM system must have surface burning characteristics such that they have a flame spreads of not more than 25 and a smoke developed index of not more than 450 when tested per ASTM E84.*
- *Per §1407.10.2 a gypsum wall board thermal barrier is required under the assemblies to limit heat transfer from exterior fire exposure into the building. However, per §1407.10.3 this barrier is not required when test data on full-scale assemblies that replicate actual installed conditions of material joints and thicknesses support its removal.*

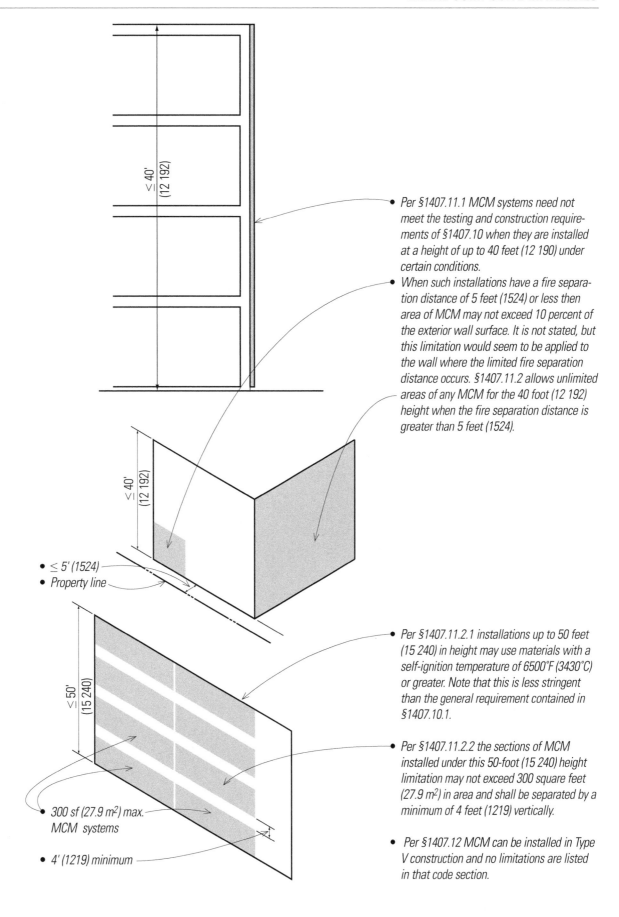

≤ 40'
(12 192)

Per §1407.11.1 MCM systems need not meet the testing and construction requirements of §1407.10 when they are installed at a height of up to 40 feet (12 190) under certain conditions.

When such installations have a fire separation distance of 5 feet (1524) or less then area of MCM may not exceed 10 percent of the exterior wall surface. It is not stated, but this limitation would seem to be applied to the wall where the limited fire separation distance occurs. §1407.11.2 allows unlimited areas of any MCM for the 40 foot (12 192) height when the fire separation distance is greater than 5 feet (1524).

≤ 40'
(12 192)

- ≤ 5' (1524)
- Property line

≤ 50'
(15 240)

Per §1407.11.2.1 installations up to 50 feet (15 240) in height may use materials with a self-ignition temperature of 6500°F (3430°C) or greater. Note that this is less stringent than the general requirement contained in §1407.10.1.

Per §1407.11.2.2 the sections of MCM installed under this 50-foot (15 240) height limitation may not exceed 300 square feet (27.9 m²) in area and shall be separated by a minimum of 4 feet (1219) vertically.

Per §1407.12 MCM can be installed in Type V construction and no limitations are listed in that code section.

- 300 sf (27.9 m²) max. MCM systems

- 4' (1219) minimum

15
Roof
Assemblies

Chapter 15 of the International Building Code establishes requirements for the design and construction of roofing assemblies for buildings. It also establishes requirements for rooftop structures, those that are built on or projecting from building roofs.

§1502 contains definitions that describe different types of roofing. The most common types are illustrated in this chapter to accompany the requirements for roof coverings.

GENERAL REQUIREMENTS

A "roof assembly" includes the weather protection elements and the elements to support design roof loads. Such assemblies are built up from several elements. These may be combined in certain systems, or appear in differing order from inside the building to outside, but the systems must accomplish the functions of each of the separately listed elements. A built-up roof with integral rigid insulation is shown for illustration:

- The roof deck
- Vapor retarder
- Substrate or thermal barrier
- Insulation
- Vapor retarder (listed twice in the text of the definition of "Roof Assembly" §1502)
- Roof Covering

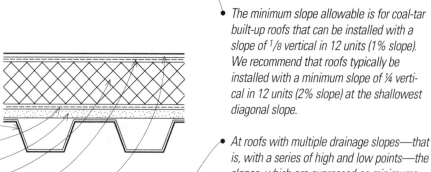

Minimum Slope
The roofing criteria include minimum slope requirements for the installation of various roof coverings. In no case are dead-flat roofs allowed for any roofing material.

Weather Protection
The general criteria in §1503 require that roofing materials and assemblies be installed in accordance with code requirements and manufacturers' recommendations.

- *Flashing and underlayments are to be installed to provide a tight weather-resistant covering and to prevent moisture from entering the wall through joints or at intersections between building elements.*
- *Drainage is to be installed per the International Plumbing Code.*
- *Attic spaces at roof assemblies are to be ventilated as required by §1203.2.*

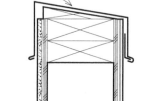

- *The tops of parapet walls are to be topped with coping of noncombustible weatherproof materials. The width of this material (typically sheet metal) is to be no less than the width of the parapet.*

- *The minimum slope allowable is for coal-tar built-up roofs that can be installed with a slope of $1/8$ vertical in 12 units (1% slope). We recommend that roofs typically be installed with a minimum slope of ¼ vertical in 12 units (2% slope) at the shallowest diagonal slope.*

- *At roofs with multiple drainage slopes—that is, with a series of high and low points—the slopes, which are expressed as minimums, should be measured at the shallowest slope of the roof. This usually occurs along a diagonal intersection of two slopes.*

- *If minimum slopes are measured along the fall line of intersecting slopes, the roof geometry will dictate that the slope of the longer intersection will be less than the allowable minimum. This is to be avoided.*

- *Flashings are to be installed at wall and roof intersections, at gutters, at changes in roof slope or direction, and around roof openings. Where flashings are metal the metal is to be corrosion resistant and have a thickness of not less than 0.019 inch (.483), which is the equivalent of No. 26 galvanized metal sheet.*

- *Gutters and drains are to be provided to conduct water off the roof and prevent ponding and leaking. Gutters at the outside of buildings other than private garages, Group R-3 occupancies or buildings not of Type V construction are to be of non-combustible construction or a minimum of Schedule 40 plastic pipe. This could be construed to mean that for healthcare occupancies in other than Type V construction, vinyl gutters may not be used.*

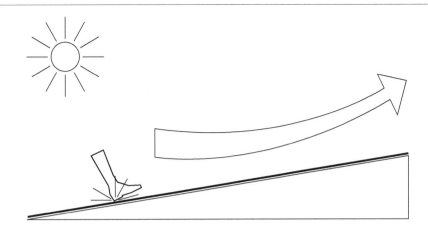

Performance Requirements

§1504 requires roofing to be securely fastened to the building to prevent damage in wind exposure. Roofs are to resist wind loads per Chapter 16 of the Code and per requirements spelled out in test criteria for specific types of roofing materials.

- *When low-slope roofs (defined as having a roof slope of less than 2:12) are installed, the roof materials must be tested to verify that they will maintain their physical integrity over the life of the roof. The tests are to cover physical properties such as sun and wind exposure, flexure of membranes in wind-load conditions and impact-resistance of the roof coverings.*

- *Gravel or stone is not to be used on roofs of buildings in hurricane-prone regions as defined in §1609.2, or an any building with certain mean roof heights when measured against anticipated wind speeds and exposure categories per Table 1504.8.*
  - *For example, a building in Exposure Category B (as described in §1609.4) and in a location where anticipated wind speeds are 100 miles per hour cannot have gravel or stone ballast on a roof over 55 feet (16 764) in height per the table.*

Excerpted from Table 1504.8

Basic Wind Speed From Figure 1609 (mph)	Maximum Mean Roof Height (per 1609.2)		
	B	C	D
85	170'	60'	30'
100	55'	15'	Not permitted
110	30'	Not permitted	Not permitted
120	15'	Not permitted	Not permitted
>120	Not permitted	Not permitted	Not permitted

Fire Classification

§1505 divides roofing assemblies into fire-resistance classifications: A, B, C and Nonclassified in descending order of effectiveness against fire exposure. Nonclassified roofing is material not listed as Class A, B or C and must be approved by the building official. The ratings are determined based upon the performance of roof coverings in fire tests. Classified roofs are to be tested and identified as to their classification.

- *The minimum fire classification of roof coverings is related to building construction type in Table 1505.1.*

- *Class A roofing assemblies are the most effective materials against severe fire test exposure. They are permitted in all types of construction.*
  - *While not tested non-combustible materials are considered as Class A roof assemblies. This includes those with coverings of brick, masonry, slate, clay or concrete roofing tiles, exposed concrete and ferrous or copper shingles or sheets.*

- *Class B or better roofing assemblies are required for all types of construction other than IIB, IIIB and VB.*

Fire Classifications of Roof Coverings

	Type of Construction								
	IA	IB	IIA	IIB	IIIA	IIIB	IV	VA	VB
Allowable	A	A	A	A	A	A	A	A	A
Minimum (italics)	B	B	B	B	B	B	B	B	B
				C		C			C

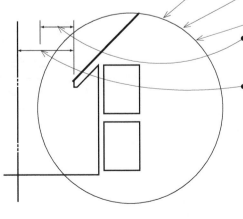

- *There is a general exception for R-3 occupancies when the roof edges are separated from other buildings by at least 6' (1829).*

- *There is also an exception for buildings that normally require a Class C roof, allowing the use of No. 1 cedar or redwood shakes and shingles on buildings two stories or less in height with a roof area of less than 6,000 sf (557.4 m²) and with a minimum 10' (3048) fire-separation distance to a lot line on all sides, except for street fronts or public ways.*

REQUIREMENTS FOR ROOF COVERINGS

Roof coverings are to be installed in accordance with the general requirements of §1506 and the more specific requirements of §1507 and in accordance and the manufacturer's installation instructions. The building official may require testing for unusual roofing assemblies where standards are unclear or the suitability of installation criteria may be in question for a proposed application. Materials are to be compatible with each other and with the building or structure to which they are applied.

Asphalt Shingles

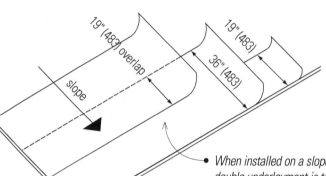

- Minimum roof slope for asphalt shingles is 2:12.

19" (483) overlap

slope

19" (483)

36" (483)

19" (483)

- When installed on a slope of less than 4:12, double underlayment is to be applied as shown in accordance with §1507.2.8.
- For slopes over 4:12 a single layer of underlayment is acceptable.

- §1507.2 specifies that asphalt shingles are to be installed over solidly sheathed decks.

- Typical shingles are to have at least 2 nails per individual shingle or 4 per shingle strip unless more nailing is called for in the manufacturers' installation instructions.
- Special fastening methods with increased numbers of fasteners are required for very steep roofs or in areas where high basic wind speeds are noted in Chapter 16 of the Code.

- Asphalt shingles are to have self-seal strips or be interlocking.
- They are to be nailed with galvanized steel, stainless-steel, aluminum or copper 12-gauge headed roofing nails that must penetrate into the sheathing at least ¾" (19.1 mm), or through sheathing of lesser thickness.

- Eaves and gables of shingle roofs are to receive drip edge flashings extending below the roof sheathing and extending back under the roof.

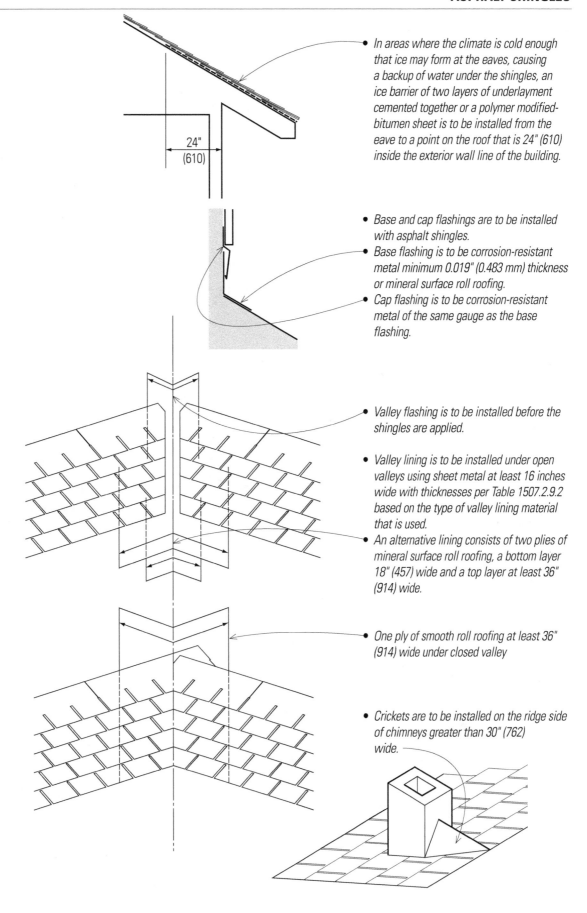

- In areas where the climate is cold enough that ice may form at the eaves, causing a backup of water under the shingles, an ice barrier of two layers of underlayment cemented together or a polymer modified-bitumen sheet is to be installed from the eave to a point on the roof that is 24" (610) inside the exterior wall line of the building.

24"
(610)

- Base and cap flashings are to be installed with asphalt shingles.
- Base flashing is to be corrosion-resistant metal minimum 0.019" (0.483 mm) thickness or mineral surface roll roofing.
- Cap flashing is to be corrosion-resistant metal of the same gauge as the base flashing.

- Valley flashing is to be installed before the shingles are applied.

- Valley lining is to be installed under open valleys using sheet metal at least 16 inches wide with thicknesses per Table 1507.2.9.2 based on the type of valley lining material that is used.
- An alternative lining consists of two plies of mineral surface roll roofing, a bottom layer 18" (457) wide and a top layer at least 36" (914) wide.

- One ply of smooth roll roofing at least 36" (914) wide under closed valley

- Crickets are to be installed on the ridge side of chimneys greater than 30" (762) wide.

CLAY AND CONCRETE TILES

Clay tiles and concrete tiles are to meet strength and durability requirements supported by testing. Concrete tiles are also required to be tested for freeze-thaw resistance, absorption and transverse breaking strength.

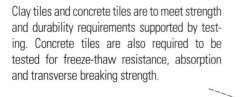

- The slope of the roof deck is to be at least 2½ vertical in 12 units horizontal (21% slope).
- When the slope is less than 4:12, double underlayment is to be installed per §1507.3.3.1.
- At slopes greater than 4:12, one layer of interlayment installed shingle fashion per 1507.3.3.2 is acceptable.

- Fasteners for these tiles are to be corrosion-resistant and of a length to penetrate at least ¾" (19.1 mm) into or through the sheathing.
- The number and configuration of fasteners is spelled out in Table 1507.3.7. The fastening criteria are dependent on basic wind speed, mean roof height, and roof slope.
- Tile is to be installed in accordance with the manufacturers' installation instructions based on climatic conditions, roof slope, underlayment system and the type of tile being installed.

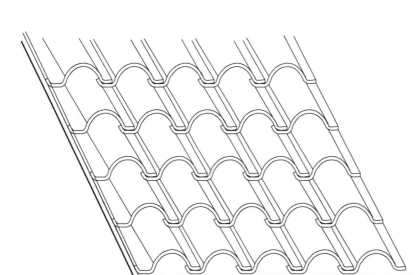

- §1507.3.1 requires concrete and clay tiles to be installed over solid or spaced board sheathing.

- Flashings are to be corrosion resistant and at least 28 galvanized sheet gage (0.019-inch).
- Flashings are to be installed at the juncture of roof tiles to vertical surfaces, and at valleys.
- Flashing to extend 11" (280) to each side of valley centerline and have a splash diverter rib 1" (25.4 mm) high
- Overlap flashing 4" (102) at ends.
- 36" (914) wide underlayment for roof slopes of 3:12 and over

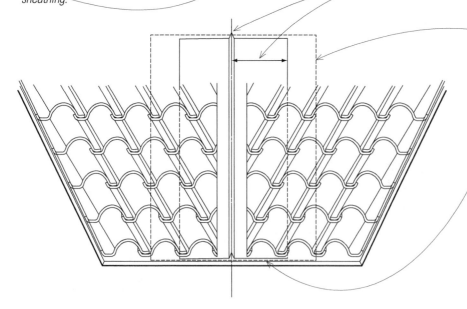

- In cold climates where there is a possibility of ice forming along the eaves and causing backups of water and the roof is under 7:12 slope, the metal valley flashing underlayment is to be solidly cemented to the roofing underlayment.

§1502 defines metal roof panels as being interlocking metal sheets having a minimum installed weather exposure of at least 3 sf (0.28 m²) per sheet. Smaller panels are defined as being metal roof shingles.

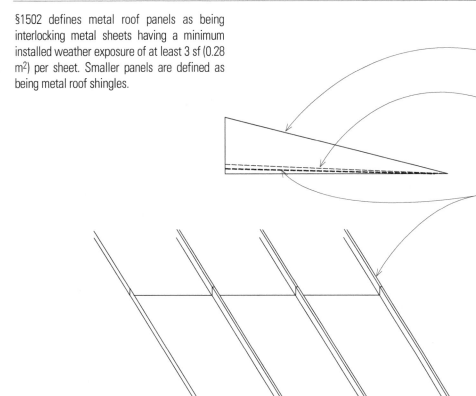

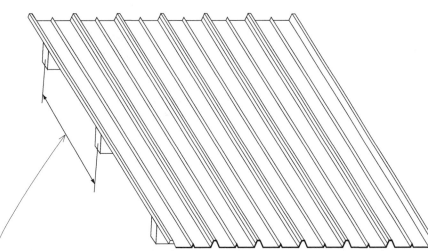

- *The minimum slope for lapped, nonsoldered seam metal roofs without applied lap sealant is to be 3:12.*
- *With lap sealant the minimum slope is ½ vertical in 12 units horizontal (4% slope).*
- *For standing seam roofs, the minimum slope is ¼ vertical in 12 units horizontal (2% slope).*

- *When the roof covering incorporates supporting structural members, the system is to be designed in accordance with Chapter 22.*
- *Metal-sheet roof coverings installed over structural decking are to comply with Table 1507.4.3(1). The table sets testing standards, application rates and thicknesses for various types of metal roofing system ranging from galvanized steel to copper, aluminum, and terne-coated stainless steel. Metal roofs must be of naturally corrosion resistant or provided with corrosion resistance per Table 1507.4.3(2).*

- *Fastenings must match the type of metal to avoid corrosion caused by galvanic electrical activity between dissimilar metals.*
- *Panels are to be installed per the manufacturers' instructions. In the absence of such instructions, the fastenings are to be galvanized fasteners for galvanized roofs and hard copper or copper alloy for copper roofs. Stainless-steel fasteners are acceptable for all types of metal roof.*

- *Note that underlayment requirements are not spelled out in the Code for metal roof panel installations. Many manufacturers recommend the use of underlayment. This is a reminder that the Code is only the minimum standard for construction quality, not the maximum standard.*

- *Metal roof panels are to be installed over solid or closely spaced decking unless specifically designed to be installed over spaced supports.*

METAL ROOF SHINGLES

Metal roof shingles are smaller in size than metal roof panels, being by definition less than 3 sf (0.28 m²) in weather exposure.

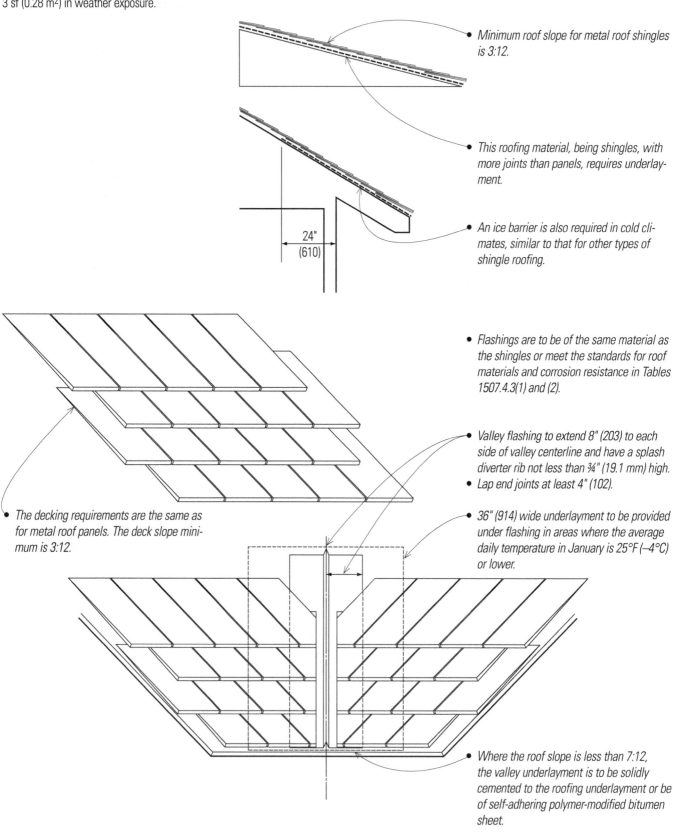

• Minimum roof slope for metal roof shingles is 3:12.

• This roofing material, being shingles, with more joints than panels, requires underlayment.

• An ice barrier is also required in cold climates, similar to that for other types of shingle roofing.

24"
(610)

• Flashings are to be of the same material as the shingles or meet the standards for roof materials and corrosion resistance in Tables 1507.4.3(1) and (2).

• Valley flashing to extend 8" (203) to each side of valley centerline and have a splash diverter rib not less than ¾" (19.1 mm) high.
• Lap end joints at least 4" (102).

• 36" (914) wide underlayment to be provided under flashing in areas where the average daily temperature in January is 25°F (−4°C) or lower.

• The decking requirements are the same as for metal roof panels. The deck slope minimum is 3:12.

• Where the roof slope is less than 7:12, the valley underlayment is to be solidly cemented to the roofing underlayment or be of self-adhering polymer-modified bitumen sheet.

Mineral-Surfaced Roll Roofing

- Mineral-surfaced roll roofing is to be applied only over solidly sheathed roofs.
- The roof slope must be at least 1:12.
- A single layer of underlayment is typically required.
- Two layers are required to serve as an ice barrier in cold climates.

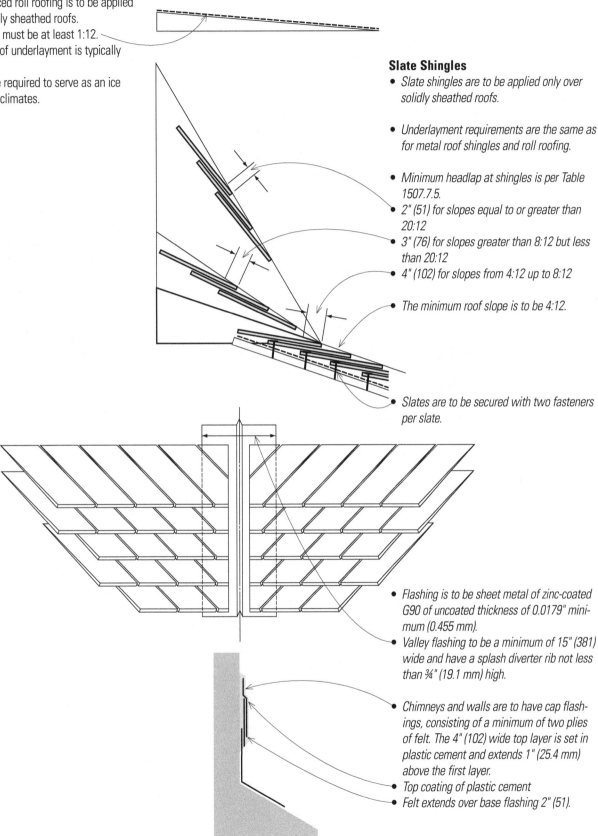

Slate Shingles

- *Slate shingles are to be applied only over solidly sheathed roofs.*

- *Underlayment requirements are the same as for metal roof shingles and roll roofing.*

- *Minimum headlap at shingles is per Table 1507.7.5.*
- *2" (51) for slopes equal to or greater than 20:12*
- *3" (76) for slopes greater than 8:12 but less than 20:12*
- *4" (102) for slopes from 4:12 up to 8:12*

- *The minimum roof slope is to be 4:12.*

- *Slates are to be secured with two fasteners per slate.*

- *Flashing is to be sheet metal of zinc-coated G90 of uncoated thickness of 0.0179" minimum (0.455 mm).*
- *Valley flashing to be a minimum of 15" (381) wide and have a splash diverter rib not less than ¾" (19.1 mm) high.*

- *Chimneys and walls are to have cap flashings, consisting of a minimum of two plies of felt. The 4" (102) wide top layer is set in plastic cement and extends 1" (25.4 mm) above the first layer.*
- *Top coating of plastic cement*
- *Felt extends over base flashing 2" (51).*

WOOD SHINGLES

The installation requirements for both wood shakes and shingles are summarized in Table 1507.8.

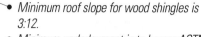

- Wood shingles may be installed over either spaced or solid sheathing.
- Spaced sheathing shall not be less than 1×4 (nominal) boards, spaced equal to exposure.
- Solid sheathing is required in areas where the average daily temperature in January is 25°F (–4°C) or lower.

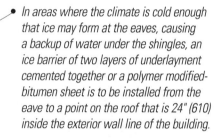

24"
(610)

- Minimum roof slope for wood shingles is 3:12.
- Minimum underlayment is to be per ASTM D 226, Type 1.

- In areas where the climate is cold enough that ice may form at the eaves, causing a backup of water under the shingles, an ice barrier of two layers of underlayment cemented together or a polymer modified-bitumen sheet is to be installed from the eave to a point on the roof that is 24" (610) inside the exterior wall line of the building.

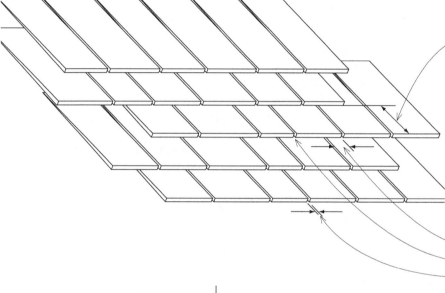

- Shingles are to have exposures in relation to roof slopes per Table 1507.8.6.
- Corrosion-resistant fasteners should penetrate at least ¾" (19.1 mm) into or through the sheathing.

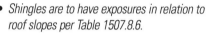

- 1½" (38) minimum sidelap
- Joints in alternate courses should not align.
- ¼" to ³⁄₈" inches (6.4 to 9.5 mm) spacing between shingles

- Flashings are to be corrosion-resistant and at least 26 galvanized sheet gage (0.019" or 0.48 mm).
- Flashing to extend 11" (280) to each side of valley centerline and have a splash diverter rib 1" (25.4 mm) high.
- Overlap flashing 4" (102) at ends.
- 36" (914) wide underlayment for roof slopes of 3:12 and over
- In cold climates where there is a possibility of ice forming along the eaves and causing backups of water and the roof is under 7:12 slope, the metal valley flashing underlayment is to be solidly cemented to the roofing underlayment.

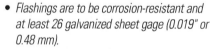

- *While wood shingles are sawn, wood shakes are formed by splitting a short log into a number of tapered radial sections, resulting in at least one texture face.*
- *Wood shakes may be installed over either spaced or solid sheathing.*
- *Spaced sheathing shall not be less than 1×4 (nominal; 25 × 102) boards, spaced equal to weather exposure.*
- *Solid sheathing is required in areas where the average daily temperature in January is 25°F (-4°C) or lower.*

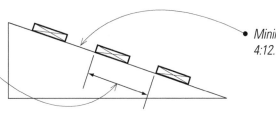

- *Minimum roof slope for wood shingles is 4:12.*

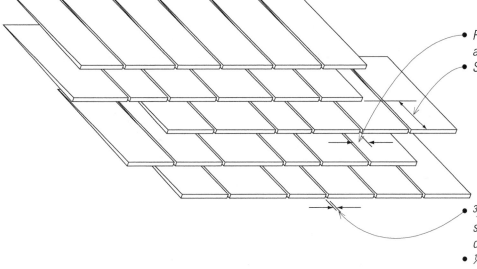

- *Fastening, sidelap and flashing requirements are similar to those of wood shingles.*
- *See below for weather exposure.*

- *³⁄₈" to ⁵⁄₈" (9.5 to 15.9 mm) spacing between shakes and tapersawn shakes of naturally durable wood.*
- *¼" to ³⁄₈" inches (6.4 to 9.5 mm) spacing between preservative tapersawn shakes.*

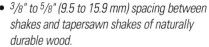

- **Shingles** *are to have exposures in relation to roof slopes per Table 1507.8.6.*
- *7½" (191) exposure for 18" (457) No. 1 grade shakes and 10" (254) exposure for 24" (610) No. 1 grade shakes.*

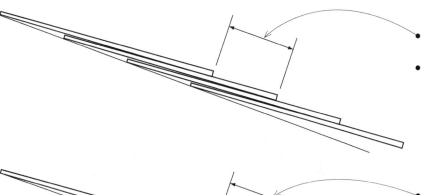

- **Shakes** *are to have exposures in relation to roof slopes per Table 1507.9.7.*
- *5½" (140) exposure for 18" (457) No. 1 grade shingles and 7½" (191) exposure for 24" (610) No. 1 grade shingles.*

BUILT-UP ROOFS

Built-up roof coverings are defined as two or more layers of felt cemented together and topped with a cap sheet aggregate or similar surfacing material. Although not stated in the Code, the intent of this section is that materials for such roofs must be complementary and work together in accordance with the manufacturer's written installation instructions. The designer should verify the compatibility of the various components of such roofs to be certain they work together chemically and mechanically in accordance with the test criteria and standards cited in §1507.10.2 and §1507.11.2.

- Material standards are specified in Table 1507.10.2.

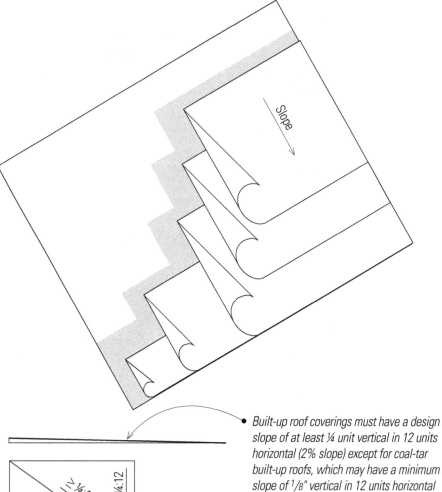

Slope

Modified Bitumen Roofing

Modified Bitumen Roofing consists of one or more layers of polymer modified asphalt sheets.

- Built-up roof coverings must have a design slope of at least ¼ unit vertical in 12 units horizontal (2% slope) except for coal-tar built-up roofs, which may have a minimum slope of ¹/₈" vertical in 12 units horizontal (1% slope).
- Minimum slopes should be maintained along the shallowest slope at intersecting valleys. Thus typical roof slopes will be slightly greater than 1/4:12.

≥ ¼:12

> ¼:12

> ¼:12

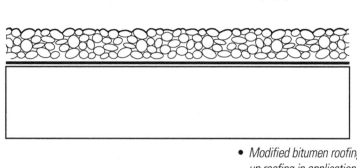

- Modified bitumen roofing is similar to built-up roofing in application. The sheets are fully adhered or mechanically attached to the substrate or held in place with a layer of ballast.
- These roofing materials are to be installed with a minimum slope of at least 1/4:12.

Single-ply roofing membranes are field applied using one layer of a homogeneous or composite material rather than multiple layers.

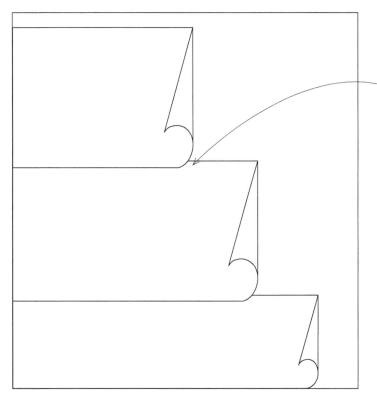

- The membrane material is seamed together with heat or adhesives or a combination of seaming methods.

- §1507.12 and 1507.13 specify the applicable ASTM standards for thermoset and thermoplastic single-ply roofing.

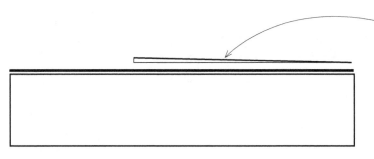

- Both require the same minimum design slope of ¼ unit vertical in 12 units horizontal (2% slope).

Sprayed Polyurethane Foam Roofing

This type of roof covering is applied by spraying a layer of polyurethane foam onto the roof deck and then applying a liquid-applied protective coating over the foam membrane after it has been chemically cured.

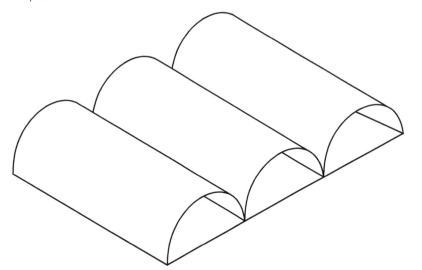

- *The design slope is to be at least ¼ unit vertical in 12 units horizontal (2% slope).*
- *§1507.14 requires that the roof assembly meet Class A, B or C fire-rating criteria per ASTM E 108 of UL 790.*
- *Application is to be per the manufacturers' installation instructions. The foam plastic materials must also comply with the code requirements for plastics contained in Chapter 26. Note that the only portion of the chapter that seems to apply is 2603.6, which is contained in the section pertaining to foam plastic insulation and not roofing as such.*

Liquid-Applied Coatings

Liquid-applied coatings are not defined in this chapter; however, there are installation requirements for this type of roof covering. Typically such coatings form a membrane by sealing the surface to which they are applied after the liquid congeals or dries.

- *Liquid-applied roofing must have a minimum design slope of ¼ unit vertical in 12 units horizontal (2% slope).*

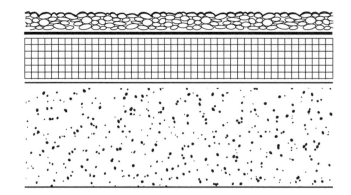

Roof Insulation

- *Thermal insulation may be installed above the roof decking if it is covered by a roof covering complying with the fire-resistance ratings of FM 4450 or UL 1256.*
- *Foam plastic insulation shall conform to the requirements of Chapter 26.*
- *Cellulosic fiber insulation shall comply with the applicable requirements for wood contained in Chapter 23.*

Penthouses

Rooftop structures include such items as penthouses for elevators, water tanks, cooling towers and spires.

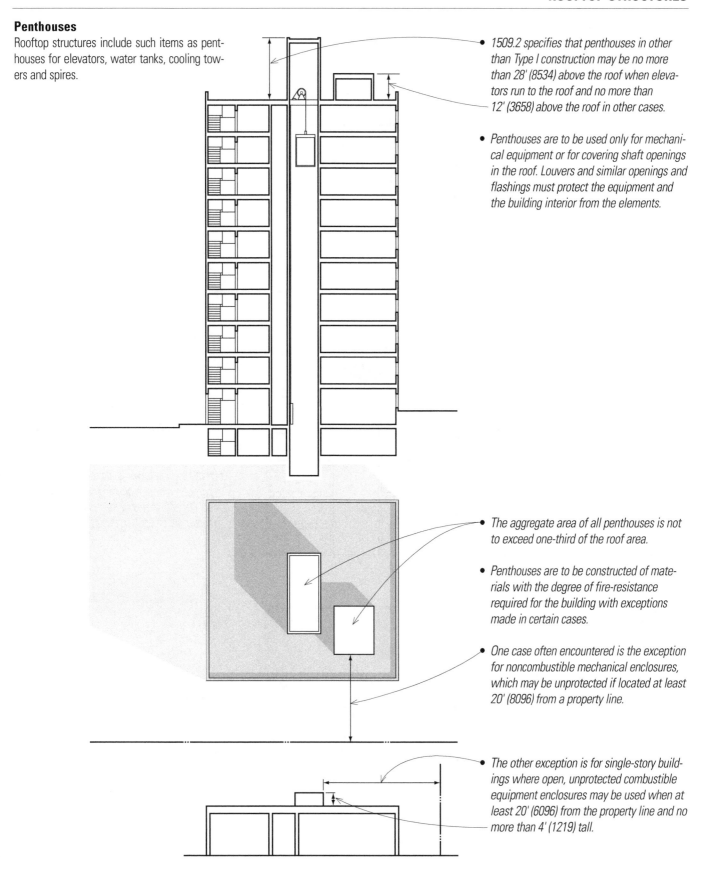

● 1509.2 specifies that penthouses in other than Type I construction may be no more than 28' (8534) above the roof when elevators run to the roof and no more than 12' (3658) above the roof in other cases.

● Penthouses are to be used only for mechanical equipment or for covering shaft openings in the roof. Louvers and similar openings and flashings must protect the equipment and the building interior from the elements.

● The aggregate area of all penthouses is not to exceed one-third of the roof area.

● Penthouses are to be constructed of materials with the degree of fire-resistance required for the building with exceptions made in certain cases.

● One case often encountered is the exception for noncombustible mechanical enclosures, which may be unprotected if located at least 20' (8096) from a property line.

● The other exception is for single-story buildings where open, unprotected combustible equipment enclosures may be used when at least 20' (6096) from the property line and no more than 4' (1219) tall.

ROOFTOP STRUCTURES

Tanks

Tanks on the roof are often used for supplying water for fire sprinklers or fire fighting. When having a capacity larger than 500 gallons, they must be supported on masonry, reinforced concrete or steel, or be of Type IV construction.

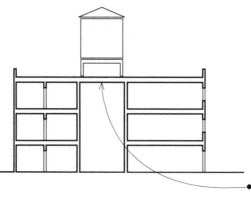

• *When such tanks are above the lowest story, supports are to be of Type I-A construction. Thus one could construe §1509.3 to say that all roof-mounted tanks are to be on supports of Type I-A construction.*

Cooling Towers

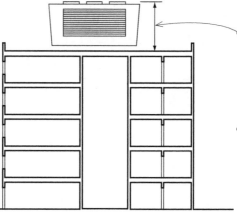

• *Cooling towers that are larger than 250 sf (23.2 m²) in area or over 15' (4572) high and located on buildings more than 50' (15 240) high are to be of noncombustible construction.*

• *If the cooling tower is located on a lower part of a roof of a building that is more than 50' (15 240) in height, this provision could be construed to apply in this case as well.*

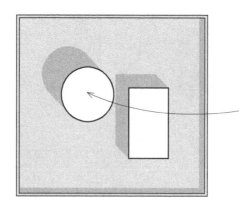

• *Cooling towers, as for penthouses, may not exceed one-third of their supporting roof area.*

Towers, Spires, Domes and Cupolas

Towers, spires, domes and cupolas are considered to be rooftop structures that contain or support no mechanical equipment and are not occupied. §1509.5 requires these structures to at least match the fire-resistance of the building supporting them, with the following exceptions:

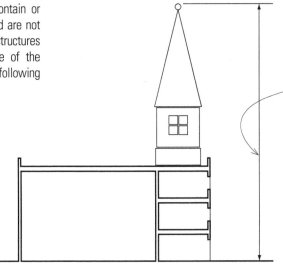

- Where any such structure exceeds 60' (18 288) in height above grade, the structure and any supporting construction must be of Type I or II construction if it exceeds 200 sf (18.6 m²) in area and is other than a belfry or architectural embellishment.
- If the structure does not meet all these criteria, it would seem to fall back to the requirement for the construction to match that of the supporting building.

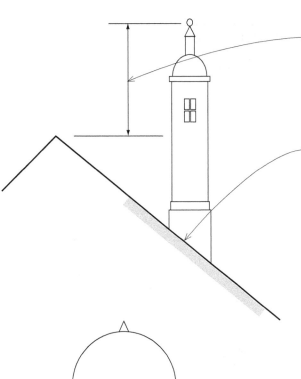

- Where the structure is higher than 25' (7620) above the highest point of the roof at which it comes in contact, exceeds 200 sf (18.6 m²) in horizontal section, or is other than a belfry or architectural embellishment, it must be constructed entirely of noncombustible materials.
- These structures must be separated from the building below by a minimum of 1½-hour fire-resistive construction and openings protected with a minimum 1½-hour fire-protection rating.

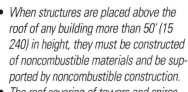

- When structures are placed above the roof of any building more than 50' (15 240) in height, they must be constructed of noncombustible materials and be supported by noncombustible construction.
- The roof covering of towers and spires are to be of the same class as required for the roof of the rest of the structure.

REROOFING

Roofs are typically replaced to comply with the requirements for new roofing materials. When more than 25% of the roof is replaced within any 12-month period, the entire roof covering is to be made to conform to new roofing requirements.

- *Per the exception in §1510.1, where a roof has positive roof drainage (i.e., a slope sufficient to ensure drainage within 48 hours of precipitation, taking into consideration all loading deflections), a reroofing need not meet the minimum design slope requirement of ¼ unit vertical in 12 units horizontal (2% slope).*

Reroofing requirements are concerned with not overloading existing roof structures by allowing multiple roof layers of heavy or combustible materials. Roof structures must be capable of supporting the loads for the new roofing system along with the weight of any existing materials that may remain. Also, the roof structure must be able to support the loads imposed by materials and equipment used during the reroofing process.

Applying new roofing without removing the old roof is allowed in limited circumstances. Existing roofing must be removed when any of the following conditions occur:

1. *The roof or roof covering is water-soaked or deteriorated such that the existing roof is not an adequate base for added roofing.*
2. *If the existing roof covering is wood shakes, slate or tiles.*

  - *Note that an exception allows metal or tile roofs to be applied over wood shakes if combustible concealed spaces are avoided by applying gypsum board, mineral fiber or glass fiber over the shakes.*

3. *Where there are already two or more roof applications of any type of roof covering. This is to avoid overloading the roof structure with multiple layers of roofing.*

Existing durable materials such as slate or clay tiles may be reapplied if in good condition and undamaged. Aggregate surfacing materials are not to be removed and reinstalled. Flashing may be reinstalled if in good condition.

- *When 25% or less of a roof covering is replaced, only the new portion must comply with the requirements of Chapter 15.*

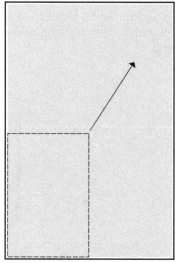

- *Entire roof covering must comply with the requirements for new roofing when more than 25% of roof is removed or replaced.*

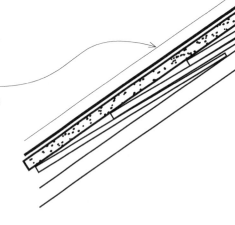

16

Structural Provisions

This chapter is a summary of structural design requirements contained in Chapters 16, 17 and 18 of the Code. Chapter 16 of the code sets forth general design criteria for structural loads to be accommodated by the structural system of a building. Detailed criteria for building materials are contained in the code chapters devoted to specific materials, such as wood (Chapter 23), concrete (Chapter 19) and steel (Chapter 22). Chapter 17 of the code governs the testing and inspection of construction materials. Chapter 18 of the code applies to the requirements for soils, site grading and foundation design.

STRUCTURAL DESIGN

The structural design requirements contained in Chapter 16 apply to all building and structures. The chapter focuses on the engineering principles that underlie the requirements and design of structural systems to accommodate anticipated loads, such as the weight of the building, the weight of occupants and materials in the building, and loads imposed by nature such as wind, snow and earthquakes.

Chapter 16 contains numerous tables and criteria that are to be applied in specific situations based on building use, occupancy, construction type and geographic location. Detailed analysis must be undertaken by the designer in concert with appropriate engineering consultants as needed to supplement the designer's training and expertise to prepare a code-compliant design. The fundamental underlying document for Chapter 16 is *ASCE 7, Minimum Design Loads for Buildings and Other Structures*, published as a separate reference standard by the American Society of Civil Engineers. The requirements of ASCE 7 are a part of the Code except where modified specifically in the IBC.

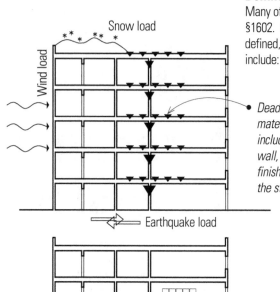

Snow load

Wind load

Earthquake load

Definitions

Many of the terms in Chapter 16 are defined in §1602. Other basic structural terms may not be defined, but also need to be understood. Terms include:

- *Dead Load: The weight of construction materials incorporated into a building, including architectural elements, such as wall, floors, roofs and ceilings, stairs and finishes, as well as equipment attached to the structure.*

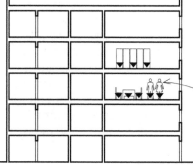

- *Live Loads: Loads produced by the use and occupancy of a building and not including dead loads or environmental loads such as wind, snow, flood or earthquake loads.*

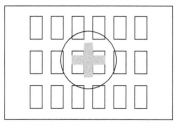

- *Essential Facilities are defined in §1602 as buildings and other structures intended to remain operational in the event of extreme environmental loading, such as from wind, snow, flood or earthquakes. Certain healthcare facilities, such as hospitals, are often considered to be essential facilities. Local requirements and designations of such facilities should be reviewed with the Authorities Having Jurisdiction during the early stages of design as structural requirements for such facilities may be much higher than for non-essential facilities.*

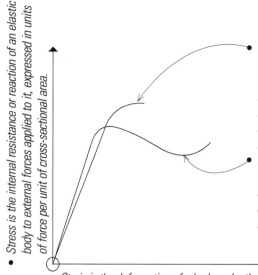

- *Stress is the internal resistance or reaction of an elastic body to external forces applied to it, expressed in units of force per unit of cross-sectional area.*

- *Nonductile Element: An element having a failure mode that results in an abrupt loss of resistance when the element is deformed beyond the deformation corresponding to its nominal strength. Such elements lose their strength when deformed beyond their load ranges.*

- *Ductile Element: An element capable of sustaining large cyclic deformation without any significant loss of strength. Such elements can deform and still maintain structural strength.*

- *Strain is the deformation of a body under the action of an applied force, equal to the ratio of the change in size or shape to the original size or shape of a stressed element.*

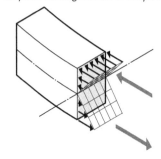

Construction Documents

Construction documents must have sufficient information to allow the AHJ to review the documents for code compliance. The documents are to show the size, section dimensions and relative locations of structural members. Design loads are to be indicated to verify compliance with structural design requirements for various types of loading. Light-frame buildings, such as houses and small commercial buildings constructed under the conventional framing provisions of §2308, have a separate, shorter set of design requirements for floor and roof live loads, ground snow load, basic wind speed, seismic design category and site class.

The construction documents are to show the loads and information for specified items. The detailed requirements for these items will be elaborated upon later in this chapter.

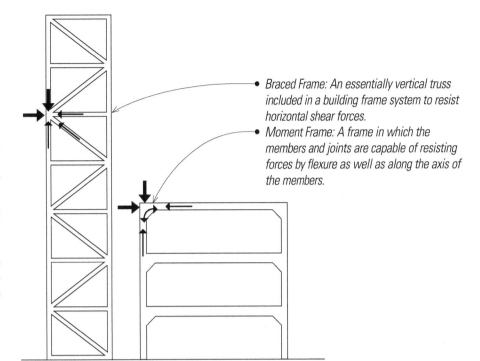

- *Braced Frame: An essentially vertical truss included in a building frame system to resist horizontal shear forces.*
- *Moment Frame: A frame in which the members and joints are capable of resisting forces by flexure as well as along the axis of the members.*

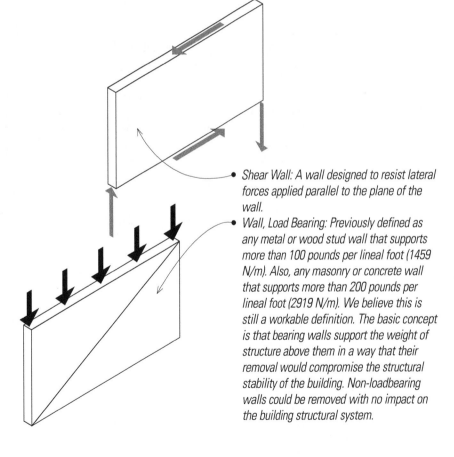

- *Shear Wall: A wall designed to resist lateral forces applied parallel to the plane of the wall.*
- *Wall, Load Bearing: Previously defined as any metal or wood stud wall that supports more than 100 pounds per lineal foot (1459 N/m). Also, any masonry or concrete wall that supports more than 200 pounds per lineal foot (2919 N/m). We believe this is still a workable definition. The basic concept is that bearing walls support the weight of structure above them in a way that their removal would compromise the structural stability of the building. Non-loadbearing walls could be removed with no impact on the building structural system.*

STRUCTURAL DESIGN

General Design Requirements

Buildings are to be designed in accordance with one of several defined and approved structural design methods, such as the strength design method, the load and resistance factor design method, the allowable stress method, and so forth. The structure is to support the factored loads in load combinations as defined in the Code.

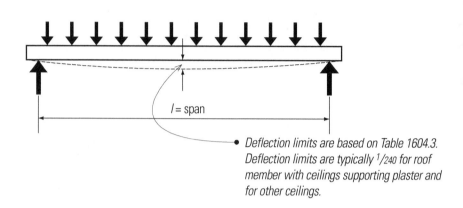

Serviceability

The structure is to be designed to limit deflections and lateral drift under anticipated loading. Load effects on the structure are to be determined by application of a rational analysis taking into account equilibrium of the structure, general stability, geometric compatibility, short-term and long-term material properties.

• *Deflection limits are based on Table 1604.3. Deflection limits are typically $^1/_{240}$ for roof member with ceilings supporting plaster and for other ceilings.*

Occupancy Category

Table 1604.5 classifies buildings into occupancy categories by importance. Factors noted in the structural chapters are for typical buildings. Values for loads and strengths are to be increased by the factors based on the importance of the structure.

• *Buildings with low occupancy loads and with little hazard to human life in the event of failure such as agricultural buildings or minor storage facilities fall into Occupancy Category I.*
• *Typical buildings, not classified as any other category, are classified as Category II.*
• *Category III buildings are those where structural failures would be a substantial hazard to human life. These include structures such as assembly areas having more than 300 occupants, schools, healthcare facilities and detention facilities.*
  • *This category includes healthcare facilities with 50 or more resident patients, but without surgery or emergency treatment facilities. This could include convalescent care facilities, rehabilitation centers, or nursing homes.*
• *Essential facilities, such as hospitals with surgery or emergency facilities, fire stations, police stations, water and power stations which need to be relied upon in emergencies, have the highest importance, Category IV.*

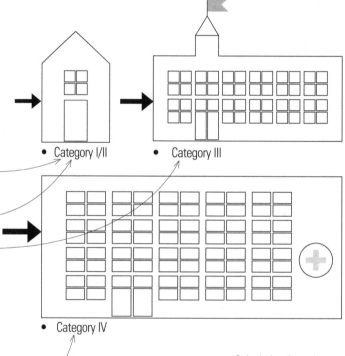

Calculation formulas apply different factors for each category, taking into account the relative importance of each facility. Thus "essential facilities" may have factors of up to 1.5 applied to loading criteria to provide additional strength to such structures in the event of such events as hurricane, earthquakes or floods. Note that these factors are applied to external loading, not other effects such as fires or loads imposed by use. These other load factors are addressed in live load tables.

Anchorage

Roofs are to be anchored to walls and columns, and walls and columns are to be anchored to foundations, to resist the anticipated uplift and sliding forces that result from the application of the prescribed loads, whether from dead loads or live loads.

Structural members including all of their components are to be designed to resist forces due to both earthquake and wind. Consideration is to be made for overturning, sliding and uplift forces. Continuous load paths are to be provided for transmitting these forces to the foundation. Seismic detailing requirements for lateral-force-resisting systems prescribed by the Code and by ASCE 7 are to be met even when wind loads are greater than seismic load effects.

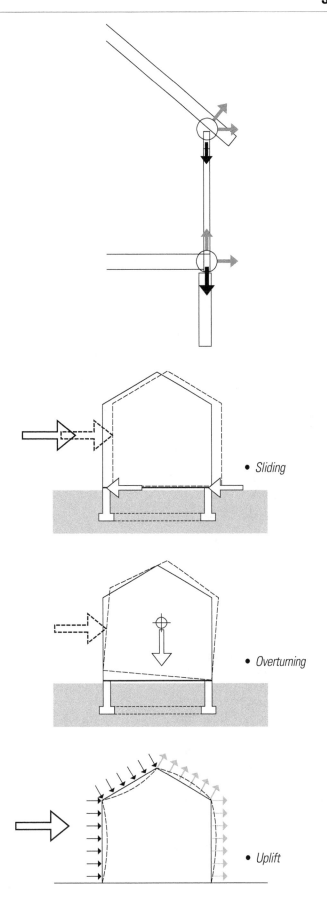

- Sliding

- Overturning

- Uplift

STRUCTURAL DESIGN

Load Combinations

Various combinations of dead loads, live loads, seismic loads and wind loads are to be applied in the design of structural systems. The load factors for each combination depend on the type of analysis used. Various combinations of loading are to be examined and the design is to resist the most critical effects of the combinations specified. There are two basic methodologies for structural design. Each is often described in various ways so the two methods make references §1602 to other descriptions found in common usage. The two methodologies are:

- "Allowable Stress Design" (also known as "Working Stress Design"), which is a method of proportioning structural members such that stresses produced in the members by nominal loads do not exceed specified allowable stresses. This is the more traditional method of structural design.
- "Load and Resistance Factor Design" (LRFD as used in steel and word design) also known as "Strength Design" (when used for concrete and masonry design) is a method of proportioning structural members and their connections such that no computed forces produced in the members by factored loads do not exceed the member design strength. These are more contemporary design methods, relying more on the power of computing than on older calculation methods.

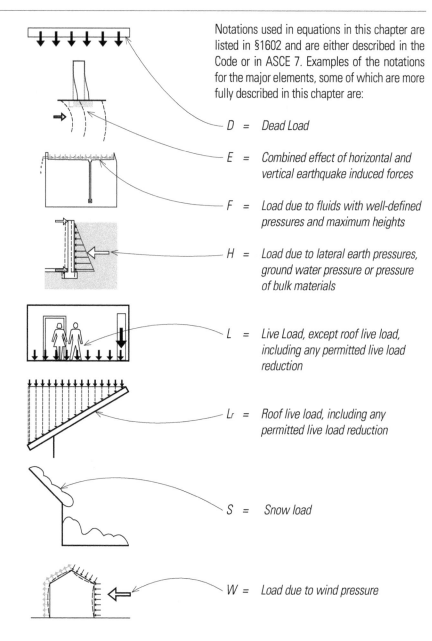

Notations used in equations in this chapter are listed in §1602 and are either described in the Code or in ASCE 7. Examples of the notations for the major elements, some of which are more fully described in this chapter are:

D = Dead Load

E = Combined effect of horizontal and vertical earthquake induced forces

F = Load due to fluids with well-defined pressures and maximum heights

H = Load due to lateral earth pressures, ground water pressure or pressure of bulk materials

L = Live Load, except roof live load, including any permitted live load reduction

L_r = Roof live load, including any permitted live load reduction

S = Snow load

W = Load due to wind pressure

Basic Load Combinations

To the right, for comparison purposes, are several basic formulae for each of the two structural design systems. The formulae take into account the way each calculation methodology accounts for safety factors. The comparisons below show roof life loads without rain or snow loads where choices between roof live loads, rain or snow loads appear in a formula. The design is to take into account the most critical effects resulting from the combination of loads.

Equation Numbers For Comparison	Combination Loads Using LRFD or Strength Design	Combination Loads Using Allowable Stress Design	Alternative Basic Load Combinations*
16-1 / 16-8	1.4 (D + F)	(D + F)	
16-2 / 16-9 / 16-16	1.2 (D + F + T) + 1.6(L + H) + 0.5 (Lr)	D + F + T + L + H	D + L + Lr
16-6 / 16-14	0.9D + 1.6W + 1.6H	0.6D + W + H	

* For allowable stress design stress increases allowed in the Code chapters on specific materials may be taken when alternate load combinations that include wind or seismic loads. Only one example of several is shown in the table.

Dead Loads

Actual weights of materials and construction are to be used to determine dead loads. Fixed service equipment is to be considered as dead load. These fixed elements include plumbing, electrical feeders, HVAC systems and fire sprinklers.

Live Loads

As defined in §1602, live loads are those produced by the use or occupancy of the building. Table 1607.1 defines minimum uniformly distributed live loads and concentrated loads for various occupancies. Per §1607.2, live loads not designated in Table 1607.1 shall be determined in accordance with a method approved by the building official. Thus live loads not noted in the table must be evaluated by an approved method. Note that the live loads listed in Table 1607.1 may be reduced in accordance with the provisions of §1607.9.

Concentrated loads are to be determined either by the table or the approved methods, and the design is to use the method producing the greatest load effect. Concentrated loads are to be applied over an area of 2½ feet square [6 ¼ sf (0.58 m²)] located so as to produce the maximum load effect on the structural members being designed.

Where partitions may be installed and later moved, a uniformly distributed partition live load of 15 psf (0.74 kN/m²) is to be assumed unless the specified floor live load exceeds 80 psf (3.83 kN/m²).

To the right is a summary of examples from Table 1607.1. The numbers opposite the occupancy correspond to the location of the examples in the complete table.

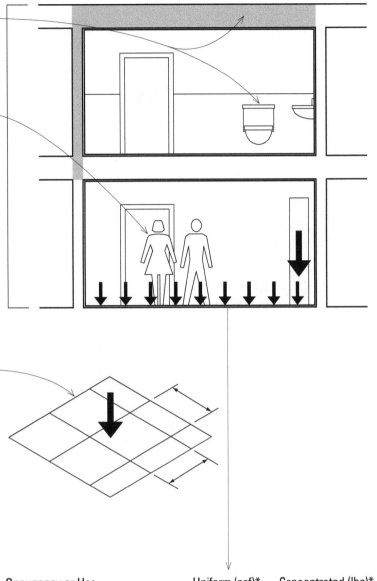

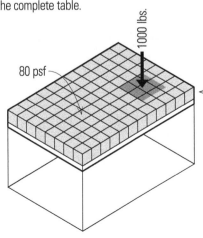

Occupancy or Use	Uniform (psf)*	Concentrated (lbs)*
4. Assembly areas—fixed seat	60	-
10. Dining rooms and restaurants	100	-
13. Corridors, except as otherwise indicated	100	-
21. Hospital		
Corridors above first floor	80	1,000
Operating rooms, laboratories	60	1,000
Patient rooms	40	1,000
26. Office building—offices	50	2,000
28. Residential, multi-family		
Private rooms and corridors to them	40	-
Public rooms and corridors to them	100	-

*1 pound per square foot (psf) = 0.0479 kN/m²; 1 pound = 0.004448 kN

STRUCTURAL DESIGN

Loads on Handrails and Guards

Handrails and guards are to be designed to resist a load of 50 pounds per lineal foot (0.73 kN/m) applied in any direction at the top and to transfer this load through the supports to the structure supporting the rail or guard.

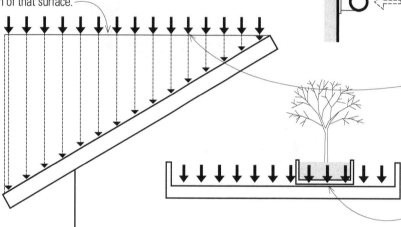

Handrail assemblies are also to be able to resist a single concentrated load of 200 pounds (0.89 kN) applied in any direction at any point along the top. This load can be considered independently of the uniform load noted previously.

Intermediate rails—that is, those not the handrail—are to be able to resist a load of 50 pounds (0.22 kN) on an area not to exceed 1 sf (0.09 m²) including the openings and spaces between rails. These are not required to be superimposed with the other railing loads noted above.

Roof Loads

Roof loads acting on a sloping surface are to be assumed to act vertically on the horizontal projection of that surface.

Grab bars are to be designed to resist a single concentrated load of 250 pounds (1.11 kN) applied in any direction at any point.

Minimum roof live loads are determined based upon roof slopes per §1607.11.2.1 through 1607.11.2.4.

Landscaped roofs are to have a uniform design live load of 20 psf (0.958 kN/m²). The weight of the landscaping materials is to be considered as dead load, and the weight is to be calculated based on saturation of the soil.

Interior Walls and Partitions

Walls that exceed 6' (1829) in height are to be designed to resist loads to which they are subjected, but not less than a horizontal load of 5 psf (0.240 kN/m²).

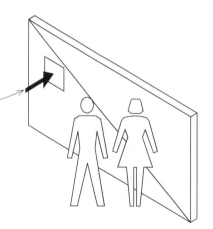

Snow Loads

Snow loads are to be determined in accordance with a reference standard: Chapter 7 of ASCE 7. Snow loads in the contiguous United States are shown in Figure 1608.2. Snow loads are determined based upon historical data and are correlated to geographic location and to elevations.

- For example, the snow load in north-central Kansas, is 25 psf (1.19 kN/m²).
- In north-east Arizona, the load varies from zero up to the 3000' (914 m) elevation; 5 psf up to the 4500' elevation (0.24 kN/m² load up to the 1372 m elevation); 10 psf up to the 5400' elevation (0.48 kN/m² load up to the 1645 m elevation) and 15 psf up to the 6300' elevation (0.72 kN/m² load up to the 1920 m elevation).
- In heavy snow areas, such as the Sierra Nevada and the Rocky Mountains, the snow load is to be determined by case studies that are based on 50-year recurrence data and must be approved by the building official.

Roofs are to be designed in accordance with ASCE 7 to accommodate snow loads under varying conditions, such as:

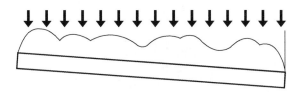

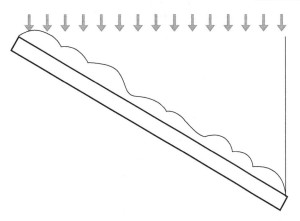

- *Snow loads on flat roofs (≤ 5°; 0.09 rad), taking into account such factors as exposure and rain-on-snow surcharge*

- *Snow loads on sloped roofs (> 5°; 0.09 rad)*

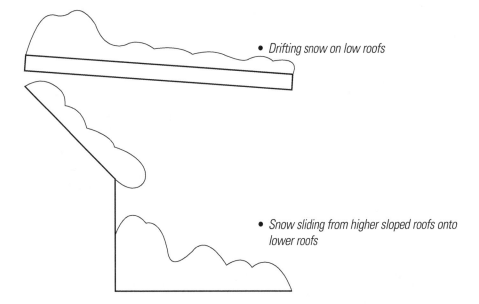

- *Drifting snow on low roofs*

- *Snow sliding from higher sloped roofs onto lower roofs*

STRUCTURAL DESIGN

Wind Loads

Buildings and portions of buildings are to be designed to withstand, at a minimum, the wind loads included in the code in accordance with Chapter 6 of ASCE 7. The wind is assumed to come from any horizontal direction, and no reduction is to be taken for the effect of shielding by other structures. This is in keeping with the principle that the Code applies to the building in question and is affected neither positively nor negatively by adjacent buildings. There are, however, portions of §1609 where adjacent site and topographic conditions may impact wind loads. See §1609.4.

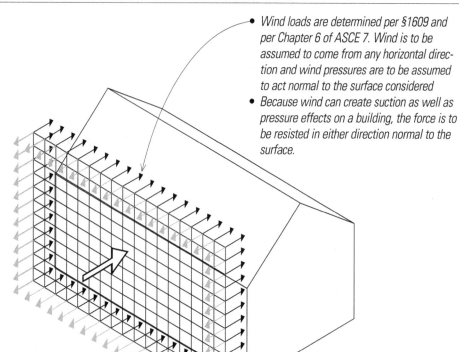

- Wind loads are determined per §1609 and per Chapter 6 of ASCE 7. Wind is to be assumed to come from any horizontal direction and wind pressures are to be assumed to act normal to the surface considered
- Because wind can create suction as well as pressure effects on a building, the force is to be resisted in either direction normal to the surface.

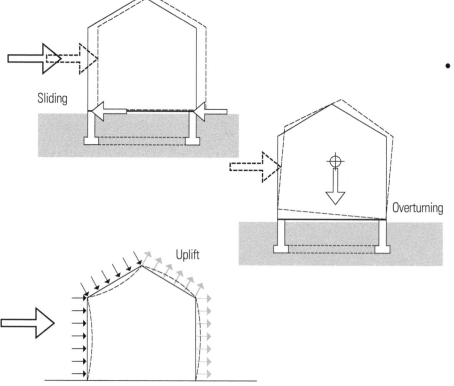

Sliding

Overturning

Uplift

- §1604.9 requires structural members and systems as well as building cladding to be anchored to resist overturning, uplift or sliding caused by wind forces. Continuous load paths for these forces are to be provided to the foundation.

Basic Wind Speed

Wind speeds for design are indicated in Figure 1609. The basic wind speed is based upon the speed for a three-second gust of wind. The wind speeds vary from 85 mph (38 m/s) in the West Coast of the U.S. to a wind speed of 150 mph (67 m/s) in hurricane-prone areas in southern Florida.

Exposure Category

The exposure category reflects the how ground surface irregularities affect design wind pressure. The exposure conditions vary from the most protected to the least protected wind exposures. Where buildings have multiple exposures, the condition that results in the highest wind force shall apply. The exposures are determined by applying a "surface roughness" category to wind calculations over each 45-degree (0.79 rad) sector from which wind can impact the building. The factors roughly increase with each surface roughness category:

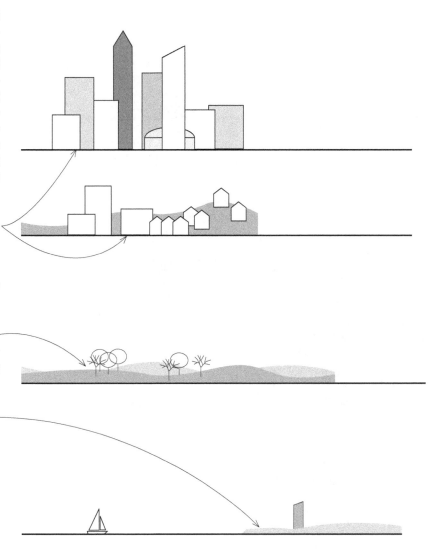

- *Surface Roughness B in urban and suburban residential areas; this is the assumed basic exposure. Exposure B occurs when the surface roughness prevails in the upwind direction for a distance of at least 2,600 feet (792 m) or 20 times the height of the building, whichever is greater.*
- *Surface Roughness C in open terrain with scattered obstructions including areas in flat open country, grasslands and water surfaces in hurricane-prone regions. Exposure C occurs when Exposure B does not apply.*
- *Surface Roughness D in flat, unobstructed areas and exposed to wind flowing over open water for a distance of at least a mile (excluding Surface Roughness C shoreline conditions in hurricane prone regions). Exposure D occurs when surface roughness D prevails upwind for at least 5,000 feet (1524 m) or 20 times the height of the building, whichever is greater. Exposure D extends inland from the shoreline for 600 feet (183 m) or 20 times the height of the structure, whichever is greater.*

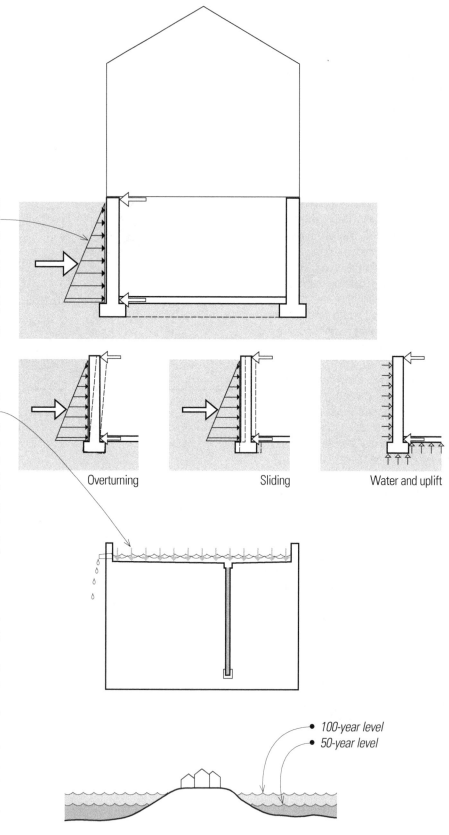

Soil Lateral Load

§1610 requires basement walls and retaining walls to be designed to resist lateral loads imposed by soils behind the walls. These walls are to be designed to be stable against overturning, sliding, excessive foundation pressure and water uplift. The soil loads are to be per Table 1610.1 and range from 30 psf per foot of depth (4.7 kPa/m) for gravels to 60 psf per foot of depth (9.4 kPa/m) for relatively dense inorganic clay soils.

Rain Loads

§1611 specifies that roofs are to be designed to accommodate the load of accumulated water when roof drains are clogged. Roofs with a slope less than a ¼ unit vertical in 12 units horizontal (2% slope) must be analyzed to determine if ponding will result in progressive deformation of the roof members leading to potential roof instability or failure in accordance with Section 8.4 of ASCE 7. The anticipated depth of water is based on the difference in elevation between the normal roof drain system and the outlet of the overflow system.

Flood Loads

§1612 requires that, in flood hazard areas established under the Federal Energy Management Agency Flood Insurance Study Program, all new buildings as well as major improvements or reconstruction projects must be designed to resist the effects of flood hazards or flood loads. Determination of whether a building falls under this requirement is based on locally adopted flood-hazard maps. These identify the anticipated flooding areas and elevations of flood waters for given anticipated return periods such as 50 or 100 years. The elevation and location of the building site must be compared to the flood-hazard maps to determine if this section is applicable.

Overturning

Sliding

Water and uplift

• *100-year level*
• *50-year level*

Earthquake Loads

§1613 contains the provisions for the seismic design of building structures. Earthquake design must be investigated for every structure and included to varying degrees based on the location of the building and the anticipated seismicity of the location. Earthquake design can be quite complex and involve detailed calculations. Certain basic types of structure, notably those wood-frame residences and light commercial buildings using the Conventional Light-Frame Construction provisions of §2308, are deemed to comply with the seismic requirements of the Code. Other more complex buildings must undergo seismic analysis based on ASCE 7. This analysis takes into account several basic factors. While we will not go into the design calculations in detail, it is worth understanding the basic criteria that are to be addressed by seismic analysis and design. The 2006 edition of the code greatly shortened the earthquake section of Chapter 16 by adopting references to ASCE 7. Our discussion is conceptual, with the assumption that those seeking greater detail will refer to the design criteria contained in ASCE 7.

Chapter 16 requires that all structures be designed and constructed to resist the effects of earthquake motions and be assigned a Seismic Design Category based on anticipated earthquake acceleration and the occupancy category of the building as outlined in §1613.5.6 and Tables 1613.5.6 (1 and 2).

Seismic forces are produced in a structure by ground motions that cause a time-dependent response in the structure. The response generated by the ground motions depends on:

- the magnitude, duration and harmonic content of the ground motions
- the dynamic properties of the structure (size, configuration and stiffness)
- the type and characteristics of the soil supporting the structure

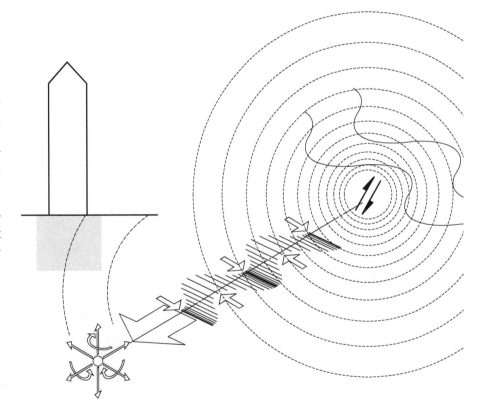

STRUCTURAL DESIGN

Site Ground Motion

§1613 provides procedures for determining design earthquake ground motions. Table 1613.5.2 defines a site-specific procedure required for sites having liquefiable soils, sensitive clays or weakly cemented soils.

The magnitude of earthquake ground motions at a specific site depends on the proximity of the site to the earthquake source, the site's soil characteristics and the attenuation of the peak acceleration. The dynamic response of a structure to earthquake ground motions can be represented by a graph of spectral response acceleration versus period.

Figures 1613.5 (1) through (14) map the maximum spectral response accelerations for the United States and its territories at short (0.2 second) and longer 1-second periods. The spectral response accelerations are given in percentages of gravity (g), assuming Site Class B (rock).

Structural Design Criteria

All structures require lateral-force-resisting and vertical-force-resisting systems having adequate strength, stiffness and energy dissipation capacity to withstand the anticipated or design earthquake ground motions.

- These ground motions are assumed to occur along any horizontal direction of a structure.
- Continuous load paths are required to transfer forces induced by earthquake ground motions from points of application to points of resistance.

Seismic Design Categories

The seismic design section requires that each structure be assigned a Seismic Design Category (A, B, C, D, E or F) based on the occupancy category and the anticipated severity of the earthquake ground motion at the site. This classification is used to determine permissible structural systems, limitations on height and irregularity, which components must be designed for seismic resistance, and the type of lateral force analysis required. The criteria for determining the seismic design categories are dependent on the seismic spectral response accelerations measured at various periods. For a detailed description of assignment of seismic design categories see §1613.5.6.

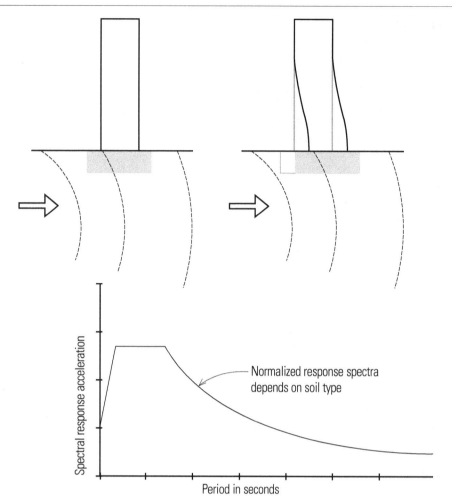

Normalized response spectra depends on soil type

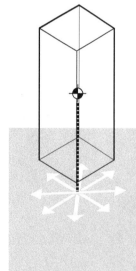

Seismic Design Category (short period used for illustration except as noted)*

| | Occupancy Category | | |
	I or II	III	IV
Short period < 0.167 g	A	A	A
0.167 g up to 0.33 g	B	B	C
0.33 g up to 0.50 g	C	C	D
> 0.50 g	D	D	D
*for 1-second period S1 ≥ 0.75 g for noted occupancy category, set by §1613.5.6	E*	E*	F*

* Seismic Design Categories from Tables 1613.5.6 (1) [for illustration]

*E If 1-second period lateral acceleration greater than or equal to 0.75 g for Occupancy Categories I, II, III

*F If 1-second period lateral acceleration greater than or equal to 0.75 g for Occupancy Category IV

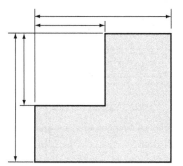

Building Configuration

Perhaps of greatest interest to architects and designers is the idea of building configuration, which classifies buildings into regular and irregular configurations. Irregularity in a building, either in its plan or its section configuration, can impact its susceptibility to damage in an earthquake.

Plan Irregularities

Plan irregularities include:

- *Torsional irregularity existing when the maximum story drift at one end of a structure is 120–140% greater than the average of the story drifts at the two ends of the structure*

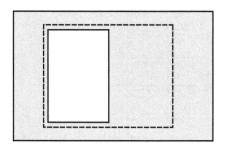

- *Reentrant corners where the projections are greater than 15% of the plan dimension in the given direction*

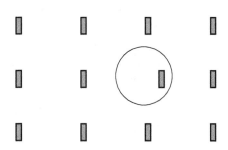

- *Discontinuous diaphragms, especially when containing cutouts or open areas greater than 50% of the gross enclosed diaphragm area*

- *Out-of-plane offsets creating discontinuities in lateral-force-resisting paths*

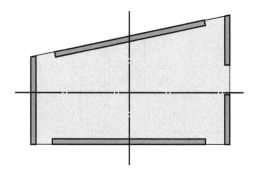

- *Nonparallel systems in which the vertical lateral-force-resisting systems are not parallel or symmetric about the major orthogonal axes of the lateral-force-resisting systems*

Vertical Irregularities
Vertical or sectional irregularities include:

- *Soft story having a lateral stiffness significantly less than that in the story above*

- *Weight or mass irregularity caused by the mass of a story being significantly heavier than the mass of an adjacent story*

- *Geometric irregularity caused by one horizontal dimension of the lateral-force-resisting system that is significantly greater than that of an adjacent story*

- *In-plane discontinuity in vertical lateral-force-resisting elements*

- *Weak story caused by the lateral strength of one story being significantly less than that in the story above*

Earthquake Loads:
Minimum Design Lateral Force and
Related Effects

The design criteria defines the combined effect of horizontal and vertical earthquake-induced forces as well as the maximum seismic load effect. These load effects are to be used when calculating the load combinations of §1605.4.

Redundancy

Redundancy provides multiple paths for a load to travel from a point of application to a point of resistance. The design criteria assign a redundancy coefficient to a structure based on the extent of structural redundancy inherent in its lateral-force-resisting system.

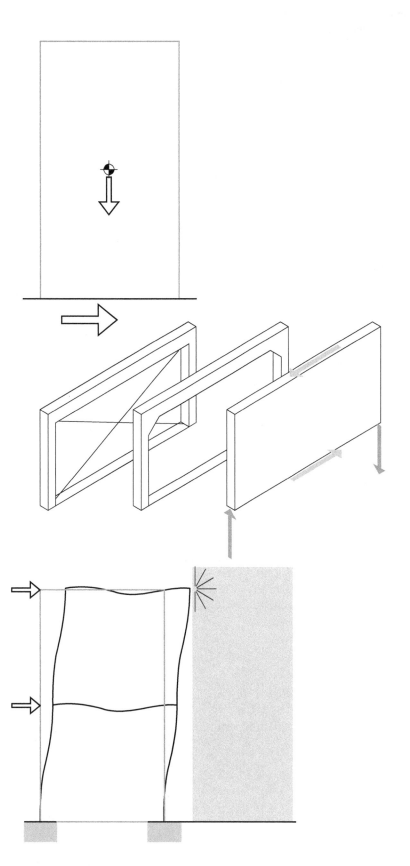

Deflection and Drift Limits

The design criteria that the design story drift not exceed the allowable story drift obtained from the criteria specifications. All portions of a building should act as a structural unit unless they are separated structurally by a distance sufficient to avoid damaging contact when under deflection.

STRUCTURAL DESIGN

Equivalent Lateral Force Procedure

The equivalent lateral force procedure for the seismic design of buildings assumes that the buildings are fixed at their base.

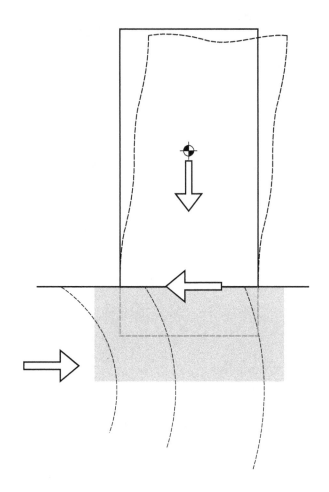

Seismic Base Shear

- *The basic formula for determining seismic base shear (V) is:*

$$V = C_s W$$

where:
C_s *= the seismic response coefficient determined from the design criteria and W = the effective seismic weight (dead load) of the structure, including partitions and permanent mechanical and electrical equipment.*

- *The seismic response coefficient is equal to a design spectral response coefficient amplified by an occupancy importance factor and reduced by a response modification factor based on the type of seismic-force-resisting system used.*

$$\text{Seismic response coefficient} = \frac{\text{Design spectral response coefficient}}{\text{(Response modification factor/Occupancy importance factor)}}$$

Vertical Distribution of Seismic Forces

The design criteria specify how the seismic base shear is to be distributed at each story level.

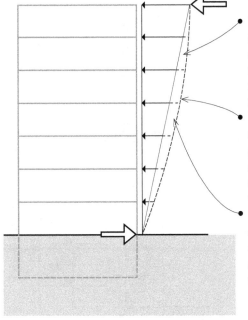

- The seismic base shear for buildings having a fundamental period not exceeding a specified period (around 0.5 second) are distributed linearly along the height with a zero value at the base and the maximum value at the top.
- The seismic base shear for buildings having a fundamental period exceeding a specified period (around 2.5 seconds) is distributed in a parabolic manner along the height with a zero value at the base and the maximum value at the top.
- The seismic base shear for buildings having a fundamental period between the specified low and high limits is distributed linear interpolation between a linear and a parabolic distribution.

Horizontal Shear Distribution

Seismic design story shear is the sum of the lateral forces acting at all levels above the story. The design criteria specify how the seismic design story shear is distributed according to the rigidity or flexibility of horizontal diaphragms and torsion.

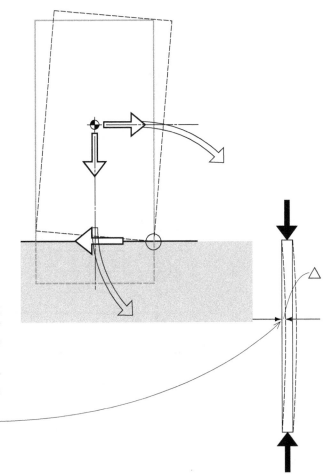

Overturning

A structure must be able to resist the overturning moments caused by the lateral forces determined to impact the structure.

Drift and P-delta Effects

Frames and columns are to be designed to resist both brittle fracture and overturning instability when building elements drift. They are also to resist the second-order effects on shear, axial forces and moments introduced during displacement. These forces are defined as the "P-delta effects."

STRUCTURAL DESIGN

Dynamic Analysis Procedure for the Seismic Design of Buildings

The static analysis contained in the design criteria may be used only for buildings that are assigned a higher Seismic Design Category because of regular configuration and height limitations. There are basically three types of dynamic analysis procedures that may be used for the seismic design of all buildings: modal response spectra analysis, linear time-history analysis and nonlinear time-history analysis.

Seismic-Force-Resisting Systems

There are several basic types of seismic-force-resisting systems:

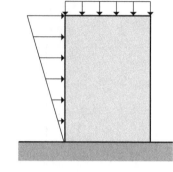

- *Bearing wall systems*

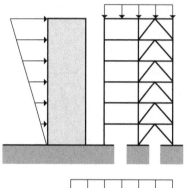

- *Building frame systems*

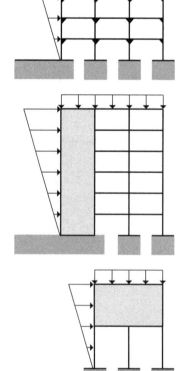

- *Moment-resisting frame systems*

- *Dual systems with special moment frames*
- *Dual systems with intermediate moment frames*

- *Inverted pendulum systems*

Detailing of Structural Components

There are requirements for the design and detailing of the components making up the seismic-force-resisting system of a building.

After a building is analyzed by calculations to determine its dynamic and static responses to seismic loads, it must be detailed to implement the design requirements of its seismic-force-resisting system.

This analysis must determine the worst-case forces based on the direction of seismic load, and that maximum force is to be used as the design basis.

Among details to be considered are:

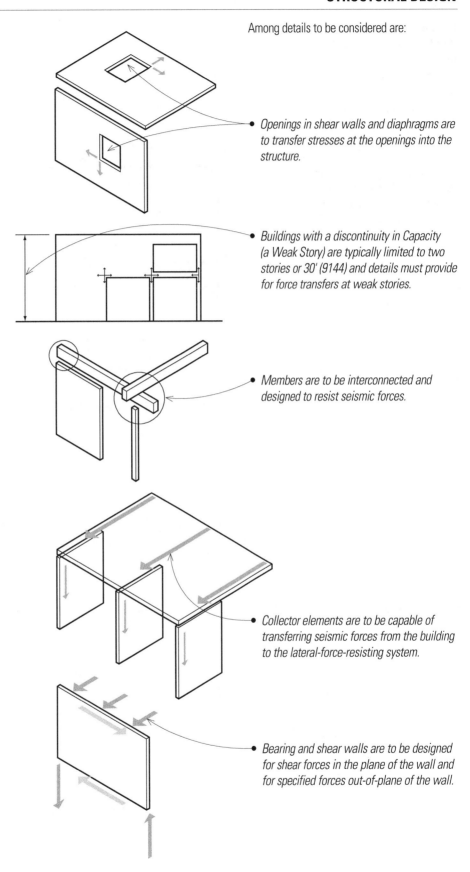

• *Openings in shear walls and diaphragms are to transfer stresses at the openings into the structure.*

• *Buildings with a discontinuity in Capacity (a Weak Story) are typically limited to two stories or 30' (9144) and details must provide for force transfers at weak stories.*

• *Members are to be interconnected and designed to resist seismic forces.*

• *Collector elements are to be capable of transferring seismic forces from the building to the lateral-force-resisting system.*

• *Bearing and shear walls are to be designed for shear forces in the plane of the wall and for specified forces out-of-plane of the wall.*

Seismic Design of Architectural, Mechanical and Electrical Components

The seismic design of non-structural elements such as architectural, mechanical and electrical components must also be considered. These components are typically not part of the structural system either for resisting conventional gravity loads or seismic forces. However, these components, when a permanent part of the building, must be seismically restrained. Failures of these systems result in a great deal of physical damage and are often a factor in casualties from earthquakes. Also, life safety systems, such as fire sprinklers and electrical systems, need to be functional after an earthquake.

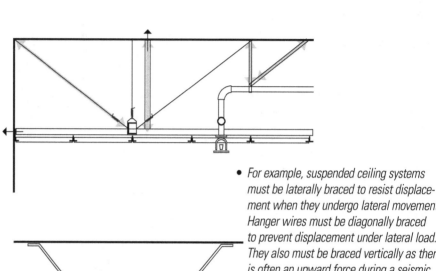

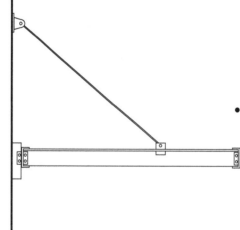

- *For example, suspended ceiling systems must be laterally braced to resist displacement when they undergo lateral movement. Hanger wires must be diagonally braced to prevent displacement under lateral load. They also must be braced vertically as there is often an upward force during a seismic event that can dislodge systems not braced for reversal of gravity loads.*

- *Such items as electrical components, mechanical equipment, ductwork, cabinetry and access floors must be seismically braced.*

- *Also, such elements as exterior architectural veneers, cantilevered exterior elements, signs and building ornament must be designed and detailed to resist seismic forces.*

17

Structural Tests and Special Conditions

Construction materials and methods of construction are subject to approval and inspection for quality, workmanship and labeling. The intent of the provisions and requirements contained in Chapter 17 is that materials should have the properties, strengths and performance that are represented and used as the basis for design. For products, these criteria are applied at the factory. For construction, these criteria are applied either during fabrication for shop-fabricated items or at the job site during construction. Testing and approval agencies must be acceptable to the building official. The building official typically recognizes accreditation by national agencies that certify approval processes and inspection protocols.

Special inspections are required for the manufacture, installation, fabrication, erection or placement of components where special expertise is required to ensure compliance with design documents and referenced standards. Such inspections may be required to be continuous or periodic depending on the critical nature of the process and whether the inspector can determine compliance without being present during the entire process. Testing and sample gathering take place to be certain that materials that are to be covered by other construction, such as soils work, footings or structural elements, are determined to meet design and code criteria before they are covered up. Special inspection does not take the place of structural observation, which is the visual observation of the construction process for the structural system. Conversely, such observation does not take the place of special inspection, as observations by their nature are periodic and not continuous. The definition of structural observation specifically states that it does not waive responsibilities for special inspections required by §1704.

Special Inspections

§1704 specifies certain types of work that requires inspection by special inspectors, employed by the owner or the responsible design professional acting as the owner's agent. The special inspector is subject to the approval of the Building Official. The special inspector is to make reports in conformance with the requirements of §1705. There are many areas of construction where special inspection may be required. Note that per Exception 3 to §1704.1 special inspections are not required in R-3 or U occupancies accessory to a residential occupancy unless otherwise required by the building official. A representative list of inspections that can be anticipated on most healthcare projects, with numbers taken from Table 1704.3, include:

	S	M	T	W	T	F	S
	1	2	3	4	5	6	7
	8	9	10	11	12	13	14
	15	16	17	18	19	20	21
	22	23	24	25	26	27	28
	29	30					

Continuous Inspection

	S	M	T	W	T	F	S
	1	2	3	4	5	6	7
	8	9	10	11	12	13	14
	15	16	17	18	19	20	21
	22	23	24	25	26	27	28
	29	30					

Periodic Inspection

Verification and Inspection	Continuous Inspection	Periodic Inspection
Steel		
1. Material verification for high strength bolts	-	
2.a. Inspection of high strength bolting with bearing type connections	-	
2.b. Inspection of high strength bolting with slip critical connections		
3. Material verification of structural steel	-	-
4. Material verification of weld filler materials	-	-
5.a. Inspection of welding structural steel	Varies	Varies
5.b. Inspection of welding of reinforcing steel	Varies	Varies
6. Inspection of steel frame joint details for compliance with construction documents	-	

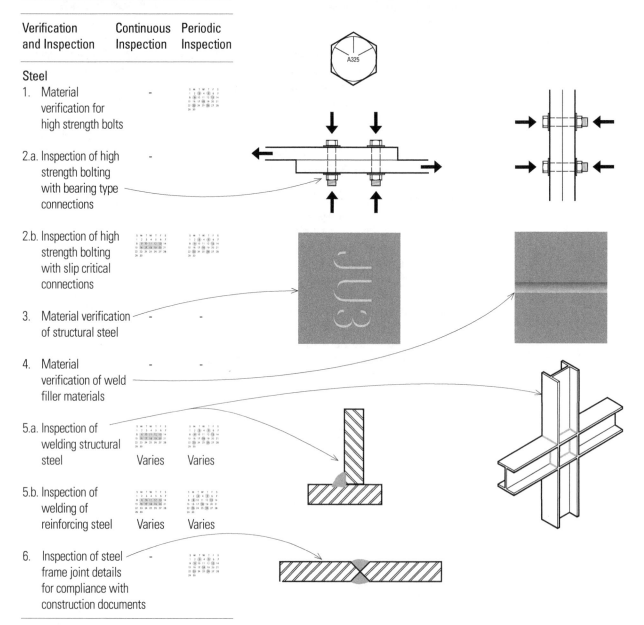

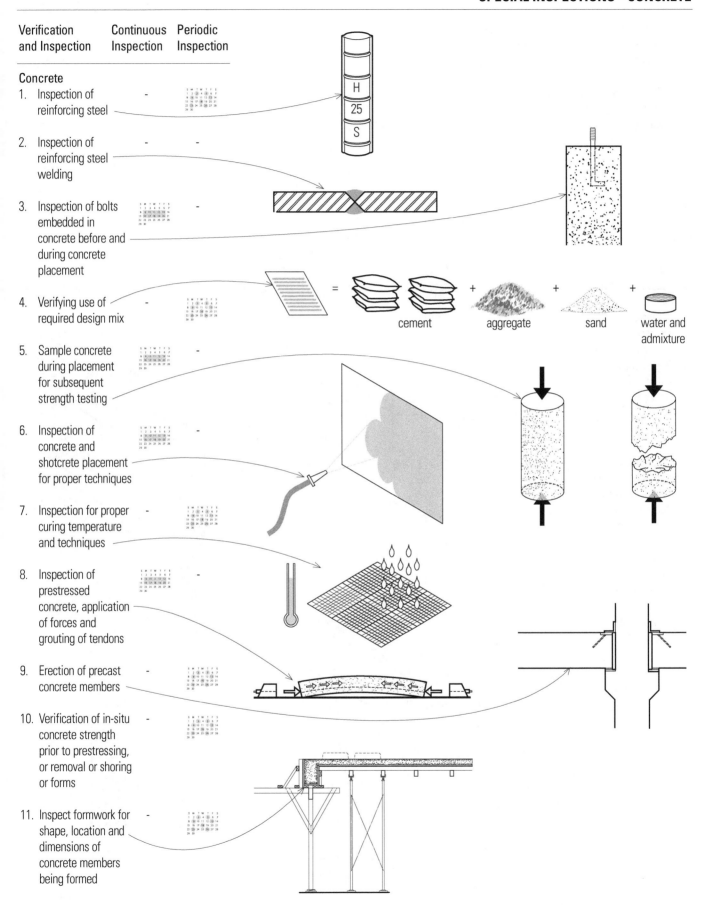

Verification and Inspection	Continuous Inspection	Periodic Inspection
Concrete		
1. Inspection of reinforcing steel	-	✓
2. Inspection of reinforcing steel welding	-	-
3. Inspection of bolts embedded in concrete before and during concrete placement	✓	-
4. Verifying use of required design mix	-	✓
5. Sample concrete during placement for subsequent strength testing	✓	-
6. Inspection of concrete and shotcrete placement for proper techniques	✓	-
7. Inspection for proper curing temperature and techniques	-	✓
8. Inspection of prestressed concrete, application of forces and grouting of tendons	✓	-
9. Erection of precast concrete members	-	✓
10. Verification of in-situ concrete strength prior to prestressing, or removal or shoring or forms	-	✓
11. Inspect formwork for shape, location and dimensions of concrete members being formed	-	✓

cement   aggregate   sand   water and admixture

Masonry construction sets two criteria for special inspections, depending on the Occupancy Category of the building. We will use a hospital with surgical facilities, Occupancy Category IV from Table 1604.5 as an example. Per §1704.5.3 such a facility would require "Level 2 Special Inspection" of the masonry work, as described in Table 1704.5.3.

Verification and Inspection	Continuous Inspection	Periodic Inspection

Masonry

1. From the beginning of masonry construction the following is to be verified to ensure compliance:

   a. Proportions of site-mixed mortar, grout and prestressing grout

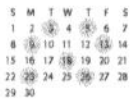

   b. Placement of masonry units and construction of mortar joints

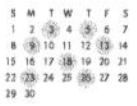

   c. Placement of reinforcement, connectors and prestressing tendons and anchors

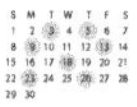

   d. Grout space prior to grouting

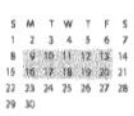

   e. Placement of grout

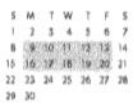

   f. Placement of prestressing grout

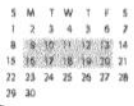

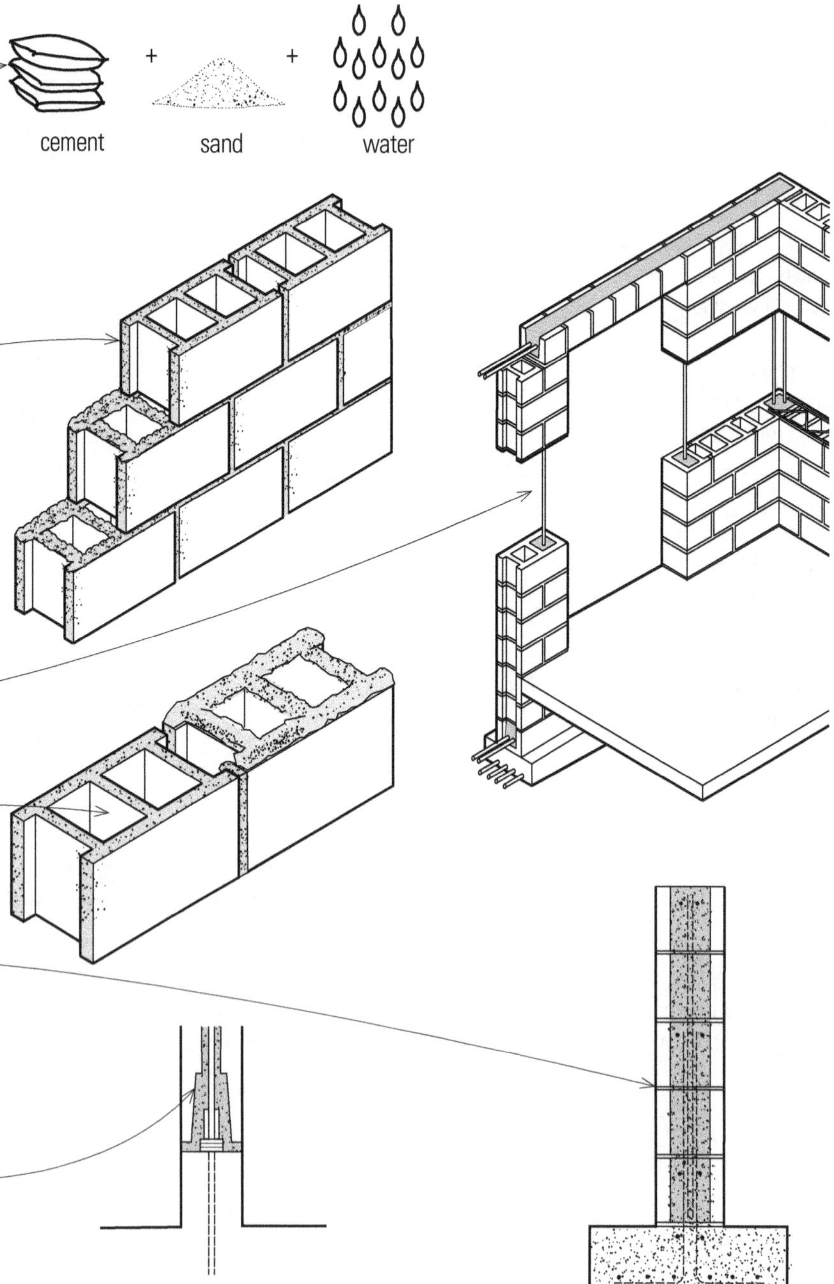

cement + sand + water

Verification and Inspection	Continuous Inspection	Periodic Inspection

Masonry

2. Inspection program to verify:

   a. Size and location of structural elements

   b. Type size and location of anchors, including anchorage of masonry to structural members, frames or other construction

   c. Specified size, grade and type of reinforcement

   d. Welding of reinforcing bars

   e. Protection of masonry during cold or hot weather

   f. Application and measurement of prestressing force

3. Preparation of any required grout or mortar specimens/prisms shall be observed by the special inspector

4. Verify compliance of special inspection provisions in the construction documents

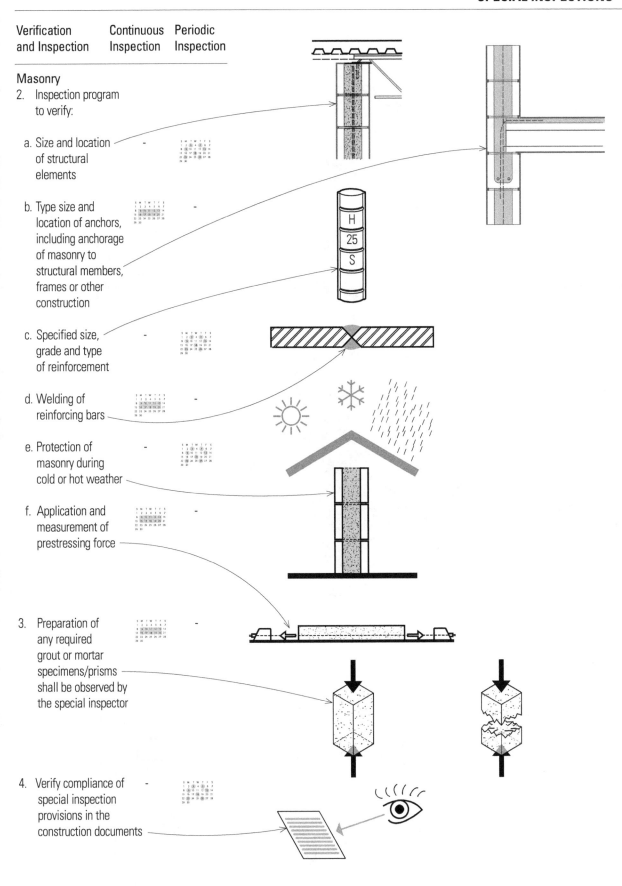

SPECIAL INSPECTIONS

Special inspections are also required for soils work (Table 1704.7), pile foundations (Table 1704.8) and pier foundations (Table 1704.9) to determine that buildings with specialized design criteria for those systems meet code requirements. In addition there are detailed special inspection requirements for the following special systems:

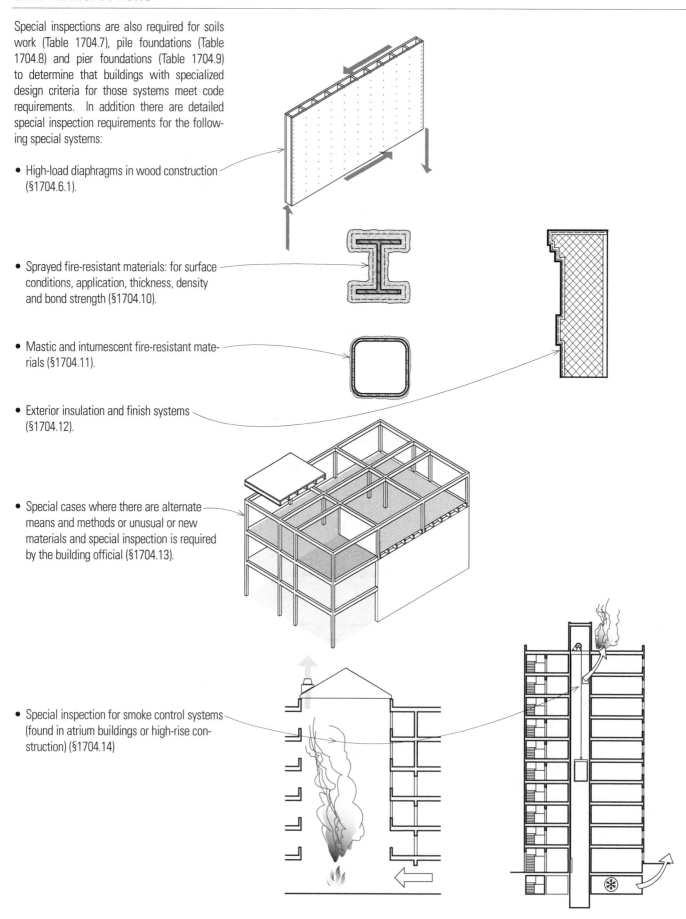

- High-load diaphragms in wood construction (§1704.6.1).

- Sprayed fire-resistant materials: for surface conditions, application, thickness, density and bond strength (§1704.10).

- Mastic and intumescent fire-resistant materials (§1704.11).

- Exterior insulation and finish systems (§1704.12).

- Special cases where there are alternate means and methods or unusual or new materials and special inspection is required by the building official (§1704.13).

- Special inspection for smoke control systems (found in atrium buildings or high-rise construction) (§1704.14)

Where special inspections are required per §1704, §1707 or §1708, then the registered design professional in responsible charge of the project is to prepare a statement of the requirements for special inspections. The statement is to be in accord with the requirements of §1705. The contractor is required by §1706 to acknowledge their awareness of special design and inspection requirements for wind and/or seismic force resisting systems.

§1707 specifies special inspection and testing requirements for seismic designs in Seismic design categories C, D, E or F. §1708 specifies testing and verification requirements for masonry materials and glass unit masonry based on the Occupancy Categories from §1604.5. Hospitals in Occupancy Category IV are to have Level 2 Quality Assurance for masonry per Table 1708.1.4.

There are also requirements in §1709 for structural observation by a registered design professional employed by the owner. Per §1709.2 these requirements come into play for buildings in Seismic Design Category D, E or F per §1613 when one or more of the following conditions exist:

1. Classified as Occupancy Category III or IV per §1604.5
2. The building height is greater than 75 feet (22 860)
3. Is in Seismic Design Category E, Occupancy Category I or II and greater than 2 stories in height
4. When such observation is required by the design professional in responsible charge
5. When such observation is specifically required by the building official

§1709.3 requires observation for wind design conditions where basic wind speeds exceed 110 mph (49 m/s) per Figure 1609. The conditions are the same as for seismic observations noted above except Item 3 is omitted.

The Code makes provisions for innovative construction materials and techniques. Alternative test procedures and load tests of actual assemblies may be used to demonstrate compliance with the intent of the code for strength and durability. The code also has provisions for in-situ testing when there is reason to believe that construction already in place may not have the stability or load-bearing capacity to carry expected loads.

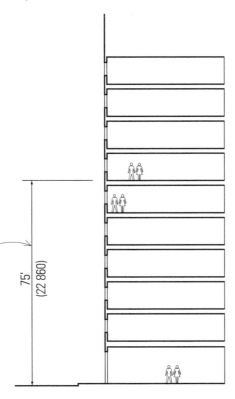

75' (22 860)

18
Soils and Foundations

Chapter 18 contains the provisions to the design and construction of buildings and foundation systems. Chapter 16 regulates buildings and foundations subject to water pressure from wind and wave action; Chapter 33 governs excavations and fills made during the course of construction.

SOILS AND FOUNDATIONS

Foundation and Soils Investigations

The building official will normally require a soils investigation to determine the stability and bearing capacity of the site soils. The height of the groundwater table should also be part of the soils investigation, as any slabs or occupied spaces below grade will require either damp proofing or waterproofing depending on the elevation of the groundwater table. The report should also classify the type of soil, recommend the type of footing, and design criteria for the footings. Where buildings are located in Seismic Design Categories C or above then §1802.2.6 or §1802.6.7 require additional investigations for potential earthquake motion-related hazards such as slope instability, liquefaction, loss of soil strength and earthquake motion induced lateral pressures.

Excavation, Grading and Fill

§1803 governs excavations, placement of back-fill, and site grading near footings and foundations.

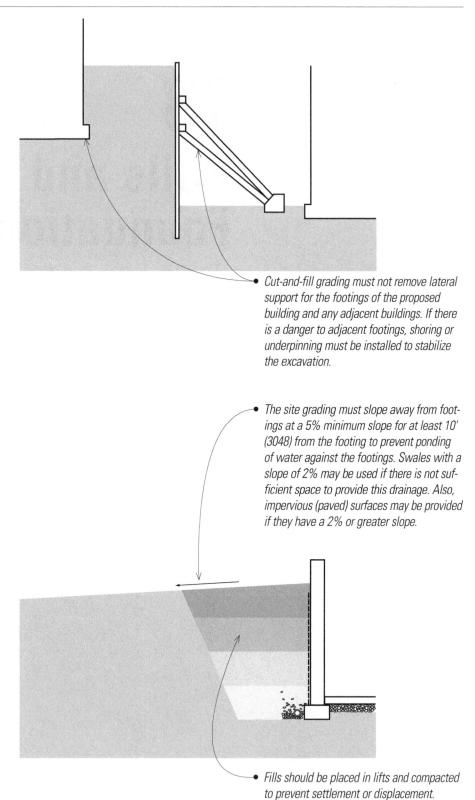

• Cut-and-fill grading must not remove lateral support for the footings of the proposed building and any adjacent buildings. If there is a danger to adjacent footings, shoring or underpinning must be installed to stabilize the excavation.

• The site grading must slope away from footings at a 5% minimum slope for at least 10' (3048) from the footing to prevent ponding of water against the footings. Swales with a slope of 2% may be used if there is not sufficient space to provide this drainage. Also, impervious (paved) surfaces may be provided if they have a 2% or greater slope.

• Fills should be placed in lifts and compacted to prevent settlement or displacement.

• Alternate systems, such as drains, may be used in conditions where the building is built into a slope.

Allowable Load-Bearing Values of Soils

Table 1804.2 contains presumptive load-bearing values for foundations and lateral pressure, based upon the observed capacities for various types of rock and soils. The use of these allowable pressures for vertical and lateral loads determines the size of the footing, based upon the weight of the structure bearing down upon them.

The allowable vertical pressures vary from 12,000 psf (575 kN/m²) for bedrock to 1,500 psf (72 kN/m²) for clay soils.

• *For example, a footing supported on clay soil will need to be eight times larger in plan area than one carried on bedrock.*

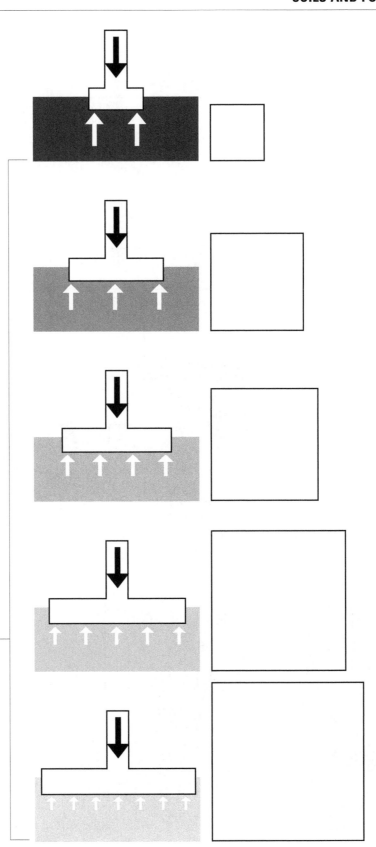

Footing and Foundations

§1805 governs the design and construction of footings and foundations.

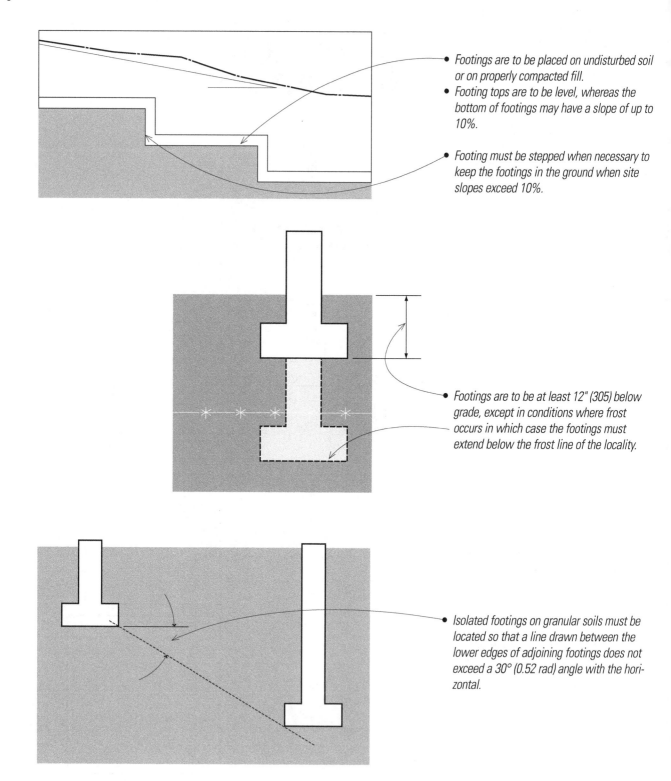

- Footings are to be placed on undisturbed soil or on properly compacted fill.
- Footing tops are to be level, whereas the bottom of footings may have a slope of up to 10%.

- Footing must be stepped when necessary to keep the footings in the ground when site slopes exceed 10%.

- Footings are to be at least 12" (305) below grade, except in conditions where frost occurs in which case the footings must extend below the frost line of the locality.

- Isolated footings on granular soils must be located so that a line drawn between the lower edges of adjoining footings does not exceed a 30° (0.52 rad) angle with the horizontal.

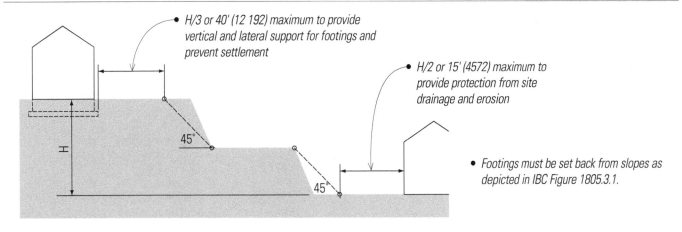

H/3 or 40' (12 192) maximum to provide vertical and lateral support for footings and prevent settlement

H/2 or 15' (4572) maximum to provide protection from site drainage and erosion

- Footings must be set back from slopes as depicted in IBC Figure 1805.3.1.

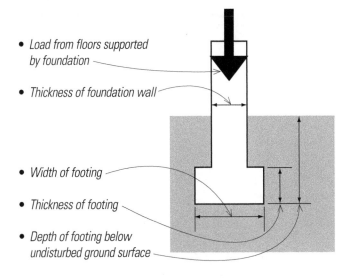

- Load from floors supported by foundation
- Thickness of foundation wall
- Width of footing
- Thickness of footing
- Depth of footing below undisturbed ground surface

While there are provisions for alternate materials for footings, most footings and foundations are made of concrete. We will focus on concrete footings and foundation designs.

- Footing concrete is have a minimum compressive strength of 2,500 psi (17.2 MPa).
- Light-frame construction footings and foundations are to be in accordance with Table 1805.4.2. The number of floors supported by the foundation determines the footing and foundation wall sizes.

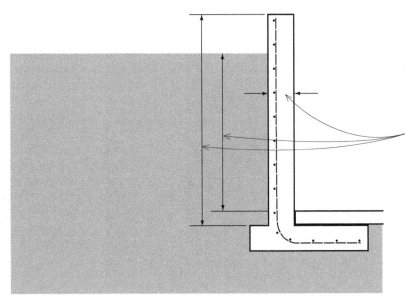

- Foundation walls for below-grade structures where earth is retained may be designed by using Tables 1805.5 (1) through (5) or by calculation.
- The tables are based on the height and the thickness of the wall. They assume steel with 60,000 psi (414 MPa) yield strength and concrete with a compressive strength of 2,500 psi (17.2 MPa). Reinforcing steel is to be placed in the wall based on the tables. Alternate reinforcing from that called for in the tables may be used as long as an equivalent cross-sectional area is maintained and the rebar spacing does not exceed 72" (1829 mm) and the rebar sizes do not exceed No. 11.

Damp-proofing and Waterproofing

§1807 requires walls that retain earth and enclose interior spaces below grade to be waterproofed or damp-proofed. The location of the water table determines whether damp-proofing or more extensive waterproofing is required.

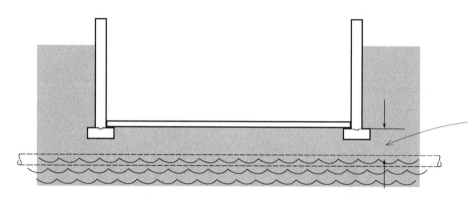

- Damp-proofing is required when there may be moisture present, but not under hydrostatic pressure.
- §1807.1.3 specifies that floors and walls may be damp-proofed when engineered means to lower ground water to not less than 6" (152) below floors and walls are provided.

Subsoil Drainage System

When conditions requiring damp-proofing occur at below-grade rooms, §1807.4 requires that additional provisions be made. Such conditions are presumed to not have hydrostatic pressures.

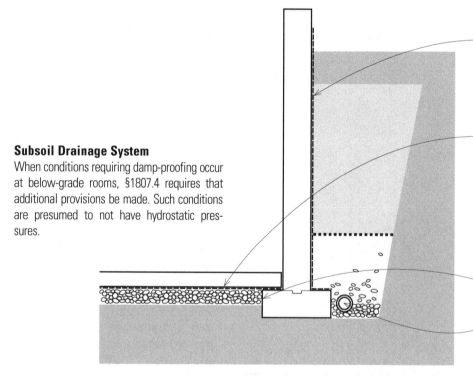

- Walls requiring damp-proofing are to be treated on the earth side of the wall. They are to receive a bituminous coating, acrylic modified cement or any of the wall waterproofing treatments noted in the Code.
- Slabs are to be protected with an underslab polyethylene membrane at least 6 mil (0.006"; 0.152 mm) in thickness with joints lapped and sealed.
- Alternately a mopped-on bitumen layer or 4-mil (0.004"; 0.102 mm) polyethylene membrane may be applied to the top of the slab when other materials will cover it.
- Floor slabs of basements are to be placed over a base course of not less than 4" (102) of gravel or crushed stone.
- A foundation drain is to be placed around the perimeter with gravel or crushed stone extending a minimum of 12" (305) beyond the outside edge of the footing.
- The drain gravel or pipe must be located with the water-flow line no higher than the floor slab.

Waterproofing is required when the site soils investigation indicates that there is water under hydrostatic pressure at the site. Waterproofing is to be installed unless a groundwater control system is installed that lowers the water table to 6″ (152) below the lowest floor level.

Walls that are to be waterproofed must be of concrete or masonry construction. They must be designed to resist the anticipated hydrostatic pressures along with expected lateral and vertical loads.

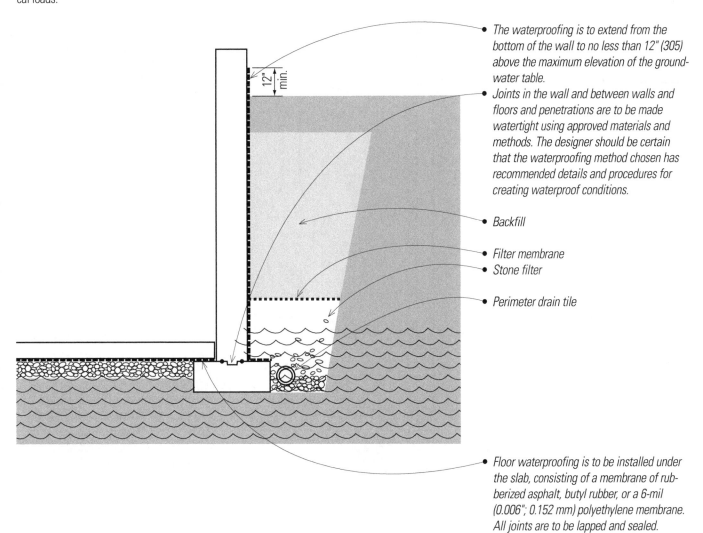

The waterproofing is to extend from the bottom of the wall to no less than 12″ (305) above the maximum elevation of the groundwater table.

Joints in the wall and between walls and floors and penetrations are to be made watertight using approved materials and methods. The designer should be certain that the waterproofing method chosen has recommended details and procedures for creating waterproof conditions.

Backfill

Filter membrane

Stone filter

Perimeter drain tile

Floor waterproofing is to be installed under the slab, consisting of a membrane of rubberized asphalt, butyl rubber, or a 6-mil (0.006″; 0.152 mm) polyethylene membrane. All joints are to be lapped and sealed.

12″ min.

Pier and Pile Foundations

When spread footings are not adequate or appropriate for a foundation system, pier and pile foundations are often used. Both systems use columnar structural elements either cast in drilled holes or driven into the ground to support foundations.

Piers

§1808.1 defines pier foundations as consisting of isolated masonry or cast-in-place concrete structural elements extending into a firm subgrade.

Piles

§1808.1 defines pile foundations as consisting of concrete, steel or wood structural elements either driven into the ground or cast-in-place.

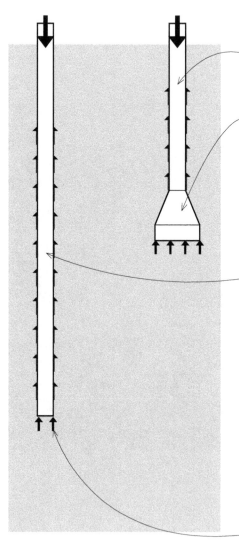

- By definition, piers are relatively short structural elements, having lengths less than or equal to 12 times their least horizontal dimension.
- Belled piers are cast-in-place concrete piers having a base that is larger than the diameter of the rest of the shaft. This enlarged base increases the load-bearing area of an end-bearing pier.

- Piles are relatively slender structural elements, having lengths exceeding 12 times their least horizontal dimension.

- §1809 covers the design and construction of driven piles, such as timber piles and precast concrete piles.
- §1810 covers the design and construction of cast-in-place concrete pile foundations, the various types of which are listed and described in §1808: augured uncased piles, caisson piles, concrete-filled steel pipe and tube piles, driven uncased piles, and enlarged-base piles.

- Both piers and piles derive their load-carrying capacity through skin friction (friction between their surfaces and the surrounding soil), and through end bearing on supporting material, or a combination of both.

Both pier and pile foundations are to be based upon the recommendations of a soil investigation. Once the capacity of the soil is determined a system design can be selected. The Code contains detailed requirements for the design and configuration of pier and pile foundations.

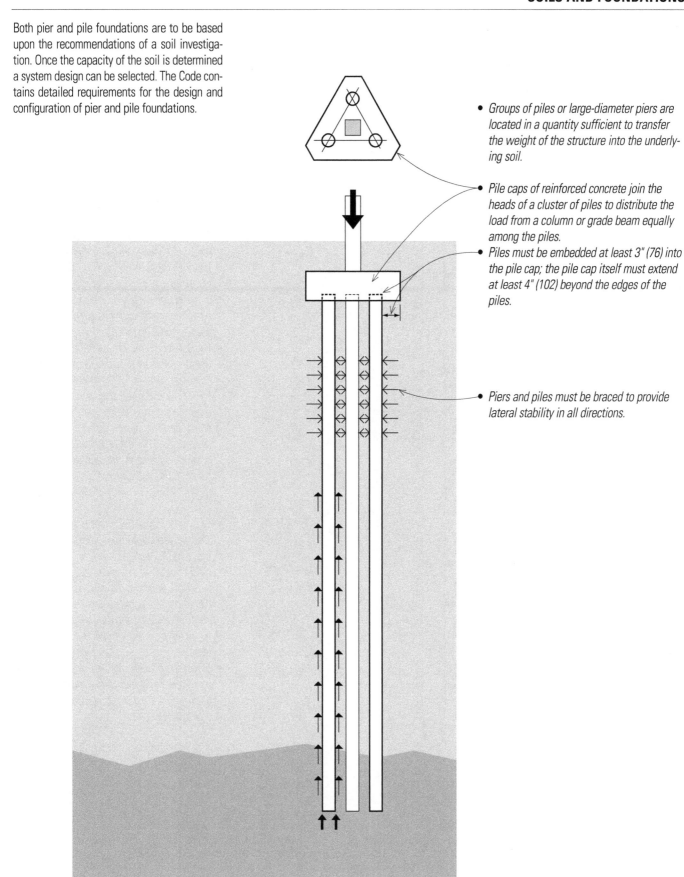

• *Groups of piles or large-diameter piers are located in a quantity sufficient to transfer the weight of the structure into the underlying soil.*

• *Pile caps of reinforced concrete join the heads of a cluster of piles to distribute the load from a column or grade beam equally among the piles.*

• *Piles must be embedded at least 3" (76) into the pile cap; the pile cap itself must extend at least 4" (102) beyond the edges of the piles.*

• *Piers and piles must be braced to provide lateral stability in all directions.*

SOILS AND FOUNDATIONS

Driven Piles

Driven piles are piles that rely on either end-bearing or friction on the surface of the pile to provide support for the building. They are inserted into the ground and driven into place by impact of a hammer, similar to driving a nail into wood. Piles may also be driven by vibratory drives, subject to verification by load tests.

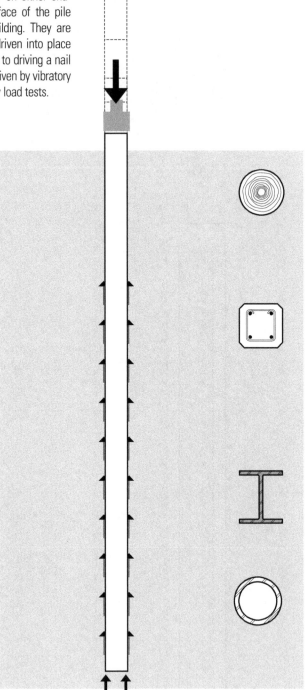

Driven piles may be of various materials:

- *Timber piles may be round or sawn. They must be preservative-treated unless the entire pile will be under water for its entire life of service. Timber piles should be capped or monitored closely as they are driven to ensure their shafts or tips are not split or shattered during driving.*
- *Precast concrete piles are to be reinforced with longitudinal rods tied in place with lateral or spiral ties. Concrete piles are to be designed to resist seismic forces as for concrete columns. Reinforcement for precast nonprestressed piles is to have a minimum cover of 2" (51). Reinforcement for precast prestressed piles should have a minimum cover of 1¼" (32) for square piles 12" (305) or smaller in size, and 1½" (38) for larger piles.*
- *Steel piles consist of either H-sections or sections fabricated from steel plates. They are to have a flange projection not exceeding 14 times the minimum thickness of either the flange or the web. Their nominal depth in the direction of the web is not to be less than 8" (203) and the flanges and webs are to have a minimum thickness of ³/₈" (9.5 mm).*
- *Steel-pipe piles driven open-ended are to be at least 8" (203) in outside diameter. Wall thickness of the pipe sections depend on the driving force used.*

Cast-in-Place Concrete Piles

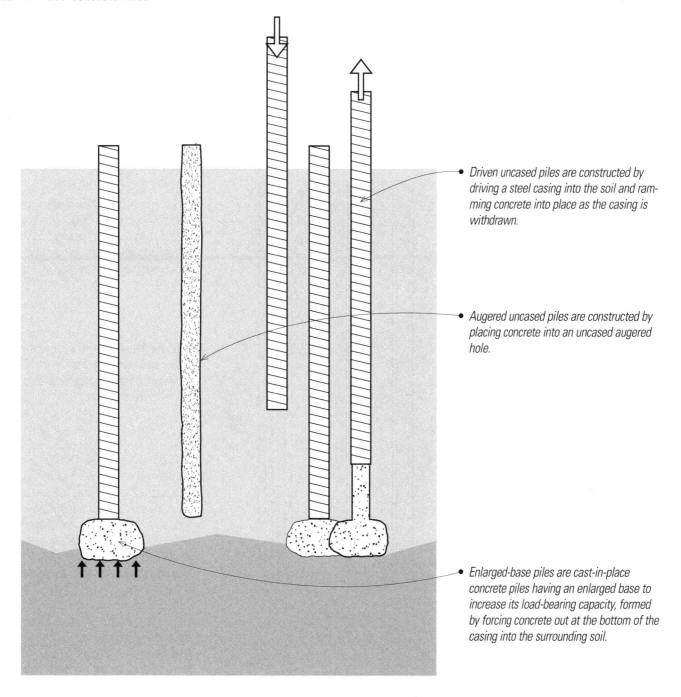

Driven uncased piles are constructed by driving a steel casing into the soil and ramming concrete into place as the casing is withdrawn.

Augered uncased piles are constructed by placing concrete into an uncased augered hole.

Enlarged-base piles are cast-in-place concrete piles having an enlarged base to increase its load-bearing capacity, formed by forcing concrete out at the bottom of the casing into the surrounding soil.

Cast-in-Place Concrete Piles

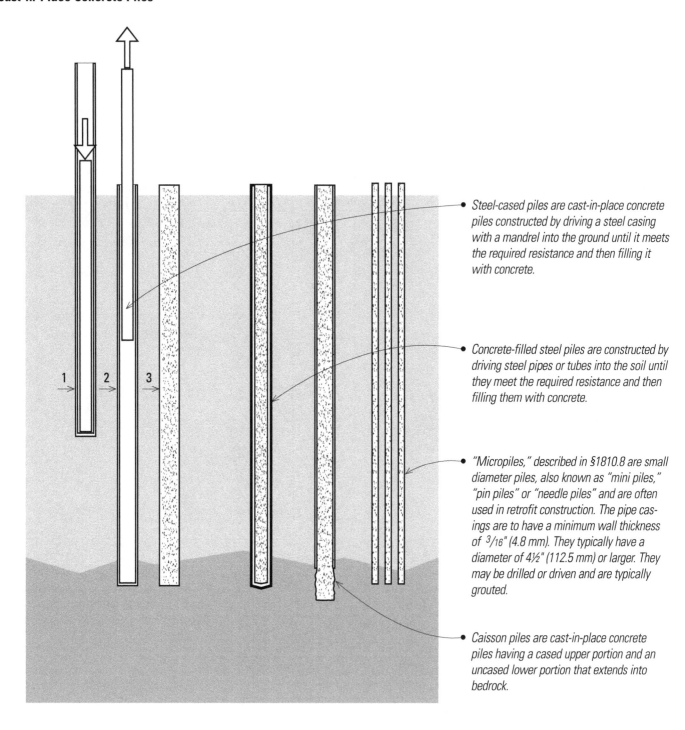

Steel-cased piles are cast-in-place concrete piles constructed by driving a steel casing with a mandrel into the ground until it meets the required resistance and then filling it with concrete.

Concrete-filled steel piles are constructed by driving steel pipes or tubes into the soil until they meet the required resistance and then filling them with concrete.

"Micropiles," described in §1810.8 are small diameter piles, also known as "mini piles," "pin piles" or "needle piles" and are often used in retrofit construction. The pipe casings are to have a minimum wall thickness of $^3/_{16}$" (4.8 mm). They typically have a diameter of 4½" (112.5 mm) or larger. They may be drilled or driven and are typically grouted.

Caisson piles are cast-in-place concrete piles having a cased upper portion and an uncased lower portion that extends into bedrock.

Pier Foundations

§1812. Piers are usually constructed by drilling a hole to a suitable bearing strata and placing concrete in the hole.

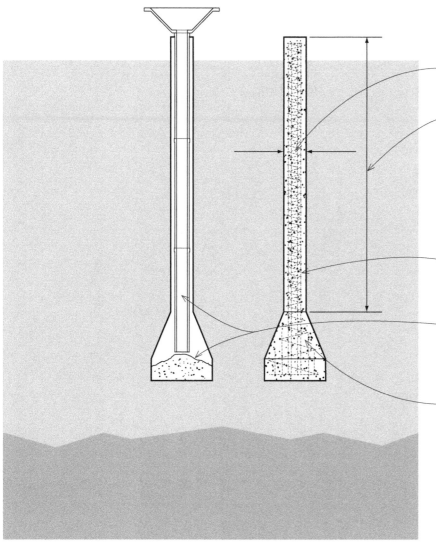

- *Single isolated piers are to have a minimum dimension of 2' (610) and, per the definition of a pier contained in §1808.1, they are to be no more than 12 times the least horizontal dimension in height.*

- *Piers are to be made of reinforced concrete with a concrete strength of at least 2,500 psi (17.2 MPa). The concrete mix is to be designed to minimize segregation during placement whether placed by funnel hopper or by pumping.*
- *Reinforcement is to be tied together as a unit and placed in the pier.*

- *Concrete is to be placed to avoid segregation using a tremie and not dropped in the pier from the top.*

- *When bottoms are belled, the edge thickness of the bell is to be that required for the edge of footings.*
- *When a steel shell is used and is considered part of the reinforcing, the steel must be protected with a coating or other method acceptable to the building official per §1807.2.17.*

- *When piers extend below the water level of the site, they must be constructed so their bottoms can be inspected and the concrete or other masonry can be placed in the dry.*

19

Building Materials and Systems

(IBC Chapters 19–30)

This chapter is a summary of design requirements contained in Chapters 19 through 33 of the Code. Chapters 19 through 23 deal primarily with the structural requirements for building materials: concrete, masonry, steel, wood and aluminum. We will touch on the implications of the structural design requirements for the physical form of the building. A detailed discussion of the mathematics of structural design is beyond the scope of this book.

Chapters 24 through 26 cover building materials used primarily for building-envelope construction and for finishes: glass and glazing, gypsum board and plaster, and plastics.

Chapters 27 through 29 cover building-code requirements for systems that are addressed in other related codes: electrical, mechanical and plumbing systems. Chapters 30 through 33 deal with miscellaneous items not readily related to other code sections: elevator and conveying systems, special construction, encroachments on the public right of way, and safeguards during construction.

Our discussion will concentrate on the code chapters related to building design. These are grouped together into sections for construction materials, finish materials and conveying systems. We will not touch on the other sections noted above, as most of the other items are addressed in other codes or by specialized constructors and are thus beyond the scope of this book.

STRUCTURAL MATERIALS

The materials chapters of the code are based upon standards developed by institutions that concentrate on developing industry standards for specific materials. These groups bring together materials experts, industry representatives, design professionals, testing agencies and building officials to develop criteria for strength, material design calculations, testing and quality assurance for various materials. These standards are developed through a consensus process similar to that for building-code development. Materials are also subjected to physical tests under controlled load or fire conditions to determine their performance and develop criteria to be incorporated into the standards. These criteria are continually updated to incorporate new knowledge learned from improvements in design techniques and analyses of building performance under actual conditions such as wind, fire and earthquake.

Chapter 19: Concrete

Chapter 19 is based upon standard ACI 318, developed by the American Concrete Institute. There are certain detailed modifications to ACI 318 made in the code-development process. These modifications are indicated by italic type in the body of the code. Because of these modifications the code should be used for reference during concrete design in conjunction with ACI 318. ACI 318 should not be used alone for design work that will be undertaken under the purview of the IBC.

§1901.2 states that Chapter 19 governs design of all concrete, whether reinforced or unreinforced, except for slabs on grade that do not transfer vertical or lateral loads from the structure to the soil. Since most, if not all, slabs on grade contained in buildings typically meet these criteria in some if not all locations, it should be assumed that Chapter 19 effectively governs the design of all concrete except for stand-alone slabs on grade such as patios or paving. As for other materials, concrete is to be designed to resist wind and seismic loads in addition to anticipated live and dead loads.

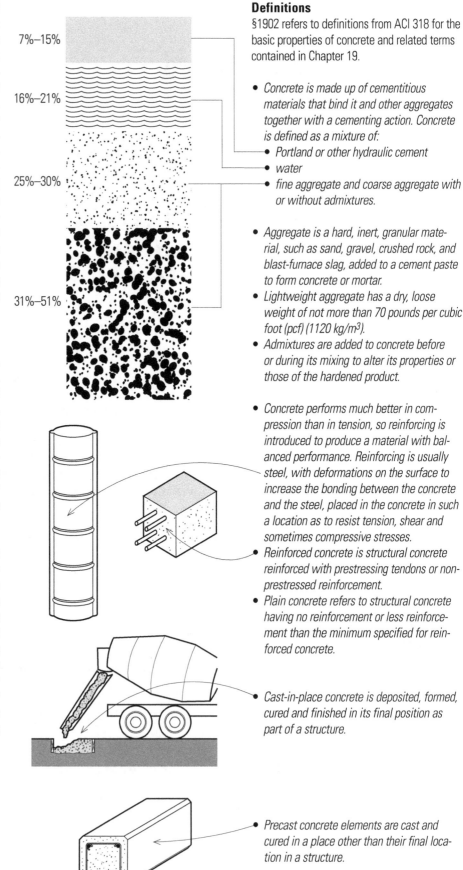

7%–15%

16%–21%

25%–30%

31%–51%

Definitions

§1902 refers to definitions from ACI 318 for the basic properties of concrete and related terms contained in Chapter 19.

- *Concrete is made up of cementitious materials that bind it and other aggregates together with a cementing action. Concrete is defined as a mixture of:*
  - *Portland or other hydraulic cement*
  - *water*
  - *fine aggregate and coarse aggregate with or without admixtures.*

- *Aggregate is a hard, inert, granular material, such as sand, gravel, crushed rock, and blast-furnace slag, added to a cement paste to form concrete or mortar.*
- *Lightweight aggregate has a dry, loose weight of not more than 70 pounds per cubic foot (pcf) (1120 kg/m³).*
- *Admixtures are added to concrete before or during its mixing to alter its properties or those of the hardened product.*

- *Concrete performs much better in compression than in tension, so reinforcing is introduced to produce a material with balanced performance. Reinforcing is usually steel, with deformations on the surface to increase the bonding between the concrete and the steel, placed in the concrete in such a location as to resist tension, shear and sometimes compressive stresses.*
- *Reinforced concrete is structural concrete reinforced with prestressing tendons or nonprestressed reinforcement.*
- *Plain concrete refers to structural concrete having no reinforcement or less reinforcement than the minimum specified for reinforced concrete.*

- *Cast-in-place concrete is deposited, formed, cured and finished in its final position as part of a structure.*

- *Precast concrete elements are cast and cured in a place other than their final location in a structure.*

Strength and Durability

There are many factors that can affect the ultimate strength of the concrete. The first of these factors is the mix design, setting the ratios of the materials and water. The second is the time that passes between the initial introduction of water and the placement of the concrete. Concrete setting is a chemical reaction that has a very definite time component. Also the mixture must be kept from segregating into its constituent parts by rotating the drum of the mixer prior to placement.

The temperature at the time of placement can also affect the strength of the concrete, with strengths declining at both high and low placement temperatures.

Air is entrained into the concrete for freeze/thaw protection based upon the anticipated weathering severity. The probability of weathering, shown in Figure 1904.2.2, places the majority of the continental U.S. in the moderate to severe weathering probability areas.

In cold weather areas where de-icing chemicals such as salts are used, steps must be taken to increase the strength of the concrete and also to protect reinforcing from corrosion to resist the long-term impacts of weathering and corrosion on the strength and durability of the concrete.

Table 1904.2.2 lists minimum concrete compressive strengths based on location in the building and on potential weather exposure. The minimum specified compressive strength is 2,500 psi, with the mix designed to achieve this strength at 28 days after placement of the wet concrete.

Cylinder Tests

Concrete is typically mixed in one location and placed in another. Because of this, concrete must be field tested to be certain that the material actually placed in the building meets the design criteria. Test cylinders are made as the material is placed. These cylinders are broken in a test machine to determine the strength of the concrete. Tests are typically done at 7 days to determine the strength trend for the concrete and at 28 days, which is the defined time when the design strength is to be achieved. Concrete mixes are proportioned to achieve workability, durability under expected exposure and development of the desired design strength.

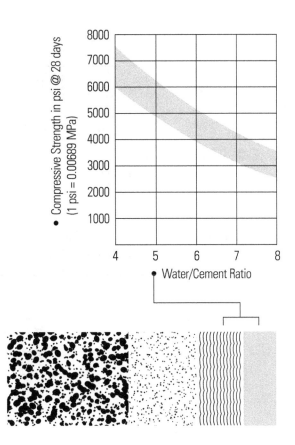

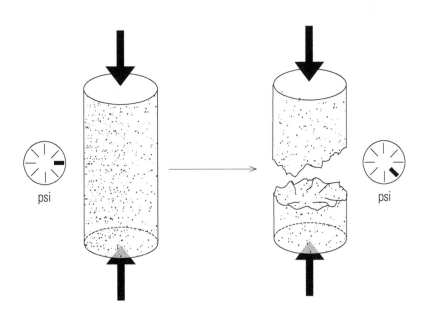

CONCRETE

Formwork

§1906.1 prescribes that the design, fabrication and erection of formwork is to comply with ACI 318. Formwork is necessary to develop the required shape and dimensions of concrete members in a structure. Forms must be substantial and tight enough to prevent leakage of cement mortar, be properly braced to maintain their proper position and shape, and be supported so as not to damage previously placed concrete.

§1906.3 provides that conduits and pipes may be embedded in concrete if they are not harmful to the concrete and comply with the limits set in ACI 318.

Joints

§1906.4 refers to ACI 318 for provisions for construction joints between two successive placements of concrete.

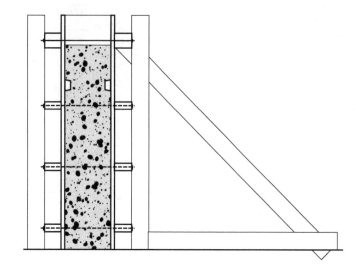

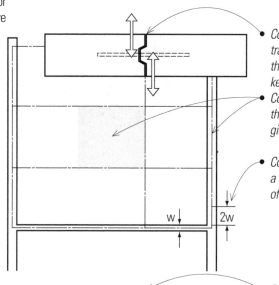

- Construction joints must provide for the transfer of shear and other forces through the joints, usually by means of mechanical keys or dowels.
- Construction joints must be located within the middle third of floor slabs, beams and girders.

- Construction joints in girders must be offset a minimum distance of two times the width of intersecting beams.

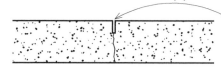

- Contraction joints are formed, sawed or tooled grooves in a concrete structure to create a weakened plane and regulate the location of cracking resulting from thermal stresses or drying shrinkage.

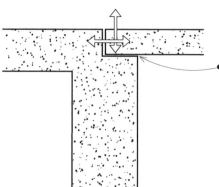

- Isolation joints separate adjoining parts of a concrete structure to allow relative movement in three directions to occur and to avoid formation of cracks elsewhere in the concrete. All or part of the bonded reinforcement may be interrupted at isolation joints, but the joints must not interfere with the performance of the structure.

Details of Reinforcement

There are several key elements of concrete and reinforcing configuration and placement as illustrated:

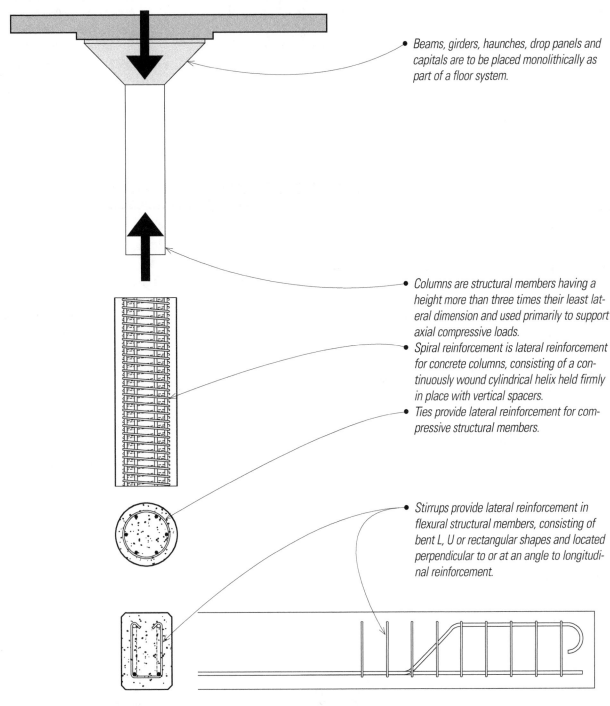

- Beams, girders, haunches, drop panels and capitals are to be placed monolithically as part of a floor system.

- Columns are structural members having a height more than three times their least lateral dimension and used primarily to support axial compressive loads.

- Spiral reinforcement is lateral reinforcement for concrete columns, consisting of a continuously wound cylindrical helix held firmly in place with vertical spacers.

- Ties provide lateral reinforcement for compressive structural members.

- Stirrups provide lateral reinforcement in flexural structural members, consisting of bent L, U or rectangular shapes and located perpendicular to or at an angle to longitudinal reinforcement.

Details of Reinforcement

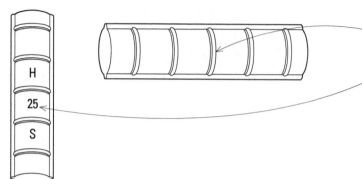

- Deformed reinforcement has surface deformations or a configuration to develop a greater bond with concrete.
- Reinforcing steel bars are marked for such things as grade of steel and manufacturer. Verification of reinforcing steel quality is often part of inspection as discussed in Chapter 17.

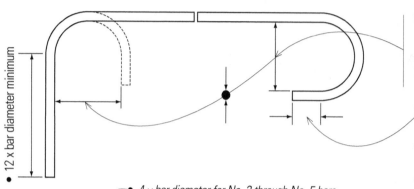

Bend Diameters
- Minimum bend diameters are regulated by §1907 and ACI 318.

- $6 \times$ bar diameter for No. 3 through No. 8 bars
- $8 \times$ bar diameter for No. 9, No 10 and No. 11 bars
- $10 \times$ bar diameter for No. 14 and No. 18 bars

- $4 \times$ bar diameter for 2½" (64) minimum

12 × bar diameter minimum

- $4 \times$ bar diameter for No. 3 through No. 5 bars
- $6 \times$ bar diameter for No. 6 through No. 8 bars

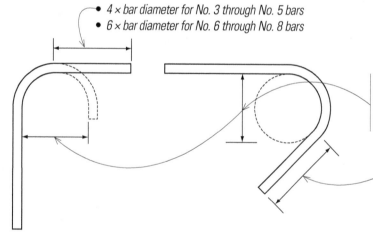

- $4 \times$ bar diameter for No. 3 through No. 5 bars
- $6 \times$ bar diameter for No. 6 through No. 8 bars

- $6 \times$ bar diameter minimum

Bar Spacing
- One bar diameter or 1" (25.4 mm) minimum for flexural members
- $1.5 \times$ bar diameter or 1½" (38) minimum for compression members

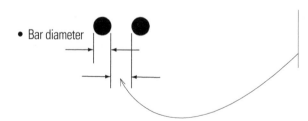

- Bar diameter

Fire Protection

The minimum depth and cover of reinforcement
are prescribed:

§1908 lists modifications made by the IBC to
ASA 318. Both documents must be used in
conjunction with each other when using the IBC
for concrete design.

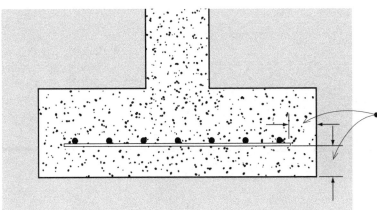

• *3" (76) minimum for concrete cast against
and permanently exposed to earth*

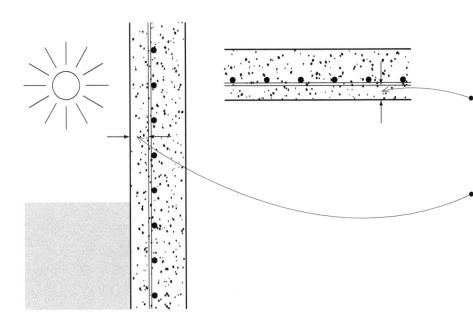

• *For concrete slabs, walls and joists not
exposed to weather or in contact with
ground. 1½" (38) for No. 14 and No. 18
bars, and ¾" (19.1 mm) for No. 11 bars and
smaller*

• *For concrete exposed to earth or weather, 2"
(51) for No. 6 through No. 18 bars, and 1½"
(38) for No. 5 bars and smaller*

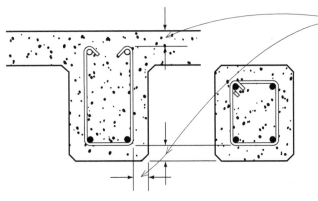

• *1½" (38) for primary reinforcement, stirrups,
ties and spirals in concrete beams and
columns*

Structural Plain Concrete

Per §1909, it is acceptable to use plain unre-inforced concrete under certain conditions, but only for continuously supported slabs, arched structures, and simple walls and pedestals. In most circumstances the use of minimal reinforc-ing to address temperature changes and shrink-age cracking is recommended practice instead of using plain concrete.

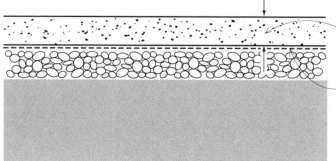

• Per §1910.1, ground-supported concrete floor slabs are to be a minimum of 3½" (89) thick, with a 6-mil (0.006"; 0.152 mm) polyethylene vapor retarder under the slab to prevent the transmission of vapor through the slab.

§1911 details requirements for the anchorage between concrete and other materials, and calls attention to the placement of embedded bolts in concrete to develop the design loads without structural failure of the concrete.

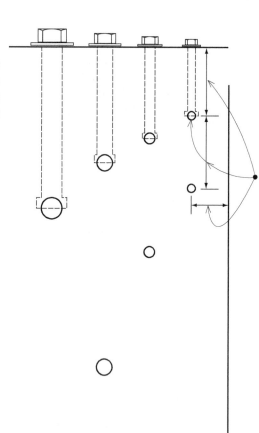

• Table 1911.2 interrelates bolt diameter, embedment length, edge distance, spacing and concrete strength.

Chapter 20: Aluminum

Chapter 20 is based completely upon *The Aluminum Design Manual, AA-94* and *Aluminum Sheet Metal Work in Building Construction, AA ASM 35*, both published by the Aluminum Association. The Code has adopted the industry standards for this material without modification. Aluminum structural design criteria are to conform to the design loads set forth in Chapter 16. §1604.3.5 refers to AA-94 for deflection limit criteria.

Aluminum members behave similarly to steel except that aluminum is more flexible and will exhibit greater deformation and deflection under the same stress and strain than will steel. Thus aluminum structural members are typically larger in thickness and/or section dimension than steel members under the same load conditions. Except under specialized conditions, where other materials may not be appropriate or durable, aluminum is used very infrequently as a primary structural material. Aluminum members are widely used in secondary structural elements such as windows, curtain walls and skylights.

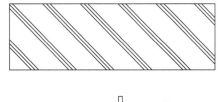

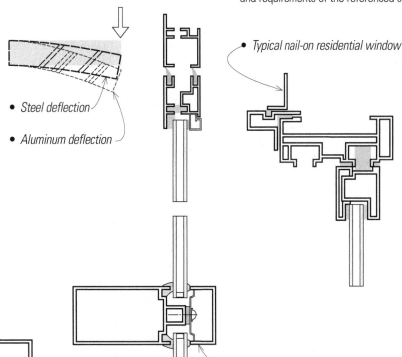

- *Steel deflection*
- *Aluminum deflection*

Aluminum structural elements may be readily made into many shapes by extrusion or fabrication. The illustrations show typical applications that are generally encountered by designers. They are consistent with the recommendations and requirements of the referenced standards.

- *Typical nail-on residential window*

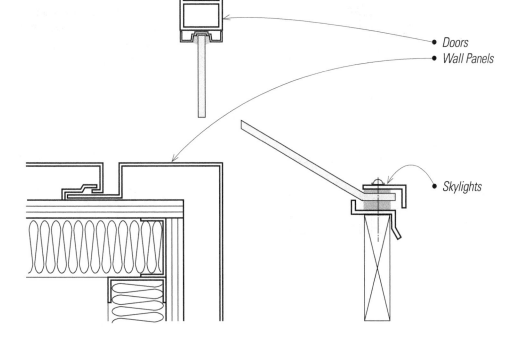

- *Store-front window system*

- *Doors*
- *Wall Panels*

- *Skylights*

Chapter 21: Masonry

Chapter 21 uses standards developed specifically for the IBC. There are references throughout the chapter to detailed standards for various materials, but there is no single standard as there are for other materials, such as concrete or aluminum. Masonry is defined as construction made up of units made of clay, shale, concrete, glass, gypsum, stone or other approved units bonded together. Ceramic tile is also classified by the code as masonry.

The units may be mortared or not, and may be grouted or not, depending upon the materials and the structural conditions. Typically, most masonry units are mortared together between units. There are some types of surface bonding mortar that hold the units together by covering the surface with a mixture of cementitious materials and glass fibers. Typical mortar is to be proportioned per Table 2103.8(1). These mortars consist of cementitious materials and fine aggregates. Cement lime mortar contains hydrated lime or lime putty.

Masonry is often grouted, especially in conditions where there are high seismic-resistive design criteria. Grout is similar in composition to concrete, but with smaller-size coarse aggregates.

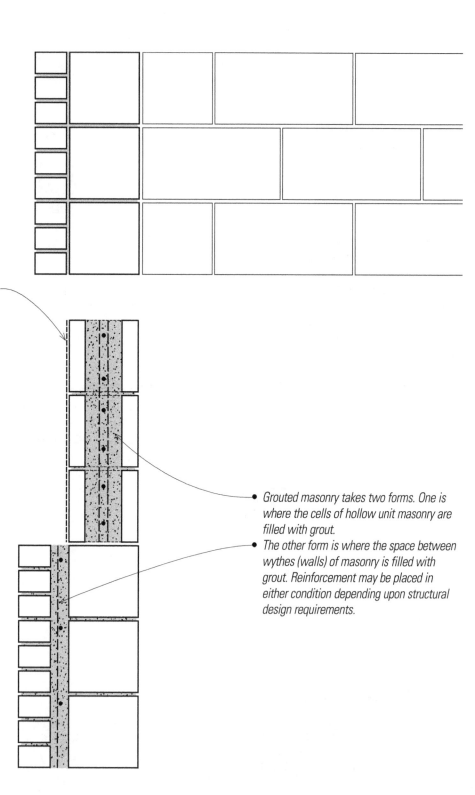

• Grouted masonry takes two forms. One is where the cells of hollow unit masonry are filled with grout.

• The other form is where the space between wythes (walls) of masonry is filled with grout. Reinforcement may be placed in either condition depending upon structural design requirements.

Construction

§2104 provides requirements for masonry construction practices. Requirements are contained both in the Code and in the referenced standard: ACI 530.1/ASCE 6/TMS 602, Masonry Structures by the American Concrete Institute.

- Bed joints are defined as the horizontal layer of mortar on which masonry units are laid.
- Head joints are the vertical mortar joints between masonry units.

- Bed and head joints are to be ³⁄₈" (9.5 mm) thick, except for starter courses over foundations, which may vary from ¼" to ¾" (6.4 mm to 19.1 mm).

- Solid units are to have fully mortared bed and head joints. Mortar is to be applied to the mortared faces by buttering before placement. Mortar is not to be forced into joints after placement.
- For hollow units, §2104.1.2.2 requires the block shells to have fully mortared bed joints over the face of the unit and the head joints to be mortared to equal the thickness of the shell.

- Note that masonry units, especially concrete masonry units, are described in nominal dimensions where the actual dimension of the unit is typically less. This is usually done to account for the thickness of joints in laying out modular dimensions.

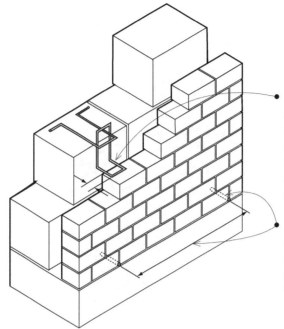

§2104.1.3 When wall ties are used, as in multiwythe or veneer construction, these ties are to be galvanized steel or stainless steel. They are to be embedded in the mortar joints to engage hollow units by at least ½" (12.7 mm) and embedded at least 1½" (38) into the mortar bed of solid units.

§2104.1.8 Weep holes are to be provided in the outside wythe of multiple wythe masonry walls at a maximum spacing of 33" (838). These weep holes are to be at least 3/16" (4.8 mm) in diameter.

§2104.3 and 2104.4 specify procedures for cold-weather and hot-weather construction. Wind factors are also included to address wind chill in cold weather and the drying effects of hot winds. The basic criteria are that the masonry units, as well as mortars and grouts, are to be kept within a temperature range of 20°F to 120°F (–6.7°C to 49°C). Also, wind speeds and temperatures must be monitored to verify that mortar does not dry out or begin to set prior to use in hot, windy weather.

§2104.5 specifies that when brick or shale units have a water-absorption rate exceeding 30 grams per 30 square inches (19.355 mm²) per minute, the units are to be wetted at the time of laying. This is to prevent the absorptive masonry from sucking water from the wet mortar and thus weakening the bond between mortar and masonry.

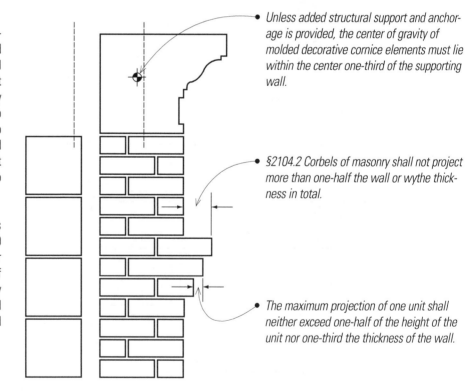

Unless added structural support and anchorage is provided, the center of gravity of molded decorative cornice elements must lie within the center one-third of the supporting wall.

§2104.2 Corbels of masonry shall not project more than one-half the wall or wythe thickness in total.

The maximum projection of one unit shall neither exceed one-half of the height of the unit nor one-third the thickness of the wall.

Seismic Design

Seismic design of masonry, as for other materials, depends on the anticipated lateral forces in the locale of the building. It is acceptable to use masonry for seismic-resistant systems, as long as the criteria for seismic design are met. The design standards are adopted by reference with some modifications made in the Code. The reference design standards are called out as "ACI 530/ASCE 5/TMS 402."

Design Methods

There are three alternative methods for design of masonry. All refer back to sections contained in ACI 530/ASCE 5/TMS 402. Two of the three methods are defined in §1602. They are:

- Allowable Stress Design, per §2107
- Strength Design, per §2108
- The third design method is Empirical Design, which is a "cookbook" approach based on best practices and field experience. It is limited to lower seismic hazard zones and may not be used in Seismic Design Categories C, E or F, or for seismic-force-resisting elements in Seismic Design Categories B or C. It also may not be used as part of lateral-force-resisting systems where basic wind speed exceeds 110 mph (79 m/s). It is described in §2109.

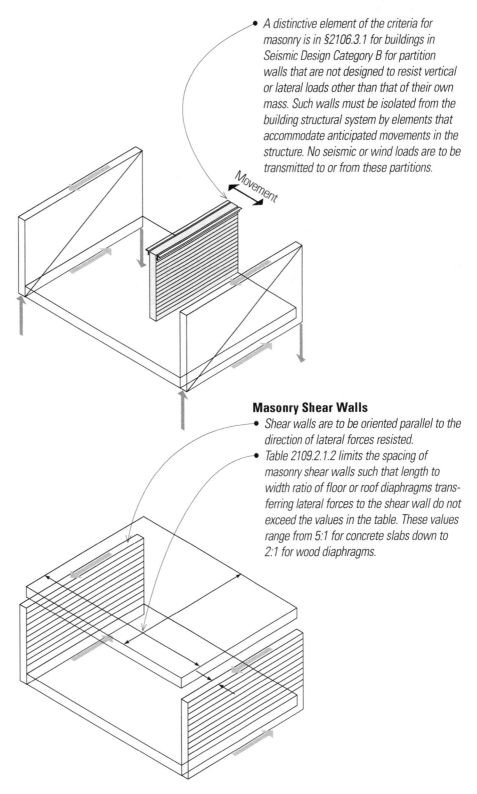

- A distinctive element of the criteria for masonry is in §2106.3.1 for buildings in Seismic Design Category B for partition walls that are not designed to resist vertical or lateral loads other than that of their own mass. Such walls must be isolated from the building structural system by elements that accommodate anticipated movements in the structure. No seismic or wind loads are to be transmitted to or from these partitions.

Masonry Shear Walls

- Shear walls are to be oriented parallel to the direction of lateral forces resisted.
- Table 2109.2.1.2 limits the spacing of masonry shear walls such that length to width ratio of floor or roof diaphragms transferring lateral forces to the shear wall do not exceed the values in the table. These values range from 5:1 for concrete slabs down to 2:1 for wood diaphragms.

MASONRY

Empirical Design

Masonry structural design may use several alternate methods for determining dimensions, connections and reinforcement. The empirical design method described in §2109 is the simplest and most direct. This method corresponds to the conventional framing provisions for wood-frame construction. It is based on customary field practices and familiar techniques. Its use is restricted to low to moderate seismic and wind-hazard areas and generally is used for buildings under 35' (10 668) in height. We will use this section to illustrate some typical provisions of masonry design. The compressive stresses for use of empirical design are contained in Table 2109.3.2. The composite strengths depend on the strengths of both the masonry units and the type of mortar used as well as whether units are grouted.

- Masonry walls must be laterally supported horizontally and vertically at intervals specified in Table 2109.4.1. This table specifies the maximum unsupported height or length of masonry walls in multiples of a wall's thickness.
- Lateral support may be provided by cross waits, pilasters or structural framing in the horizontal direction, and by floor or roof diaphragms in the vertical direction.

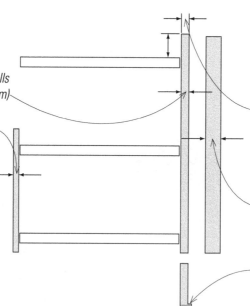

- A solid unit or fully grouted bearing wall can have an unsupported height or length 20 times its thickness. Thus an unbraced 8" (203) thick masonry bearing wall can be 13.33' (4063) in length or height [(8 × 20)/12].
- An exterior nonbearing masonry wall can have an unsupported height or length up to 18 times its thickness; an interior nonbearing masonry wall can have an unsupported height or length up to 36 times its thickness.

Thickness of Masonry

- §2109.5 states that masonry bearing walls are to be a minimum thickness of 8" (203 mm) for walls over one story high
- and 6" (152 mm) for walls of one story.

- §2109.5.4 requires unreinforced parapets to be at least 8" (203) thick and cannot extend higher than three times their thickness. Thus an 8" (203) parapet wall can be 24" (610) tall (8 × 3).

- Minimum thickness of stone rubble walls is 16" (406).

- Where a masonry wall of hollow units decreases in thickness, a course of solid masonry is required to transmit loads from face shells above to those below.

Masonry Bonding

§2109.6 requires either bond courses or wall ties to connect the two wythes of a multiple-wythe wall across the space between them. There are specific criteria for various methods:

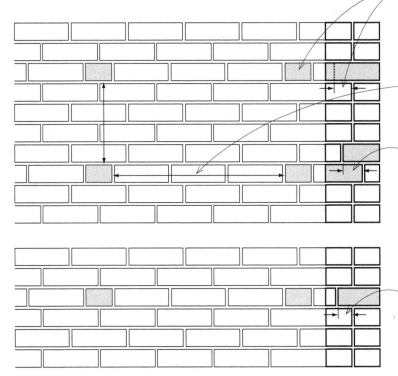

• *§2109.6.2.1 specifies that, where solid units are bonded by masonry headers, no less than 4% of the wall surface of each of the faces must be composed of headers that extend not less than 3" (76 mm) into the backing.*

• *Adjacent full-length headers can be no more than 24" (610) apart either vertically or horizontally.*

• *Where a single header does not extend through the wall, headers from the opposite side are to overlap at least 3" (76 mm) or the headers from opposite sides are to be covered by another header course overlapping the header below by at least 3" (76 mm).*

• *The design standard defines hollow masonry units.*

• *§2109.6.2.2 provides that, where hollow units make up the thickness of a wall and are bonded by masonry headers, the stretcher courses are to be bonded at vertical intervals not exceeding 34" (864) by lapping at least 3" (76) over the unit below or by lapping at vertical intervals not exceeding 17" (432) with units at least 50% greater in thickness than the units below.*

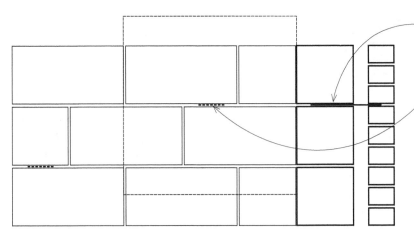

• *§2109.6.3.1 states that adjacent wythes may be bonded with wall ties of W2.8 wire (4.8 mm, MW 18) size or wire of equivalent stiffness embedded in the horizontal mortar joints.*

• *There is to be one such metal tie for every 4 ½ sf (0.42 m²) of wall area. Ties in alternate courses are to be staggered.*

• *The maximum vertical distance between ties may not exceed 24" (610 mm) and the horizontal distance may not exceed 36" (914 mm).*

• *At hollow units laid with cells vertical, rods or ties bent in a rectangular shape are to be used. In other walls the ties are to have bent ends with hooks no less than 2" (51) long. These criteria apply except when using adjustable wall ties, which require closer spacing for at least every 1.77 sf (0.164 m²) of wall area.*

Anchorage

§2109.7 requires masonry walls that intersect and depend on each other for lateral support be anchored or bonded together.

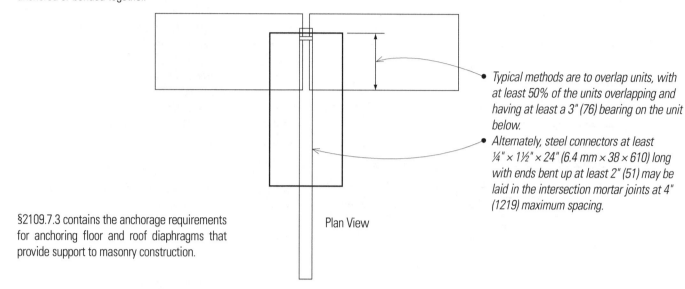

Plan View

§2109.7.3 contains the anchorage requirements for anchoring floor and roof diaphragms that provide support to masonry construction.

• Typical methods are to overlap units, with at least 50% of the units overlapping and having at least a 3" (76) bearing on the unit below.

• Alternately, steel connectors at least ¼" × 1½" × 24" (6.4 mm × 38 × 610) long with ends bent up at least 2" (51) may be laid in the intersection mortar joints at 4" (1219) maximum spacing.

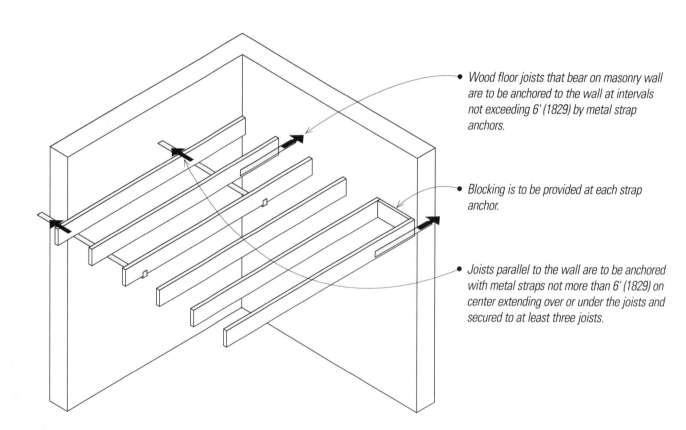

• Wood floor joists that bear on masonry wall are to be anchored to the wall at intervals not exceeding 6' (1829) by metal strap anchors.

• Blocking is to be provided at each strap anchor.

• Joists parallel to the wall are to be anchored with metal straps not more than 6' (1829) on center extending over or under the joists and secured to at least three joists.

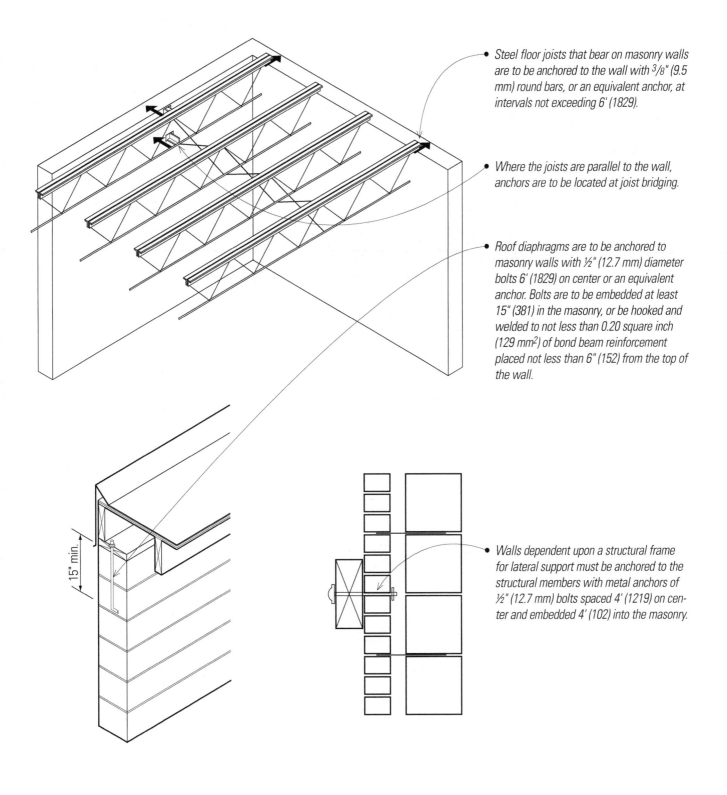

• Steel floor joists that bear on masonry walls are to be anchored to the wall with $^{3}/_{8}$" (9.5 mm) round bars, or an equivalent anchor, at intervals not exceeding 6' (1829).

• Where the joists are parallel to the wall, anchors are to be located at joist bridging.

• Roof diaphragms are to be anchored to masonry walls with ½" (12.7 mm) diameter bolts 6' (1829) on center or an equivalent anchor. Bolts are to be embedded at least 15" (381) in the masonry, or be hooked and welded to not less than 0.20 square inch (129 mm²) of bond beam reinforcement placed not less than 6" (152) from the top of the wall.

15" min.

• Walls dependent upon a structural frame for lateral support must be anchored to the structural members with metal anchors of ½" (12.7 mm) bolts spaced 4' (1219) on center and embedded 4' (102) into the masonry.

Glass Unit Masonry

§2110 covers the requirements for nonload-bearing glass unit masonry elements in both interior and exterior walls. They are not to be used in fire walls, party walls or fire partitions. The exception to this section allows glass unit masonry having a minimum fire-resistance rating of ¾ hour to be used as opening protectives or in fire partitions that are required to have a fire-resistance rating of 1 hour or less and do not enclose exit stairways or exit passageways.

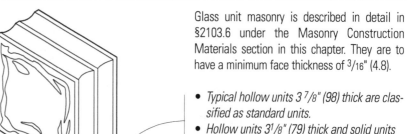

Glass unit masonry is described in detail in §2103.6 under the Masonry Construction Materials section in this chapter. They are to have a minimum face thickness of ³/₁₆" (4.8).

- *Typical hollow units 3 ⁷/₈" (98) thick are classified as standard units.*
- *Hollow units 3¹/₈" (79) thick and solid units 3" (76) thick are classified as thin units.*

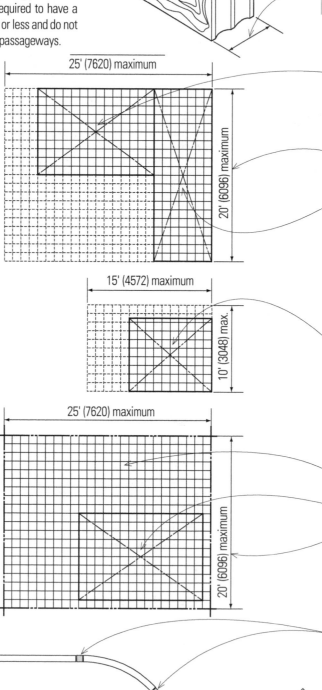

- *Exterior standard unit panel sizes are based on a 20 psf (958 N/m²) wind load and can be maximum of 144 sf (13.4 m²). Panel areas can be adjusted for other wind speeds based on Figure 2110.3.1.*
- *Maximum dimension of these panels at 25' (7620) in width of 20' (6096) in height.*

- *The overall dimension requirements must be read in concert with the panel area limitations. For example, maximizing heights will result in shorter walls to meet the area limitations. A 20' (6096) high wall can only be 7.2' (2194) high and still meet the criteria for maximum area per Figure 2110.3.1 with a 20 psf (958 N/m²) wind load.*

- *Exterior thin unit panels may be 85 sf (7.9 m²) maximum with a maximum width of 15' (4572) and a maximum height of 10' (3048). These panels may not be used in places where the wind pressure exceeds 20 psf (958 N/m²).*

- *The maximum size for interior standard-unit panels is 250 sf (23.2 m²).*
- *The maximum area for interior thin-unit panels is 150 sf (13.9 m²).*
- *The maximum dimension between structural supports is 25' (7620) in width or 20' (6096) in height. There is no differentiation between unit types for these dimensions.*

- *Where glass masonry units are laid up in curved patterns, the individual curved areas must conform to the area criteria for the appropriate unit type. However, additional structural supports are required where a curved section joins a straight section and at inflection points in multicurved walls.*

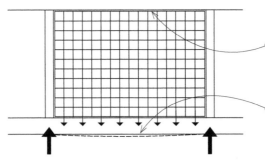

- Units panels are to be isolated from the adjoining structure so that in-plane loads are not transferred to the panels.

- Also, unit panels are to be supported on structural members having a deflection of less than $^1/_{600}$.

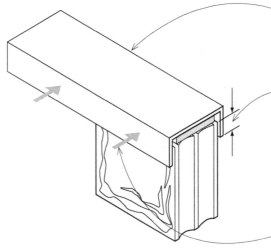

- Glass unit masonry panels are to be laterally supported along their tops and sides.
- Support is to be by panel anchors spaced not more than 16" (406) on center or by continuous channels with at least 1" (25.4 mm) overlap over the unit.

- The channels are to be oversize to allow expansion material in the opening and packing or sealant between the face of the unit and the channel. Panels are to have expansion joints along the top and sides of panels with at least $^3/_8$" (9.5 mm) allowance for expansion.
- Channels are to resist applied loads or a minimum of 200 pounds per lineal foot (2919 N/m) of panel, whichever is greater.

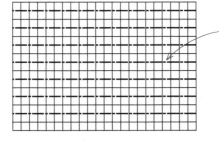

- Units are to be mortared together with mortar complying with §2103.8.
- Unit panels are to have reinforcing in the horizontal mortar bed joints at not more than 16" (406) on center. Reinforcing is also required above and below openings in the panel. The reinforcing is to be a ladder type with two parallel wires of size W1.7 and welded cross wires of W 1.7, and are to be lapped a minimum of 6" (152) at splices.

Chapter 22: Steel

Chapter 22 divides steel construction materials into three broad classes. The first is cold-formed steel, made up of members bent from steel sheets or strips, including roof decks, floor and wall panels, studs, floor joists and other structural members. The design of cold-formed members is governed by the American Iron and Steel Institute (AISI) Specifications for the Design of Cold-Formed Steel Structural Members.

§2210 governs the design of light-framed cold-formed steel walls. These requirements are analogous to those for conventional wood-framed construction in that the Code sets design and configuration criteria for typical installations that are deemed to be code compliant when they meet the prescriptive criteria.

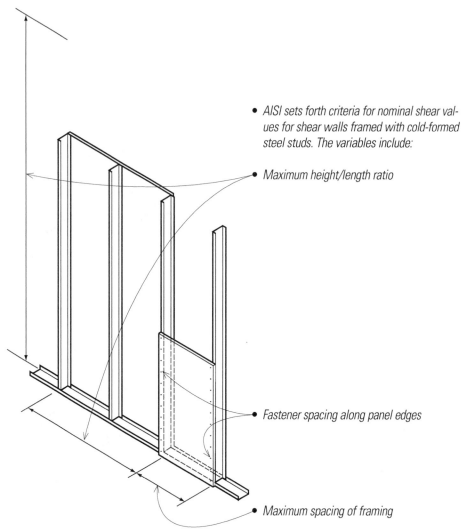

- AISI sets forth criteria for nominal shear values for shear walls framed with cold-formed steel studs. The variables include:

- Maximum height/length ratio

- Fastener spacing along panel edges

- Maximum spacing of framing

The second category is for steel joists, made of combinations of hot-rolled or cold-formed solid or open-web sections. Design of steel joists is governed by specifications from the Steel Joist Institute (SJI).

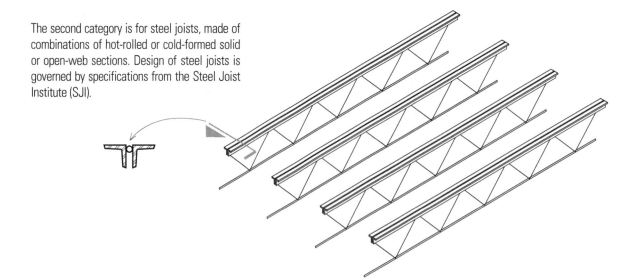

The third broad category is for structural steel members, which are rolled steel shapes, usually hot rolled, and not otherwise falling into the other two categories. The design of these members falls under the specifications of the American Institute of Steel Construction (AISC) 360. There are separate sets of specifications for structural steel design depending upon the design method used, load and resistance factor design, or allowable stress design.

Bolts and welding of connections are governed by the applicable specifications for each type of steel construction. As for other materials, steel structures are to comply with seismic and wind design requirements where applicable.

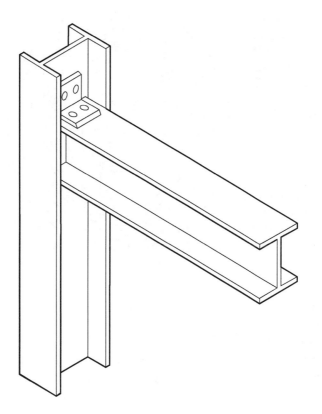

Chapter 23: Wood

Chapter 23 governs the materials, design and construction of wood members and their fasteners. Wood structures are to be designed using one of the approved methods: allowable stress design per §2304, §2305, and §2306; load and resistance factor design per §2304, §2305 and §2307; or using conventional light-frame construction provisions per §2304 and §2308. Wood-frame construction is very prevalent, especially in home-building in the United States. The Code contains a prescriptive approach for wood-frame construction as it is typically practiced. These criteria are termed as conventional light-frame wood construction. This is defined as a system where the primary structure is made up of repetitive wood framing members. We will use these conventional construction provisions for our illustrations of code-compliant wood-frame construction. We have not duplicated the illustrations contained in the code, as they are self explanatory. Note that our discussion does not include the exceptions to the prescriptive requirements. Code-compliant design must take all of the code requirements and exceptions under consideration for the specific conditions of the project.

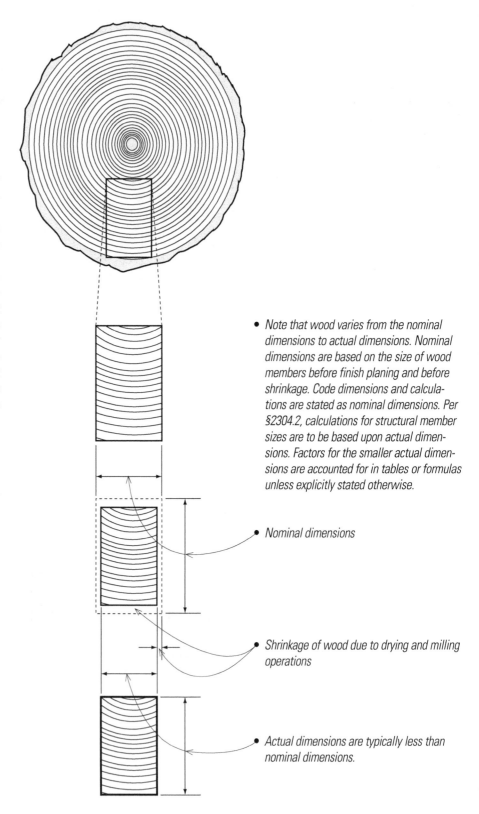

- *Note that wood varies from the nominal dimensions to actual dimensions. Nominal dimensions are based on the size of wood members before finish planing and before shrinkage. Code dimensions and calculations are stated as nominal dimensions. Per §2304.2, calculations for structural member sizes are to be based upon actual dimensions. Factors for the smaller actual dimensions are accounted for in tables or formulas unless explicitly stated otherwise.*

- *Nominal dimensions*

- *Shrinkage of wood due to drying and milling operations*

- *Actual dimensions are typically less than nominal dimensions.*

Minimum Standards and Quality

§2303 specifies minimum standards and quality for lumber, sheathing, siding and other wood-based products. Lumber is to be graded per the standards of the Department of Commerce (DOC) American Softwood Lumber Standard, PS 20.

Lumber is to be stamped to indicate that it has been reviewed and graded according to species and grade. Calculations are based upon the stress capabilities of the wood depending upon the grade and the lumber species.

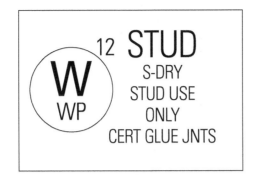

Wood is also classified as to its natural durability. The heartwood for species such as redwood and cedar are classified as decay-resistant. Redwood and eastern red cedar are also classified as termite-resistant. The Code requires the use of naturally durable woods or pressure-treated woods in certain situations, especially in conditions where wood members are close to adjacent soil.

Each of the criteria is specific to the type of wood materials they cover. Wood structural panels are to be per requirements of standards DOC PS 1 or PS 2. Structural glued laminated timbers are to conform to ASTM D 5055.

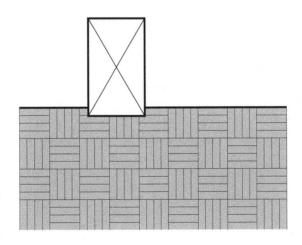

WOOD

General Construction Requirements

§2304 contains the construction requirements for structural elements and systems built of wood or wood-based products.

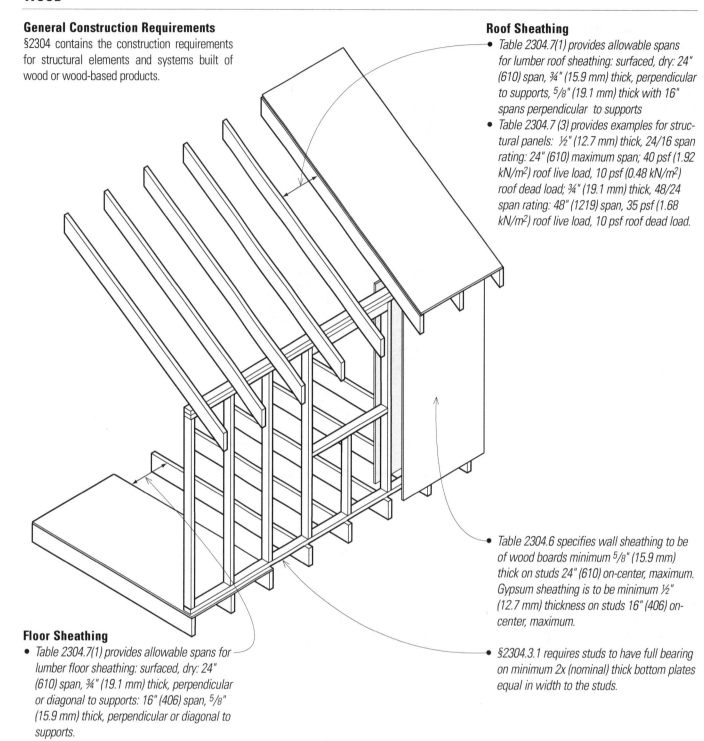

Roof Sheathing

- Table 2304.7(1) provides allowable spans for lumber roof sheathing: surfaced, dry: 24" (610) span, ¾" (15.9 mm) thick, perpendicular to supports, ⁵/₈" (19.1 mm) thick with 16" spans perpendicular to supports
- Table 2304.7 (3) provides examples for structural panels: ½" (12.7 mm) thick, 24/16 span rating: 24" (610) maximum span; 40 psf (1.92 kN/m²) roof live load, 10 psf (0.48 kN/m²) roof dead load; ¾" (19.1 mm) thick, 48/24 span rating: 48" (1219) span, 35 psf (1.68 kN/m²) roof live load, 10 psf roof dead load.

- Table 2304.6 specifies wall sheathing to be of wood boards minimum ⁵/₈" (15.9 mm) thick on studs 24" (610) on-center, maximum. Gypsum sheathing is to be minimum ½" (12.7 mm) thickness on studs 16" (406) on-center, maximum.

- §2304.3.1 requires studs to have full bearing on minimum 2x (nominal) thick bottom plates equal in width to the studs.

Floor Sheathing

- Table 2304.7(1) provides allowable spans for lumber floor sheathing: surfaced, dry: 24" (610) span, ¾" (19.1 mm) thick, perpendicular or diagonal to supports: 16" (406) span, ⁵/₈" (15.9 mm) thick, perpendicular or diagonal to supports.
- Table 2304.7 (3) provides examples for structural panels: ½" (12.7 mm) thick, 24/16 span rating: 16" (406) span, 100 psf (4.79 kN/m²) total floor load (1/360) deflection: ¾" (19.1 mm) thick, 48/24 span rating: 24" (610) span, 65 psf (3.11 kN/m²) total floor load (1/360) deflection.

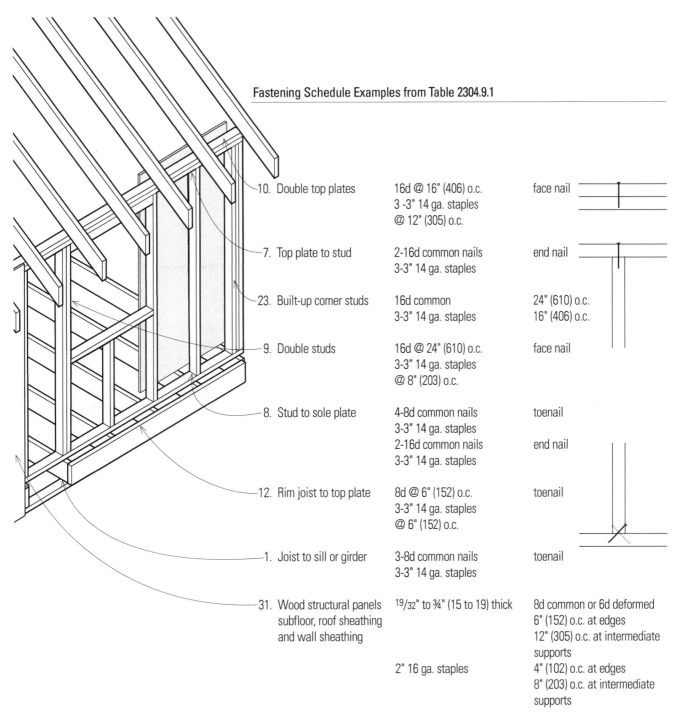

Fastening Schedule Examples from Table 2304.9.1

10.	Double top plates	16d @ 16" (406) o.c. 3 -3" 14 ga. staples @ 12" (305) o.c.	face nail
7.	Top plate to stud	2-16d common nails 3-3" 14 ga. staples	end nail
23.	Built-up corner studs	16d common 3-3" 14 ga. staples	24" (610) o.c. 16" (406) o.c.
9.	Double studs	16d @ 24" (610) o.c. 3-3" 14 ga. staples @ 8" (203) o.c.	face nail
8.	Stud to sole plate	4-8d common nails 3-3" 14 ga. staples 2-16d common nails 3-3" 14 ga. staples	toenail end nail
12.	Rim joist to top plate	8d @ 6" (152) o.c. 3-3" 14 ga. staples @ 6" (152) o.c.	toenail
1.	Joist to sill or girder	3-8d common nails 3-3" 14 ga. staples	toenail
31.	Wood structural panels subfloor, roof sheathing and wall sheathing	$19/32$" to ¾" (15 to 19) thick 2" 16 ga. staples	8d common or 6d deformed 6" (152) o.c. at edges 12" (305) o.c. at intermediate supports 4" (102) o.c. at edges 8" (203) o.c. at intermediate supports

Note that there are manufactured, tested and approved joist hangers and framing hangers that perform the functions of the fasteners noted in Table 2304.9.1. These are acceptable if they are listed and approved, and used in the manner intended.

WOOD

Protection against Decay and Termites

§2304.11 requires protection against decay and termites by the use of naturally durable or preservative-treated wood in the following conditions.

- §2304.11.2.6 requires protection for wood siding that is less than 6" (152) to exposed earth except where the material is naturally durable or preservative-treated wood.
- §2304.11.2.2 requires protection for all framing, including wood sheathing, that is less than 8" (203) from exposed earth.

§2304.11.2.5 requires a ½" (12.7 mm) air space on the top, sides and end of girders entering exterior masonry or concrete walls.

§2304.11.2.1: If joists and subflooring are closer than 18" (457) to exposed ground in crawl space, or if girders are closer than 12" (305) to exposed ground, then the floor assembly including posts, girders, joists and subfloor must be of naturally durable or preservative-treated wood.

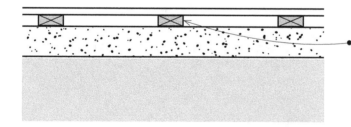

§2304.11.2.4 requires similar protection for sleepers and sills on a concrete slab that is in direct contact with the earth.

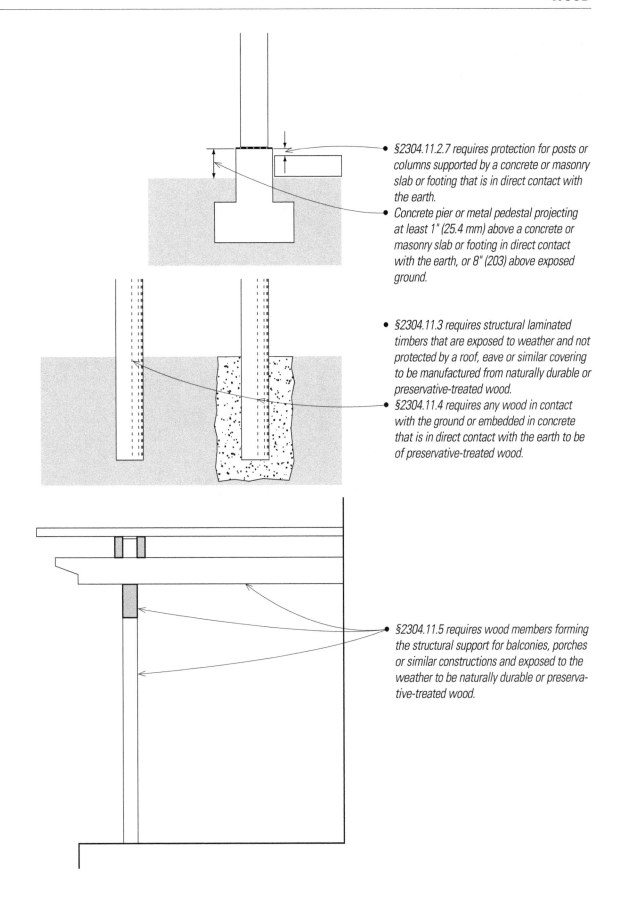

• *§2304.11.2.7 requires protection for posts or columns supported by a concrete or masonry slab or footing that is in direct contact with the earth.*

• *Concrete pier or metal pedestal projecting at least 1" (25.4 mm) above a concrete or masonry slab or footing in direct contact with the earth, or 8" (203) above exposed ground.*

• *§2304.11.3 requires structural laminated timbers that are exposed to weather and not protected by a roof, eave or similar covering to be manufactured from naturally durable or preservative-treated wood.*

• *§2304.11.4 requires any wood in contact with the ground or embedded in concrete that is in direct contact with the earth to be of preservative-treated wood.*

• *§2304.11.5 requires wood members forming the structural support for balconies, porches or similar constructions and exposed to the weather to be naturally durable or preservative-treated wood.*

Conventional Light-Frame Construction

The provisions of §2308 are intended for use in residential and light commercial construction. Buildings constructed using these provisions are typically Type V-B construction. They may also be made to conform to Type V-A requirements with the addition of fire-resistant building materials. See Table 503 for limitations on where this building type may be used in healthcare occupancies. Buildings are limited to three stories in height, bearing wall heights to 10' (3048), roof trusses and rafters are to span no more than 40' (12 192), and loads are limited. Braced walls are to be placed in the structure at maximum intervals to provide lateral bracing against lateral loads imposed by wind or seismic forces.

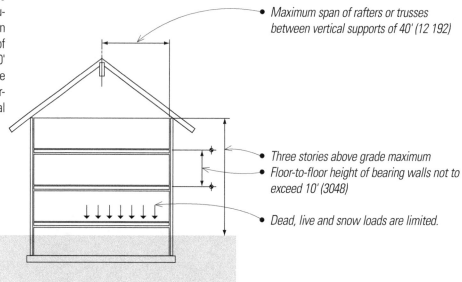

- *Maximum span of rafters or trusses between vertical supports of 40' (12 192)*

- *Three stories above grade maximum*
- *Floor-to-floor height of bearing walls not to exceed 10' (3048)*

- *Dead, live and snow loads are limited.*

Braced Wall Lines

§2308.3 requires braced wall lines as prescribed in §2308.9.3 and spaced not more than 35' (10 668) on center in both the longitudinal and transverse directions of each story.

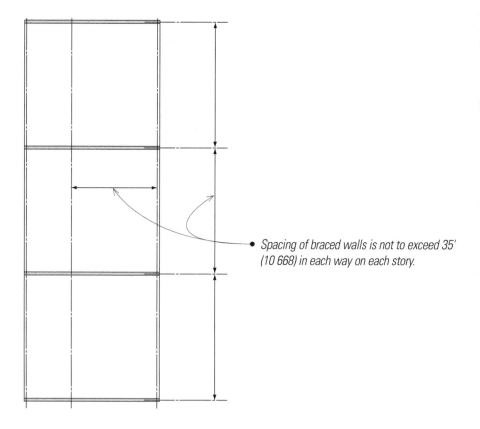

- *Spacing of braced walls is not to exceed 35' (10 668) in each way on each story.*

§2308.3.2 prescribes braced wall panel connections to transfer forces from roofs and floors to braced wall panels and from braced walls in upper stories to braced wall panels in stories below.

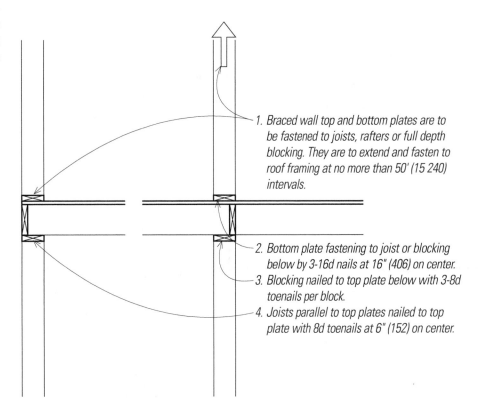

1. Braced wall top and bottom plates are to be fastened to joists, rafters or full depth blocking. They are to extend and fasten to roof framing at no more than 50' (15 240) intervals.

2. Bottom plate fastening to joist or blocking below by 3-16d nails at 16" (406) on center.
3. Blocking nailed to top plate below with 3-8d toenails per block.
4. Joists parallel to top plates nailed to top plate with 8d toenails at 6" (152) on center.

Foundation Plates or Sills

§2308.6 refers to the design and construction of foundations and footings prescribed in Chapter 18.

- Sills or plates are to be anchored to the footing with ½" (12.7 mm) diameter minimum steel bolts or approved anchors.
- Bolts are to extend 7" (178) into concrete or masonry and be spaced not more than 6' (1829) apart.
- Every plate must have at least two anchors; end anchors are to be no more than 12" (305) and no less than 4" (102) from the end of the member.

Girders

§2308.7 specifies that girders for single-story construction are to be minimum 4 × 6 (102 × 152) with a 6' (1829) maximum span on maximum 8' (2438) centers.

- Built-up girders made from 2x members are to be per Tables 2308.9.5 or 2308.9.6.

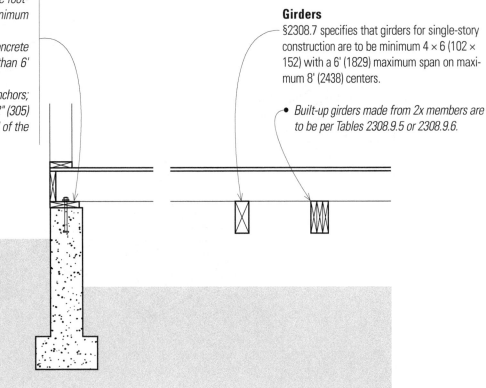

Floor Joists

§2308.8 prescribes that joists are to be per Table 2308.8 (1) or (2).

- For example, in residential areas, with 1/360 maximum deflection, a live load of 40 psf. 10 psf dead load, 16" (406) spacing, a Douglas Fir-Larch #1 2 × 10 joist will span 16'- 5" (5004).

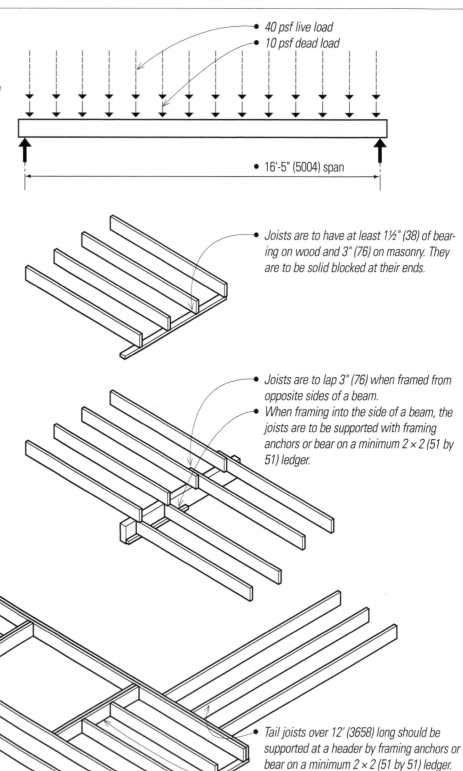

- 40 psf live load
- 10 psf dead load
- 16'-5" (5004) span

- Joists are to have at least 1½" (38) of bearing on wood and 3" (76) on masonry. They are to be solid blocked at their ends.

- Joists are to lap 3" (76) when framed from opposite sides of a beam.
- When framing into the side of a beam, the joists are to be supported with framing anchors or bear on a minimum 2 × 2 (51 by 51) ledger.

- Tail joists over 12' (3658) long should be supported at a header by framing anchors or bear on a minimum 2 × 2 (51 by 51) ledger.
- Header joists are to be doubled when the header span exceeds 4' (1219).

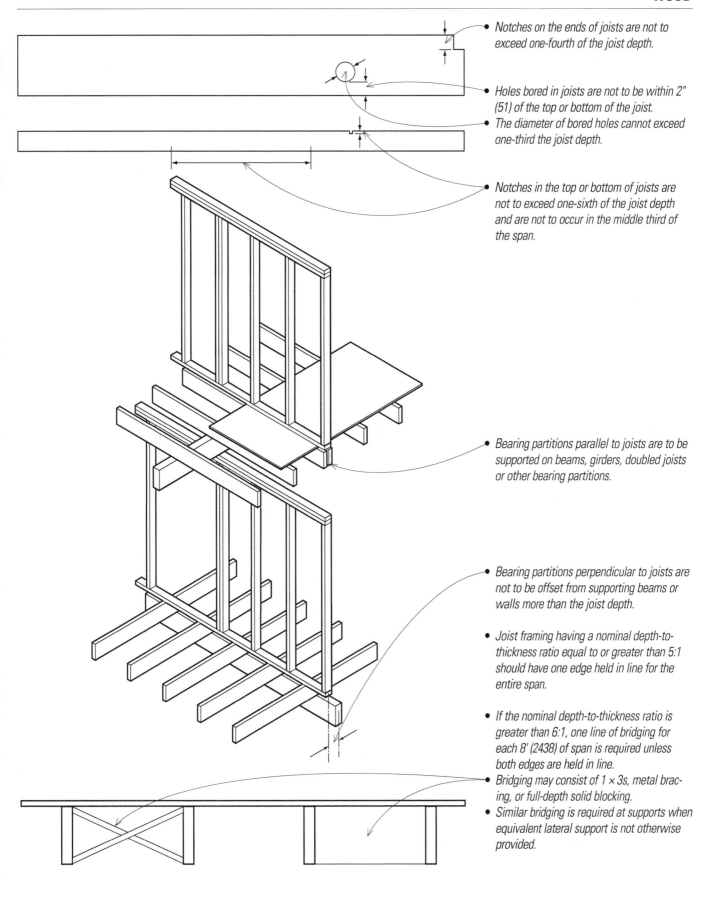

- Notches on the ends of joists are not to exceed one-fourth of the joist depth.

- Holes bored in joists are not to be within 2" (51) of the top or bottom of the joist.
- The diameter of bored holes cannot exceed one-third the joist depth.

- Notches in the top or bottom of joists are not to exceed one-sixth of the joist depth and are not to occur in the middle third of the span.

- Bearing partitions parallel to joists are to be supported on beams, girders, doubled joists or other bearing partitions.

- Bearing partitions perpendicular to joists are not to be offset from supporting beams or walls more than the joist depth.

- Joist framing having a nominal depth-to-thickness ratio equal to or greater than 5:1 should have one edge held in line for the entire span.

- If the nominal depth-to-thickness ratio is greater than 6:1, one line of bridging for each 8' (2438) of span is required unless both edges are held in line.
- Bridging may consist of 1 × 3s, metal bracing, or full-depth solid blocking.
- Similar bridging is required at supports when equivalent lateral support is not otherwise provided.

Wall Framing

§2308.9 requires studs to be placed with their wide dimension perpendicular to the wall. The size, height and spacing of wood studs are called out in Table 2308.9.1.

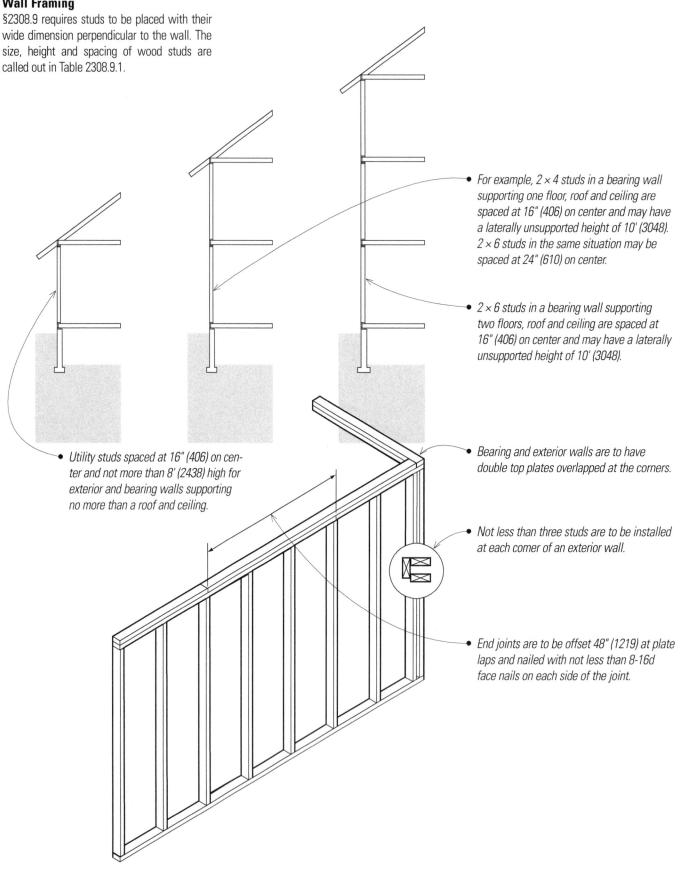

• For example, 2 × 4 studs in a bearing wall supporting one floor, roof and ceiling are spaced at 16" (406) on center and may have a laterally unsupported height of 10' (3048). 2 × 6 studs in the same situation may be spaced at 24" (610) on center.

• 2 × 6 studs in a bearing wall supporting two floors, roof and ceiling are spaced at 16" (406) on center and may have a laterally unsupported height of 10' (3048).

• Utility studs spaced at 16" (406) on center and not more than 8' (2438) high for exterior and bearing walls supporting no more than a roof and ceiling.

• Bearing and exterior walls are to have double top plates overlapped at the corners.

• Not less than three studs are to be installed at each corner of an exterior wall.

• End joints are to be offset 48" (1219) at plate laps and nailed with not less than 8-16d face nails on each side of the joint.

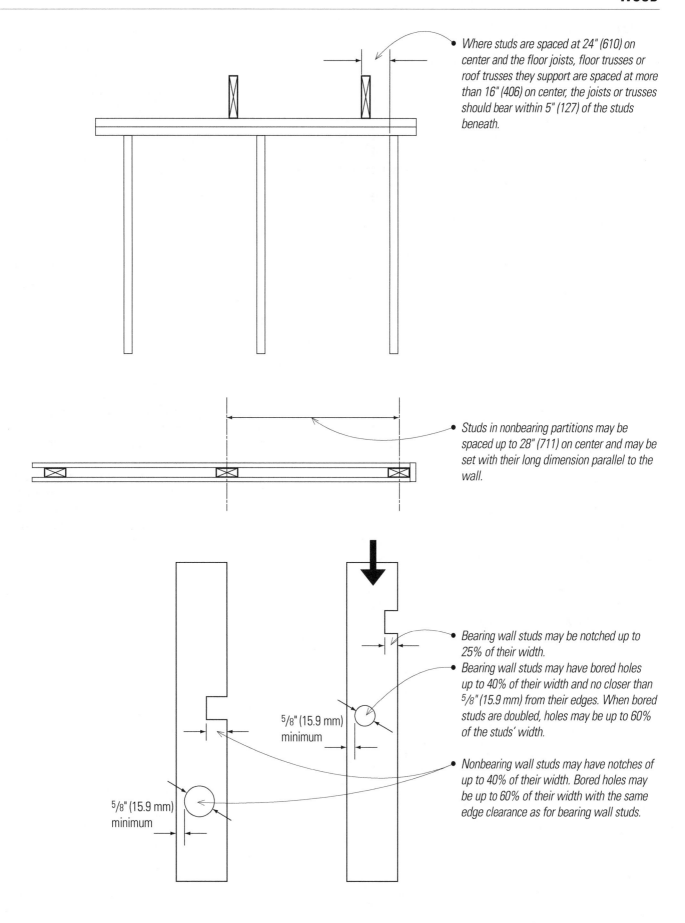

Where studs are spaced at 24" (610) on center and the floor joists, floor trusses or roof trusses they support are spaced at more than 16" (406) on center, the joists or trusses should bear within 5" (127) of the studs beneath.

Studs in nonbearing partitions may be spaced up to 28" (711) on center and may be set with their long dimension parallel to the wall.

$^5/8$" (15.9 mm) minimum

$^5/8$" (15.9 mm) minimum

Bearing wall studs may be notched up to 25% of their width.

Bearing wall studs may have bored holes up to 40% of their width and no closer than $^5/8$" (15.9 mm) from their edges. When bored studs are doubled, holes may be up to 60% of the studs' width.

Nonbearing wall studs may have notches of up to 40% of their width. Bored holes may be up to 60% of their width with the same edge clearance as for bearing wall studs.

Bracing

§2308.9.3 requires that bracing be provided per Table 2308.9.3(1), depending on the type of bracing, seismic design category and height of the building.

- *Braces are to be collinear or offset by no more than 4' (1219) across the building.*

• *Braced wall panels are to be located at each end of a building and no more than 25' (7620) on center.*

• *Braced wall panels should start no more than 8' (2438) from each end of a braced wall line.*

• *Braced wall panels are to be located at each end of a building and no more than 25' (7620) on center.*

• *In Seismic Design Category C, the total length of braced wall panels along a braced wall line in the first story of a two-story building, or the second or third story of a three-story building, should not be less than 25% of building length.*

• *In Seismic Design Category C, the total length of braced wall panels along a braced wall line in the first story of a three-story building should not be less than 40% of building length.*

§2308.9.3 provides prescriptive requirements for eight types of braced wall panels. A typical braced panel, used in many types of construction and allowable in all bracing conditions, is wood structural panel sheathing with plywood or oriented strand board. Other types of bracing include diagonal boards, fiberboard sheathing panels, gypsum board, particleboard and portland cement plaster.

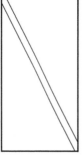
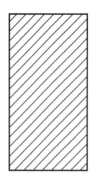

Cripple Walls

§2308.9.4 specifies that cripple walls, walls extending from the top of the foundation to the framing of the lowest floor level, are to be of the same size stud as the walls above. They must be more than 14" (356) high to use studs, or if not, must be solid blocking. If more than 14" (356) high are to be braced per Table 2308.9.3(1) as for any other braced wall panel.

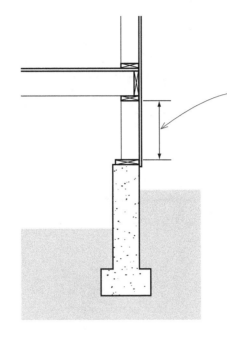

- *Nailing at the foundation plate and the top plate is not to exceed 6" (152) on center.*
- *For example, nailing for $^3/_8$" (9.5 mm) plywood per Table 2304.9.1 would be 6d nails, 6" (152) on center at the edges and 12" (305) on center in the field of the panel.*

Headers

§2308.9.5.1 requires that headers over openings in exterior walls be provided per Table 2308.9.5. This table is to be used for buildings of one or two stories.

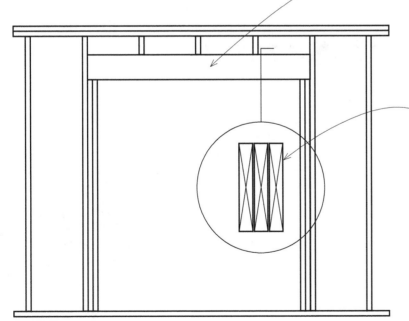

- *Headers are to be of 2 × material set on edge and nailed together per Table 2304.9.1 or lumber of equivalent size.*

- *For example, for a 36' (10973) wide building with a 30 psf (1.44 kN/m²) snow load, a header supporting the roof, ceiling and one clear span floor, a 5'-1" (1549) opening would require a header of either 2-2 × 12s with two jack studs supporting each end.*
- *Using three 2xs would require a wall depth of at least 4½" (114) to allow the jack studs to carry all three members of the built-up header.*

WOOD

Roof and Ceiling Framing

§2308.10 contains the criteria for roofs having slopes exceeding 3:12.

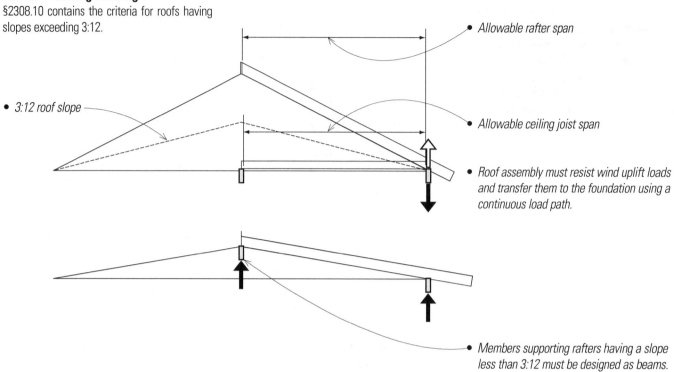

- Allowable rafter span

- 3:12 roof slope

- Allowable ceiling joist span

- Roof assembly must resist wind uplift loads and transfer them to the foundation using a continuous load path.

- Members supporting rafters having a slope less than 3:12 must be designed as beams.

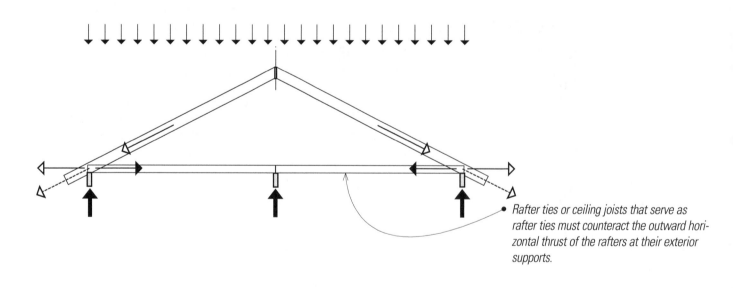

- Rafter ties or ceiling joists that serve as rafter ties must counteract the outward horizontal thrust of the rafters at their exterior supports.

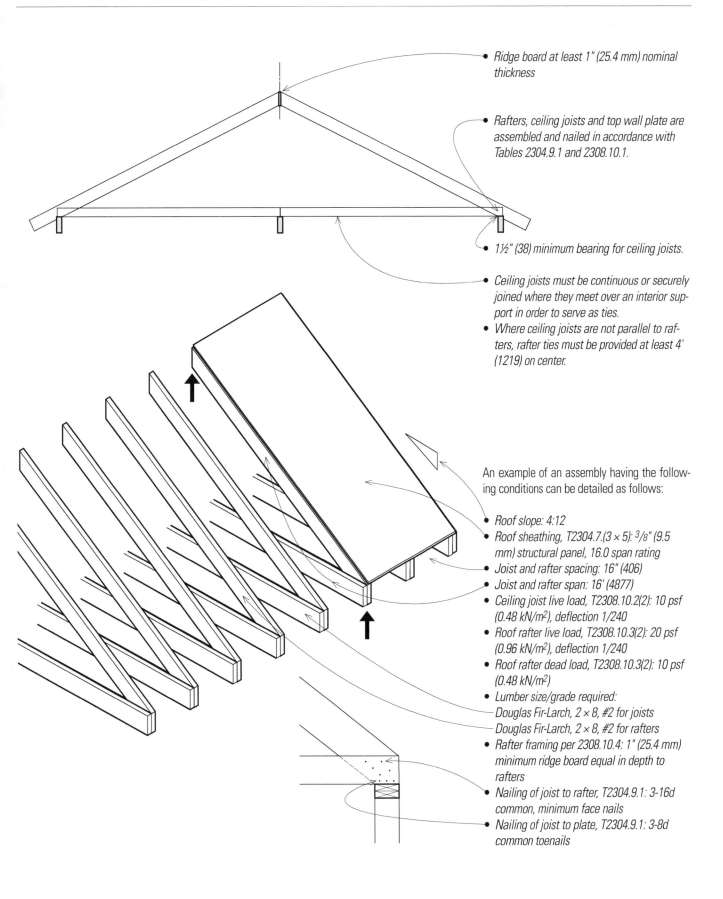

- Ridge board at least 1" (25.4 mm) nominal thickness

- Rafters, ceiling joists and top wall plate are assembled and nailed in accordance with Tables 2304.9.1 and 2308.10.1.

- 1½" (38) minimum bearing for ceiling joists.

- Ceiling joists must be continuous or securely joined where they meet over an interior support in order to serve as ties.
- Where ceiling joists are not parallel to rafters, rafter ties must be provided at least 4' (1219) on center.

An example of an assembly having the following conditions can be detailed as follows:

- Roof slope: 4:12
- Roof sheathing, T2304.7.(3 × 5): ³/₈" (9.5 mm) structural panel, 16.0 span rating
- Joist and rafter spacing: 16" (406)
- Joist and rafter span: 16' (4877)
- Ceiling joist live load, T2308.10.2(2): 10 psf (0.48 kN/m²), deflection 1/240
- Roof rafter live load, T2308.10.3(2): 20 psf (0.96 kN/m²), deflection 1/240
- Roof rafter dead load, T2308.10.3(2): 10 psf (0.48 kN/m²)
- Lumber size/grade required:
  Douglas Fir-Larch, 2 × 8, #2 for joists
  Douglas Fir-Larch, 2 × 8, #2 for rafters
- Rafter framing per 2308.10.4: 1" (25.4 mm) minimum ridge board equal in depth to rafters
- Nailing of joist to rafter, T2304.9.1: 3-16d common, minimum face nails
- Nailing of joist to plate, T2304.9.1: 3-8d common toenails

Chapter 24: Glass and Glazing

Chapter 24 contains the provisions governing the materials, design and construction of glass, light-transmitting ceramic, and light-transmitting plastic panels used for glazing. The requirements apply to vertical and sloped applications such as for windows and skylights.

§2403 requires that glass bear a manufacturer's mark on each pane designating the thickness and type of glass or glazing material.

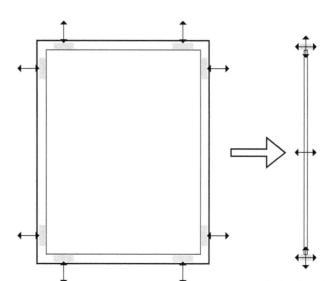

- *Glass is to be firmly supported in such a manner that the deflection of a pane of glass in a direction perpendicular to the pane must not exceed 1/175 of the glass edge length or ¾" (19.1 mm), whichever is less when subjected to either positive or negative combined loads per §1605, such as dead loads, live loads, wind loads, snow loads or seismic loads.*

- *For load calculations, §2404 assumes that the glass is supported on all four sides. Analysis or test data prepared by a registered design professional is required to verify that glass not supported on four sides will meet the Code's safety criteria. The wind loads are to be calculated in accordance with §1609. This section in turn references Chapter 6 of ASCE 7.*

 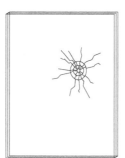 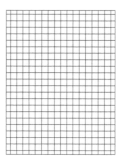

- *Tempered glass is heat-treated to strengthen it. The treatment also makes the glass panes shatter into small granular pieces that are much less sharp and dangerous than shards from untreated glass. However, since the entire pane shatters, protection or limitation of the panes' sizes is required by the Code.*
- *Laminated glass is made up of two panes of glass with a plastic layer between them. Glass fragments are contained by the plastic in the event of breakage of the glass panes, thus not having the panels fall out of their frames or rain glass pieces on the area below.*
- *Wired glass has pieces of wire embedded in the glass that act to contain glass fragments and keep the pane in place if the glass breaks.*

Sloped Glazing and Skylights

§2405 applies where glazing is installed at a slope more than 15° (0.26 rad) from the vertical plane. These requirements apply to all types of glazing materials, but the detailed requirements focus on the use of various types of glass. We will illustrate basic criteria and exceptions used in typical skylight designs glazed with glass. Because skylights often occur above walkways or occupied spaces, the Code places special safety requirements on skylight glazing to protect building occupants. Special glazing materials or protective measures are required.

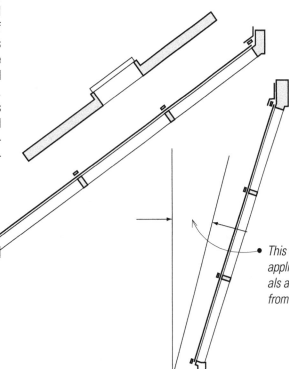

- *This section on sloped glazing and skylights applies to the installation of glazing materials at a slope greater than 15° (0.26 rad) from a vertical plane.*

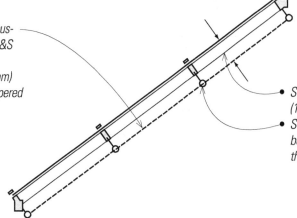

- *§2405.3 requires screening of noncombustible material not thinner than No. 12 B&S gauge [0.0808" or 2 mm] with mesh no larger than 1" by 1" (25.4 mm by 25.4 mm) below heat-strengthened and fully tempered glass in monolithic glazing systems.*

- *Screen should be located a maximum of 4" (102) below the glazing.*
- *Screening should be securely fastened and be able to adequately support the weight of the falling glass.*

- *This screening is also required where heat-strengthened glass, fully tempered glass or wired glass is used in multiple-layer glazing systems.*

- *These screens are intended to protect occupants below from falling chunks of glass, but they are not often seen in practice, due to use of allowable glazing materials or use of one of the five exceptions to §2405.3.*

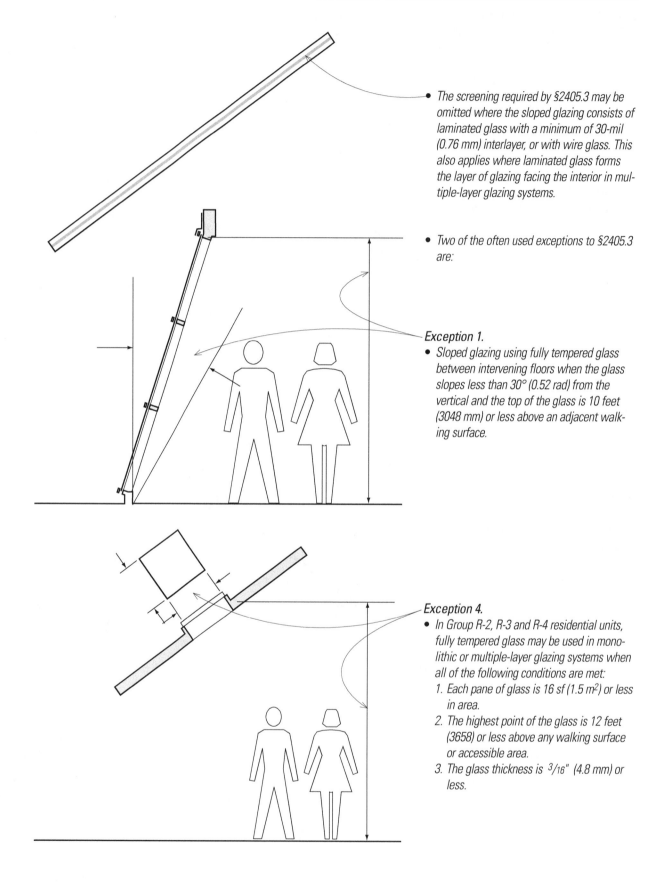

• The screening required by §2405.3 may be omitted where the sloped glazing consists of laminated glass with a minimum of 30-mil (0.76 mm) interlayer, or with wire glass. This also applies where laminated glass forms the layer of glazing facing the interior in multiple-layer glazing systems.

• Two of the often used exceptions to §2405.3 are:

Exception 1.
• Sloped glazing using fully tempered glass between intervening floors when the glass slopes less than 30° (0.52 rad) from the vertical and the top of the glass is 10 feet (3048 mm) or less above an adjacent walking surface.

Exception 4.
• In Group R-2, R-3 and R-4 residential units, fully tempered glass may be used in monolithic or multiple-layer glazing systems when all of the following conditions are met:
  1. Each pane of glass is 16 sf (1.5 m²) or less in area.
  2. The highest point of the glass is 12 feet (3658) or less above any walking surface or accessible area.
  3. The glass thickness is ³/₁₆" (4.8 mm) or less.

Safety Glazing

A major design consideration is the use of safety glazing, primarily tempered glass, laminated glass or impact-resistant plastic under certain circumstances. The basic criteria for safety glazing contained in §2406 may be summarized as requiring safely glazing for glass panels subject to human impact under normal conditions of use.

Panes of glass in showers or baths, along corridors, at storefronts next to sidewalks or in glass handrails need safety glazing. The criteria for safety glazing are set forth in the 16 CFR 1201 by the Consumer Product Safety Commission. Under most circumstances, this glazing is basically the same type as for skylights, tempered glass or laminated glass. Wired glass does not meet the CPSC criteria, but its use is accepted in fire doors where wire glass is often used, so it is acceptable per an exception. There are ten specific conditions where safety glazing is required. We will illustrate all ten, along with some common exceptions that are often applied:

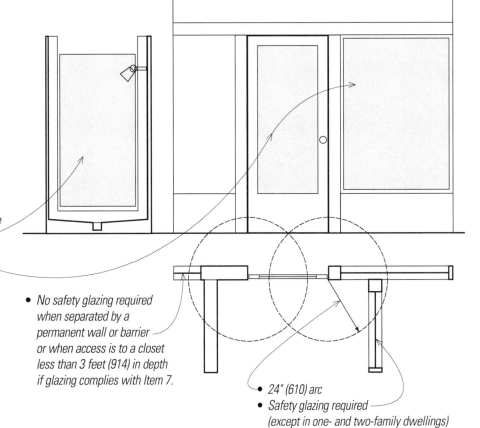

1. Glazing in swinging doors.
2. Glazing in fixed and sliding panels of sliding doors and bifold closet doors with mirrors
3. Glazing in storm doors
4. Glazing in unframed swinging doors

5. Glazing in doors for hot tubs, saunas, showers and bathtubs, as well as compartment glazing that is less than 60" (1524) above the floor of the enclosure.

6. Glazing in fixed or operable panels where the nearest edge of the glazing is within a 24" (610) arc of either vertical edge of the door in a closed position and the bottom edge of the glazing is less than 60" (1524) above the walking surface. This does not apply to glazing perpendicular to the plane of the closed door in residential dwelling units.

- No safety glazing required when separated by a permanent wall or barrier or when access is to a closet less than 3 feet (914) in depth if glazing complies with Item 7.

- 24" (610) arc
- Safety glazing required (except in one- and two-family dwellings)

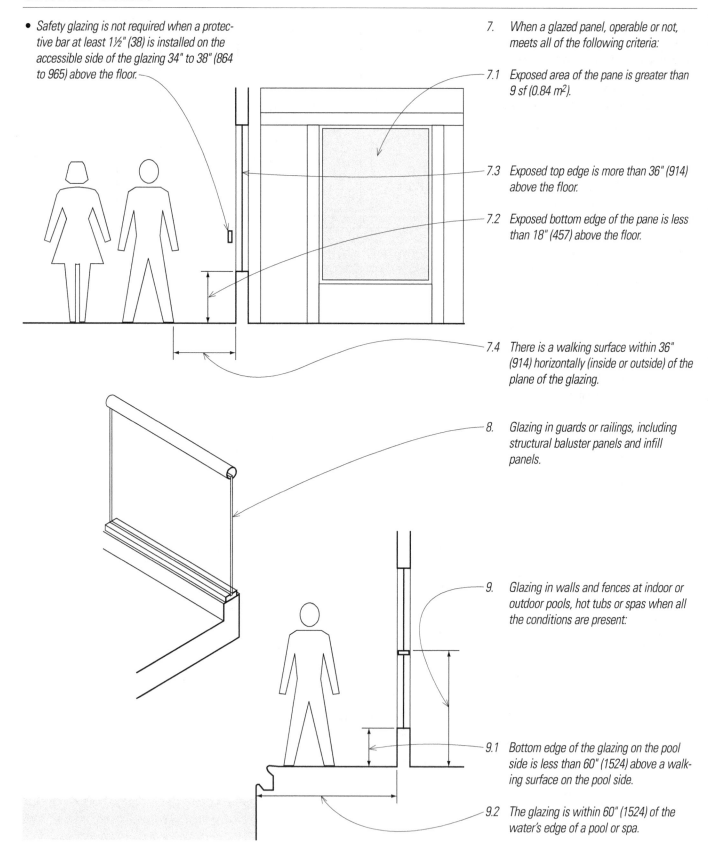

• Safety glazing is not required when a protective bar at least 1½" (38) is installed on the accessible side of the glazing 34" to 38" (864 to 965) above the floor.

7. When a glazed panel, operable or not, meets all of the following criteria:

7.1 Exposed area of the pane is greater than 9 sf (0.84 m²).

7.3 Exposed top edge is more than 36" (914) above the floor.

7.2 Exposed bottom edge of the pane is less than 18" (457) above the floor.

7.4 There is a walking surface within 36" (914) horizontally (inside or outside) of the plane of the glazing.

8. Glazing in guards or railings, including structural baluster panels and infill panels.

9. Glazing in walls and fences at indoor or outdoor pools, hot tubs or spas when all the conditions are present:

9.1 Bottom edge of the glazing on the pool side is less than 60" (1524) above a walking surface on the pool side.

9.2 The glazing is within 60" (1524) of the water's edge of a pool or spa.

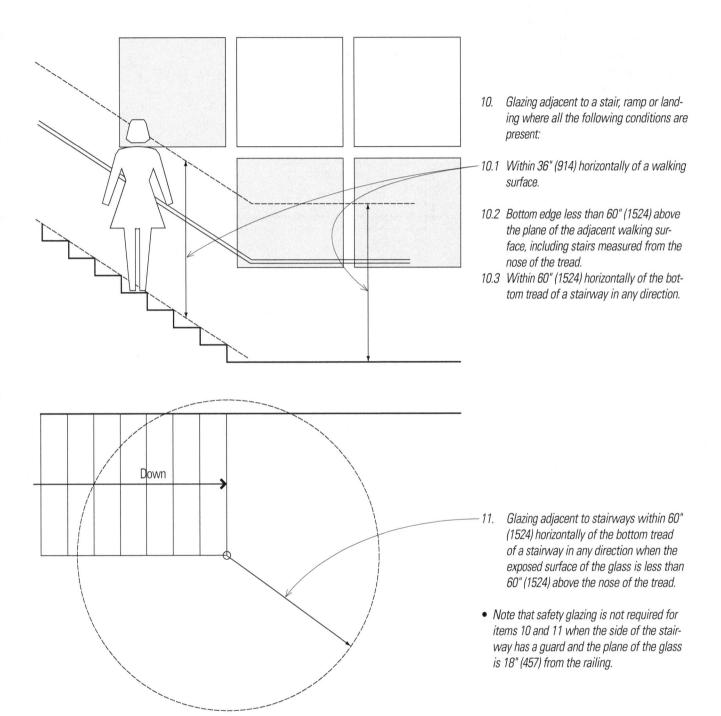

10. *Glazing adjacent to a stair, ramp or landing where all the following conditions are present:*

10.1 *Within 36" (914) horizontally of a walking surface.*

10.2 *Bottom edge less than 60" (1524) above the plane of the adjacent walking surface, including stairs measured from the nose of the tread.*

10.3 *Within 60" (1524) horizontally of the bottom tread of a stairway in any direction.*

Down

11. *Glazing adjacent to stairways within 60" (1524) horizontally of the bottom tread of a stairway in any direction when the exposed surface of the glass is less than 60" (1524) above the nose of the tread.*

• *Note that safety glazing is not required for items 10 and 11 when the side of the stairway has a guard and the plane of the glass is 18" (457) from the railing.*

Chapter 25: Gypsum Board and Plaster

Code requirements for the use of gypsum wallboard and plaster focus on durability and weather-resistance of the material. The standard in Chapter 25 relate to quality of construction, not to quality of appearance. A key measure of where standards for exterior walls are to be applied is determined by the weather exposure of the surface. Weather-exposed surfaces are defined in §2502 as obviously as one might expect, with three exceptions that are not to be considered as weather-exposed surfaces:

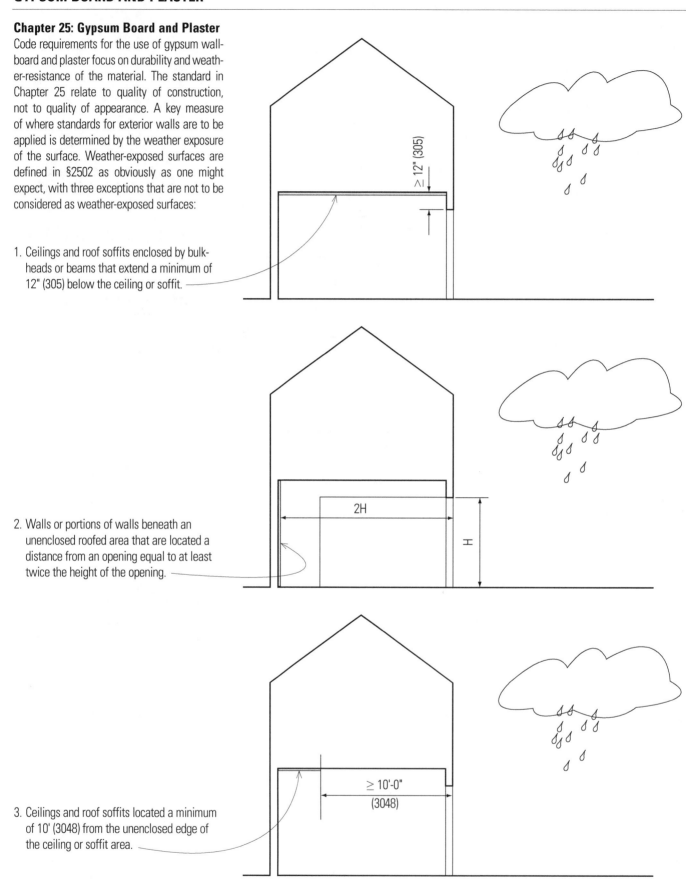

1. Ceilings and roof soffits enclosed by bulkheads or beams that extend a minimum of 12" (305) below the ceiling or soffit.

2. Walls or portions of walls beneath an unenclosed roofed area that are located a distance from an opening equal to at least twice the height of the opening.

3. Ceilings and roof soffits located a minimum of 10' (3048) from the unenclosed edge of the ceiling or soffit area.

Vertical and Horizontal Assemblies

§2504 requires wood framing, stripping and furring for lath or gypsum board is to be a minimum nominal thickness of 2" (51) in the least dimension. Wood furring strips over solid backing may be 1 × 2 (25 × 51).

Shear Wall Construction

§2505 permits gypsum board and lath-and-plaster walls to be used for shear walls. The use is dependent on the type of wall-framing materials. Wood-framed walls and steel-framed walls may be used to resist both wind and seismic loads. Walls resisting seismic loads are subject to the limitations of Section 12.2.1 of ASCE 7.

Materials

§2506 and 2507 specify that gypsum board and lath and plaster materials are to meet Code-designated ASTM standards. Materials are to be stored to protect them from the weather.

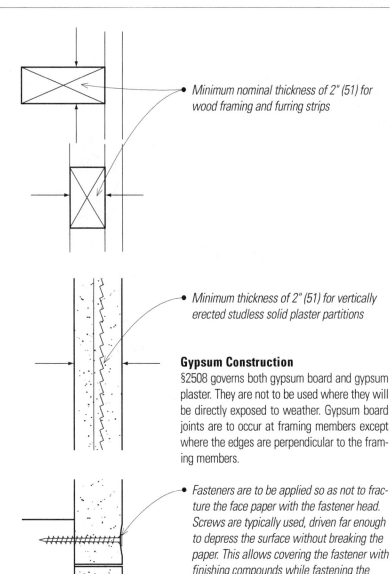

• *Minimum nominal thickness of 2" (51) for wood framing and furring strips*

• *Minimum thickness of 2" (51) for vertically erected studless solid plaster partitions*

Gypsum Construction

§2508 governs both gypsum board and gypsum plaster. They are not to be used where they will be directly exposed to weather. Gypsum board joints are to occur at framing members except where the edges are perpendicular to the framing members.

• *Fasteners are to be applied so as not to fracture the face paper with the fastener head. Screws are typically used, driven far enough to depress the surface without breaking the paper. This allows covering the fastener with finishing compounds while fastening the material securely.*
• *With few exceptions most fire-resistance-rated assemblies are to have joints and fasteners treated.*

Gypsum Board in Showers and Water Closets

§2509 governs the use of gypsum board as a base for tile or wall panels for tubs, showers and water closet compartment walls. All gypsum board used in these situations is to be water-resistant. Note that water-resistant gypsum board cannot be used in these applications if a vapor retarder is used in shower or bath compartments. Also, water-resistant gypsum backing board cannot be used in ceilings where framing exceeds 12" (305) on center for ½" (12.7 mm) gypsum backing board and 16" (406) on center for ⅝" (15.9 mm) gypsum backing board.

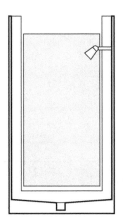
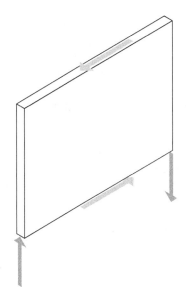

Exterior Plaster

§2512 requires cement plaster, also called stucco, to be applied in three coats over corrosion-resistant metal lath or wire fabric lath, or applied in two coats over masonry or concrete. The three coats have specific names and functions.

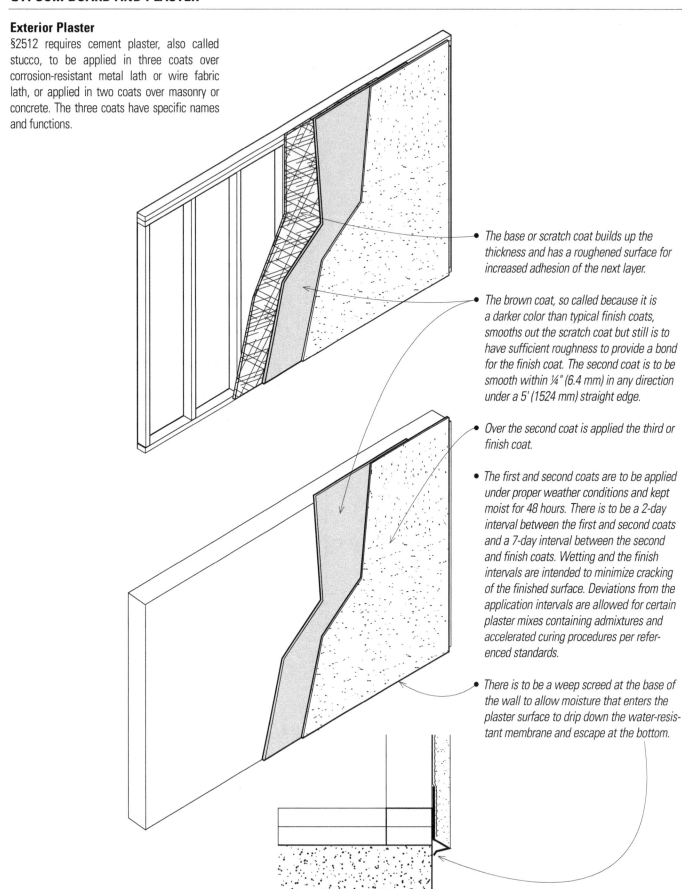

• The base or scratch coat builds up the thickness and has a roughened surface for increased adhesion of the next layer.

• The brown coat, so called because it is a darker color than typical finish coats, smooths out the scratch coat but still is to have sufficient roughness to provide a bond for the finish coat. The second coat is to be smooth within ¼" (6.4 mm) in any direction under a 5' (1524 mm) straight edge.

• Over the second coat is applied the third or finish coat.

• The first and second coats are to be applied under proper weather conditions and kept moist for 48 hours. There is to be a 2-day interval between the first and second coats and a 7-day interval between the second and finish coats. Wetting and the finish intervals are intended to minimize cracking of the finished surface. Deviations from the application intervals are allowed for certain plaster mixes containing admixtures and accelerated curing procedures per referenced standards.

• There is to be a weep screed at the base of the wall to allow moisture that enters the plaster surface to drip down the water-resistant membrane and escape at the bottom.

Chapter 26: Plastics

Chapter 26 covers a wide variety of plastic building materials that can take many diverse forms. Foam plastic boards and shapes, plastic insulation, plastic veneers, plastic finish and trim shapes and light transmitting plastics for glazing are all covered by this chapter.

Because plastics are typically flammable, the Code places restrictions on their use related to their flame spread and smoke generation. Foam plastics are typically to have a flame spread index of no more than 75 and a smoke-developed index of no more than 450 when tested at their anticipated thickness.

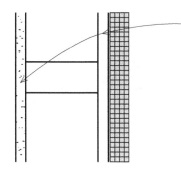

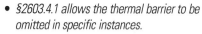

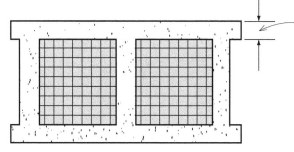

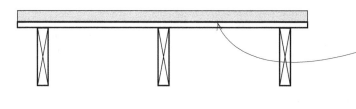

- A thermal barrier of ½" (12.7 mm) gypsum board is to separate foam plastic from the interior of a building. Equivalent materials may be used if they limit the temperature rise to a specified limit for 15 minutes.

- *§2603.4.1 allows the thermal barrier to be omitted in specific instances.*

- *The thermal barrier may be omitted if the foam plastic insulation is covered on each face by at least 1" (25.4 mm) of masonry or concrete.*

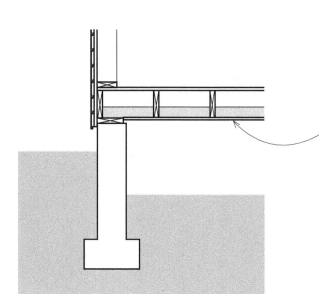

- *A thermal barrier is not required when the foam insulation is part of a Class A, B or C roof-covering assembly and if the assembly passes specified fire tests.*

- *In attics or crawl spaces "where entry is made only for service of utilities," thermal barriers are not required if the foam insulation is covered with any of the specified protective materials.*

PLASTICS

In buildings not of Type V construction, limitations are placed by §2603.5 on the use of foam plastic insulation. It is acceptable to use this material, which often occurs in exterior insulation and finish system assemblies and in manufactured exterior panels, when they meet the code criteria.

Foam plastic insulation may be used in buildings not of Type V construction in the following instances:

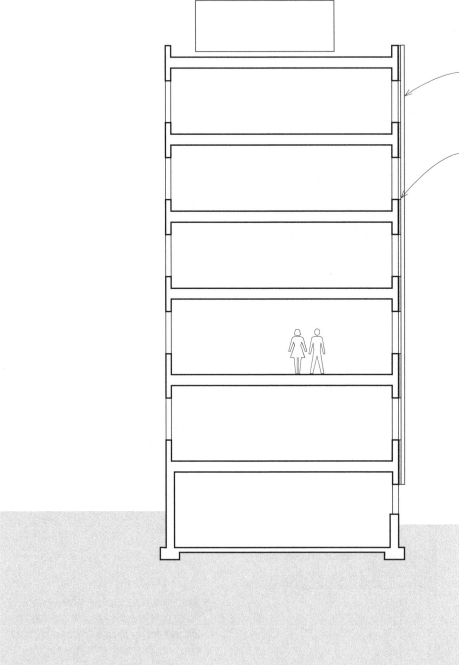

- §2603.5.1: Where the wall is to have a fire-resistance rating, the wall assembly shall demonstrate that the rating is maintained per tests conducted under ASTM E-119.

- §2603.5.2: A thermal barrier meeting the provisions of §2603.4 is to be provided except at one-story buildings. This thermal barrier is equivalent to one layer of ½" (12.7 mm) gypsum board.

- §2603.5.4: The flame-spread index is to be 25 or less and the smoke-developed index is to be 450 or less and must be less than 4 inches (102) thick.

- §2603.5.5: The wall is to be tested per NFPA 285. Note also the exception for one-story buildings complying with §2603.4.1.4, which allows use of specified foam plastics covered with aluminum or steel when the building is sprinklered.

- §2603.5.7: Exterior walls with foam plastic insulation are not to exhibit sustained flaming when tested per NFPA 268. As exceptions, the assemblies may be protected with a thermal barrier, 1" (25.4 mm) of concrete, metal panels or 7/8" (22.2 mm) of stucco.

Light-Transmitting Plastics

§2606 governs plastics used for ceiling light diffusers, glazing in skylights, glazing in walls, glazing in showers and light-transmitting plastics in light fixtures. These materials are divided into two classes based upon their burning rate. Class CC1 has a burning extent of 1" (25.4 mm) or less per ASTM D 635. Class CC2 materials have a burning rate of 2½" per minute (1.06 mm/s) or less per the same test.

Light-Transmitting Plastic Wall Panels

§2607 allows plastics to be used as light-transmitting wall panels in buildings other than Group A-1, A-2, H, I–2 and I–3 occupancies. They can be used in B occupancy clinics or healthcare provider's office buildings. There are limitations on area and panel sizes based on the location on the property and the class of the plastic material. Table 2607.4 lists allowable areas, and vertical and horizontal separations between plastic wall panels. The areas and percentages listed in the table may be doubled for buildings that are fully sprinklered. Panels are to be separated by distances as noted in the table unless a flame barrier extends 30" (762) beyond the exterior wall in the plane of the floor.

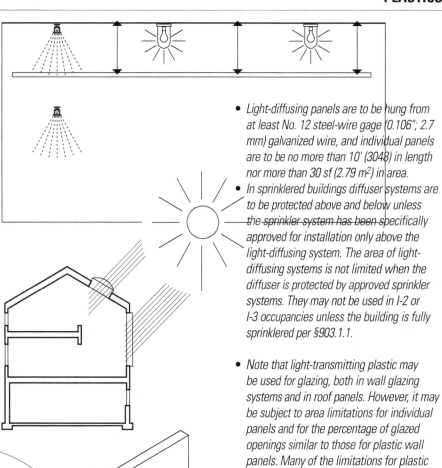

- *Light-diffusing panels are to be hung from at least No. 12 steel-wire gage (0.106"; 2.7 mm) galvanized wire, and individual panels are to be no more than 10' (3048) in length nor more than 30 sf (2.79 m²) in area.*

- *In sprinklered buildings diffuser systems are to be protected above and below unless the sprinkler system has been specifically approved for installation only above the light-diffusing system. The area of light-diffusing systems is not limited when the diffuser is protected by approved sprinkler systems. They may not be used in I-2 or I-3 occupancies unless the building is fully sprinklered per §903.1.1.*

- *Note that light-transmitting plastic may be used for glazing, both in wall glazing systems and in roof panels. However, it may be subject to area limitations for individual panels and for the percentage of glazed openings similar to those for plastic wall panels. Many of the limitations for plastic glazing are excepted for buildings that are fully sprinklered.*

2'-6" (762)

Fire Separation Distance	Class of Plastic	Maximum % of Exterior Wall	Maximum Area of Panel	Minimum Separation
Under 6' (<1829)	Plastic panels are not permitted			
6'–10' (1829–3048)	CC1	10%	50 sf (4.7 m²)	8' (2438) vertical, 4' (1219) horizontal
	CC2	Not permitted	Not permitted	
11'–30' (3353–9144)	CC1	25%	90 sf (8.4 m²)	6' (1829) vertical, 4' (1219) horizontal
	CC2	15%	70 sf (6.5 m²)	8' (2438) vertical, 4' (1219) horizontal
Over 30' (>9144)	CC1	50%	unlimited	3' (914) vertical, none horizontal
	CC2	50%	100 sf (9.3 m²)	6' (1829) vertical, 3' (914) horizontal

Electrical Systems

Chapter 27 specifies that electrical systems are to be designed and constructed in accordance with the *ICC Electrical Code*, which is in turn based on the *National Electrical Code* published by the NFPA.

The *International Building Code* contains scoping requirements for provision of emergency and standby power as defined in the *ICC Electrical Code*. Among these are:

- Emergency power for voice communications systems in Group A occupancies per §907.2.1.2.
- Standby power for smoke-control systems per §909.11.
- Emergency power for exit signs per §1011.5.3
- Emergency power for means of egress illumination per §1006.3

- Emergency and standby power for high-rise building systems:
  - Fire command center
  - Fire pumps
  - Emergency voice/alarm communications systems
  - Lighting for mechanical equipment rooms
  - Elevators

- Standby power for elevators per §3003.1

Note that the Code requires emergency and standby power systems be maintained and tested in accordance with the *International Fire Code*.

Mechanical Systems

Chapter 28 requires mechanical appliances, equipment and systems to be constructed, installed and maintained in accordance with the *International Mechanical Code* and the *International Fuel Gas Code*. Masonry Chimneys are to comply with both the IMC and Chapter 21 of the *International Building Code*.

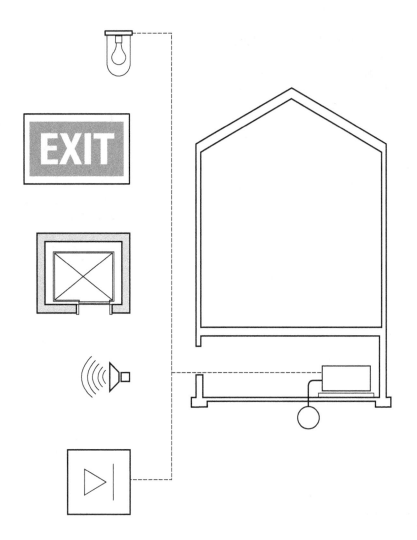

| Occupancy | Water Closets | | Lavatories | Tubs Showers | Drinking Fountains | Other |
	Male	Female				
Restaurant	1/75	1/75	1/200	—	1/500	1 service sink
Theater	1/125	1/65	1/200	—	1/500	1 service sink
Arena	1/75/ first 1,500, 1/120/ # above	1/40/ first 1,500 1/120/ # above	1/200 (m) 1/150 (f)	—	1/1,000	1 service sink
Mercantile	1/250*	1/500*	1/750	—	1/1,000	1 service sink
Business	1/25*	1/25*	1/80	—	1/100	1 service sink
Single-Family	1/dwelling		1/dwelling	1/dwelling	—	1 kitchen sink, 1 clothes washer connection/ dwelling

* Separate facilities will be required in many cases per §2902.2. See also §2902.4 for public toilet requirements

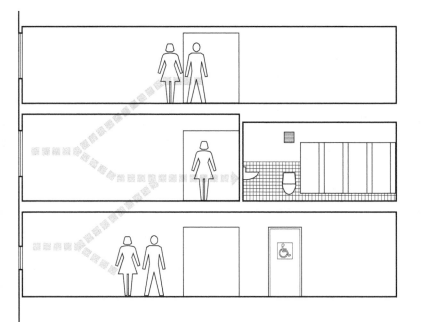

• *In public facilities, customers, patrons, visitors and employees are to be provided with public toilet facilities. These facilities are to be located not more than one story above or below the space needing the facilities. The path of travel to such facilities is not to exceed 500' (15 240). In conjunction with Chapter 11, the path of travel should be accessible or alternate accessible facilities should be provided.*

Plumbing Systems

Chapter 29 requires that plumbing systems be constructed, installed and maintained in accordance with the *International Plumbing Code*.

Minimum Plumbing Fixtures

§2902 and Table 2902.1 specifies the number of plumbing fixtures to be provided in various occupancy groups. This table has cross references to the *International Plumbing Code* for such items as the number of urinals in relation to the number of water closets. Thus the two codes must be read together when determining the number of fixtures. Many jurisdictions have not adopted the entire family of International codes. The designer should verify with the AHJ which plumbing code is adopted in the area where the building is constructed. The plumbing fixture table in the Uniform Plumbing Code is quite different from the tables in the "I" codes. It is essential to know during design which table to apply when determining the number of fixtures to be provided.

Fixture counts are determined by the occupancy type of the building and by the number of occupants as determined under the Code. Table 2902.1 sets out requirements for water closets (urinals are per the IPC, as noted above), lavatories, bathtubs/showers, drinking fountains, kitchen sinks and service sinks. The number of fixtures is based on observations of use patterns for various occupancies.

Occupants are to be presumed to be half-male and half-female unless statistical data demonstrating a different distribution of the sexes is approved by the building official. In certain occupancies, there are to be more women's fixtures than men's. This is based on use patterns where women take longer to use facilities than do men. Also, certain assembly occupancies have fixture counts based on use patterns during concentrated times, such as intermissions during theater performances.

Fixture counts for residential occupancies are expressed in terms of numbers of fixtures per dwelling unit, rather than by occupant load.

ELEVATORS AND CONVEYING SYSTEMS

Chapter 30 governs the design, construction and installation of elevators and other conveying systems, including escalators, moving walks, personnel hoists and materials. We will discuss requirements with significant design impacts.

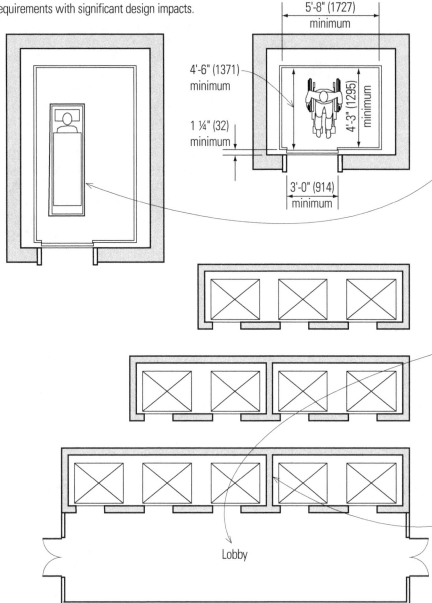

Lobby

- Where passenger elevators are required to be accessible per Chapter 11, the elevators are to comply with ICC A117.1. Typical elevator cab dimensions and controls for accessible elevators are illustrated..
- Note that per §1008.1, means of egress doors in Group I-2 occupancy used for the movement of beds shall provide a clear width not less than 41.5 inches (1054 mm).
- Certain elevators in I-2 occupancies will need to accommodate hospital beds and gurneys.

- Hoistways are to have fire-resistance ratings as required by Chapters 6 and 7. Doors in elevator shafts, including the elevator car doors, are to comply with the requirements of Chapter 7. Elevator car doors and hoistway doors have a fire rating, but they are not smoke-tight. That is why elevator lobbies are required in many circumstances where smoke migration from floor to floor may present a hazard. There are several proprietary systems that address the need to isolate elevator shafts from the rest of the building without provision of lobbies. But no matter the ultimate design solution, it is important to remember that design measures may need to be taken to address the lack of smoke protection in elevator hoistway doors.

- When four or more cars serve the same portion of a building, they are to be located in two separate hoistways. The purpose of this requirement is to minimize the chance that a fire or other emergency can disable or contaminate with smoke all of the elevators in a bank. Three elevators can be in a single enclosure, but five elevators would need a shaft division between two sets of cars.

- Elevators are typically not allowed to be used for egress unless part of an accessible means of egress per §1007.4.
- Standardized signage is to be posted near elevators not used for egress stating:

  IN FIRE EMERGENCY,
  DO NOT USE ELEVATOR.
  USE EXIT STAIRS.

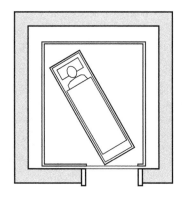

- In buildings four stories or more in height, at least one elevator is to be provided for fire-department emergency access to all floors. This elevator is to be sized to accommodate a 24" by 84" (610 by 1930) ambulance stretcher, in a horizontal and open position (that is, not partially folded up). This elevator is to be identified with the star of life, the international symbol for emergency medical services.

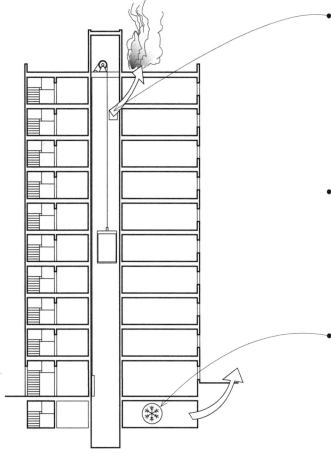

- Hoistways of elevators that penetrate more than three stories are to have a means of venting smoke and hot gases to the outer air in case of fire. Vents are to be located below the floor or floors at the top of the hoistway and must open either directly to the outer air or through noncombustible ducts to the outer air. The vent area is to be not less than 3½% of the area of the hoistway, but not less than 3 sf (0.28 m^2) per elevator car.

- Other than independent plumbing for drains or sump pumps for the elevator shaft itself, elevator shafts are not to be used for plumbing or mechanical systems.

- Elevator machine rooms for solid-state elevator controls are to have an independent ventilation or air-conditioning system to protect against overheating of the control equipment. Most elevator controls are now solid-state, so this should be assumed to be applicable in almost all installations. The system design criteria for temperature control will be determined by the elevator-equipment manufacturer's requirements.

Escalators

Per §3005.2.1. escalator floor openings are to be enclosed unless Exception 2 of §707.2 is satisfied. Note also that per §1003.7 escalators are not to be used as a component of a required means of egress.

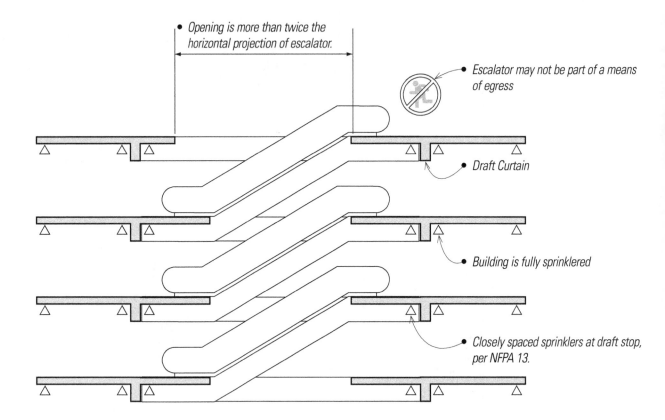

- Opening is more than twice the horizontal projection of escalator.

- Escalator may not be part of a means of egress

- Draft Curtain

- Building is fully sprinklered

- Closely spaced sprinklers at draft stop, per NFPA 13.

20
Existing Structures
(IBC Chapter 34)

This chapter contains the design requirements for work performed on existing structures. The provisions of this chapter control work related to the alteration, repair, addition or change of occupancy of existing structures. All of these activities trigger compliance with the relevant portions of Code Chapter 34. Compliance with other codes, such as the International Plumbing, Electrical and Mechanical Codes, is also required by reference in the International Building Code. This group of codes takes a coordinated approach to correlating requirements for existing buildings.

Note that a change of occupancy, whether it involves any physical work on a building or not, will trigger code-compliance review for the area of the building where the occupancy change occurs. This is not required for a change of occupancy only when the building official determines that the new use is less hazardous for life safety or fire risk than the existing one.

The International Code Council has developed a separate code that addresses existing structures. Check with your local jurisdiction to see if the IEBC has been adopted, and how it is applied to existing buildings. The IEBC includes many of the same requirements, such as IBC alternative compliance methods in Chapter 34 of the IBC. The IEBC creates a prescriptive set of requirements for repair, alterations and change of occupancy. Some jurisdictions have adopted both the criteria of Chapter 34 in the IBC and the IEBC allowing the owner and architect to make the decision as to which method to use. §101.3 of the IEBC states: "The intent of this code *[the IEBC]* is to provide flexibility to permit the use of alternative approaches to achieve compliance with minimum requirements to safeguard the public health, safe and welfare insofar as they are affected by the repair, alteration, change of occupancy, addition and relocation of existing buildings."

DEFINITIONS

§3402 contains definitions that supplement those generally applicable definitions in Chapter 2 of the Code. These definitions are primarily used in discussions regarding provision of access to existing buildings for persons with disabilities.

PRIMARY FUNCTION is the major activity for which the facility is intended, such as a banking hall in a bank or the dining area in a cafeteria. The primary function in a public service facility takes place in areas where the public is accommodated. It also occurs where the main functions of a building such as a private office building take place. Access should be provided to areas that are added or renovated unless such work is *technically infeasible*.

TECHNICALLY INFEASIBLE applies to alterations that have little likelihood of being accomplished because they would necessitate removing existing load-bearing structure or where there are other physical or site constraints that prevent modifications that would meet the criteria for new construction.

The definition of technically infeasible should be used with great caution, since it will likely lead to a lack of provision of accessibility for persons with disabilities. The Americans With Disabilities Act requires building owners to be proactive in barrier removal. Any decision to not undertake access work during other alterations must be carefully documented to justify any decision the owner makes that work is technically infeasible.

Additions, Alterations or Repairs

§3403 governs additions, alterations and repairs to any building or structure. These terms have specific definitions in Chapter 2. Additions will likely make the existing building higher or larger, but may have little impact on the existing building other than their attachment of the new part to the existing part. Alterations may involve extensive reworking of the plan, section and materials of the building, but do not result in any increase to the height or area of the building. Repairs are done on existing building fabric to maintain the use and appearance of the elements.

An understanding of the provisions of §3403 is crucial to understanding how the Code treats work in existing buildings. These requirements are different than those for change in occupancy, which is addressed in §3406. Physical construction work in areas of an existing building, whether for additions, alterations or repairs, are to conform to the Code requirements for new construction.

Additions to an existing building must result in a completed building that complies with Chapter 5 of the Code for heights and areas. However, it is essential to understand that those portions of the building not altered and not affected by the alteration, repair or addition are not required to comply with the requirements for a new structure. In other words, work on an existing building does not trigger a wholesale code-compliance review or upgrade. There are provisions in §3410, discussed below, for compliance alternatives that begin to apply a rating system to upgrading buildings in order to gradually increase the code compliance of buildings as they are altered over time. But for now, most renovation projects will be done under the prescriptive portions of Chapter 34.

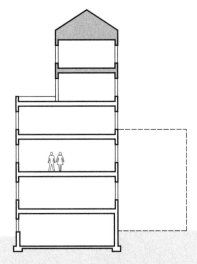

- *Addition is an extension or increase in floor area or height of a building or structure.*

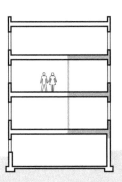

- *Repair is the reconstruction or renewal of any part of an existing building for the purpose of its maintenance.*

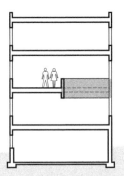

- *Alteration is any construction or renovation to an existing structure other than repair or addition.*

The exemption implied by the statement that work only needs to meet new code standards in the area of work is not a blanket exemption from examining the impact of the new work on the entire existing structure.

- Per §3403.2, while it may be possible to draw a definitive line around the renovation work in the building, the Code still requires that additions or alterations not increase the structural force in any element by more than 5% unless the element will remain in compliance with new code requirements. Also, if unsound structural elements are uncovered during addition or alteration work, those defective elements are to be made to conform to the requirements for new structures.

- The Code does allow areas of the structure with less structural live load capacity than current requirements to remain in place if certain criteria are met. Such areas must have been compliant with the codes in effect at the time of erection, the loads must be reviewed and approved by the building official, and any live load reduction must be posted with the approved load.

- Additions or alterations must also satisfy seismic criteria applied in a similar fashion to other structural criteria. Where additions or alterations increase seismic forces by more than 10% on existing members, then the structural system is to be analyzed for compliance with the seismic design criteria of ASCE 7.

- Per §3403.3, alterations or repairs that are nonstructural in nature may be made of the same materials as the original building as long as they do not impact the structure or adversely impact required fire-resistance.

- Per §3403.4, alteration or replacement of an existing stair need not comply with the requirements of §1009 for such items as rise and run when the physical conditions of space or construction will not allow a reduction in pitch or slope. Although the Code allows reuse or duplication of existing stairs, it is always prudent for the designer to try and incorporate new Code provisions into stair renovations whenever possible. Attention should be paid to such elements as regularizing tread and riser heights, even if not fully compliant with new Code standards; tread profiles; providing closed risers for accessibility, and adjusting handrail heights and graspability.

- Stair compliant with current code, 7" (178) rise, 11" (279) run per §1009.3.

- Existing stair, for example, 8" (203) rise, 10" (254) run, not compliant with current code

FIRE ESCAPES

§3404 allows continued use of fire escapes in existing buildings, but they may never be used in new buildings. New fire escapes may be added to existing buildings where site conditions do not allow any other secondary means of egress. They can be used for no more than 50% of the required number of exits or exit capacity.

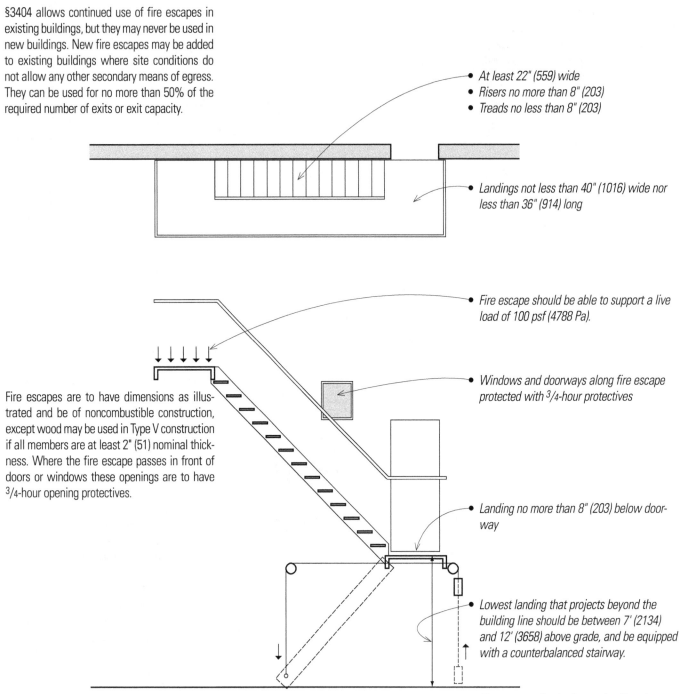

- At least 22" (559) wide
- Risers no more than 8" (203)
- Treads no less than 8" (203)

- Landings not less than 40" (1016) wide nor less than 36" (914) long

- Fire escape should be able to support a live load of 100 psf (4788 Pa).

- Windows and doorways along fire escape protected with 3/4-hour protectives

Fire escapes are to have dimensions as illustrated and be of noncombustible construction, except wood may be used in Type V construction if all members are at least 2" (51) nominal thickness. Where the fire escape passes in front of doors or windows these openings are to have 3/4-hour opening protectives.

- Landing no more than 8" (203) below doorway

- Lowest landing that projects beyond the building line should be between 7' (2134) and 12' (3658) above grade, and be equipped with a counterbalanced stairway.

- Clearance under the lowest landing above alleyways and thoroughfares less than 30' (9144) wide is not to be less than 12' (3658). This clearance requirement may be seen as a conflict with the other requirement that the landing itself be no higher than 12' (3658) as this leaves no room for structure.

When a change in the use of an existing building places that building in another occupancy category, whether in the same group or a different group, then §3406 requires that the building be made to comply with the code requirements for new construction. The building official may allow this change to occur without conforming to all the requirements of the code, provided the new or proposed use is less hazardous, based on life and fire risk, than the existing one. The only blanket exception to the requirements to meet the new Code is granted for stairways where the existing space and construction will not allow a reduction in pitch or slope.

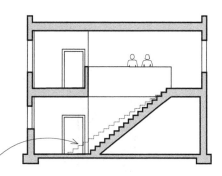

Historic Buildings

§3407 indirectly defines historic buildings as those buildings listed in or eligible for listing in the National Register of Historic Places, or designated by appropriate state or local law as historic. These buildings need not meet new construction requirements for any category of renovation or reuse, including change in occupancy, when the buildings are not judged by the building official to constitute a distinct life-safety hazard. Although not stated in the Code, it seems logical to presume that the building official's findings must be solicited and written down as part of the permit review for work in a historic structure. Note that accessibility requirements apply to historic buildings even if other new Code provisions do not.

ACCESSIBILITY FOR EXISTING BUILDINGS

§3409 governs the provision of accessibility in existing buildings, whether related to maintenance, change of occupancy, additions and alterations to existing buildings, including historic buildings. To summarize the requirements for this section, we could state that if access can be provided as part of the work on an existing building, then it should be, whenever and wherever possible.

The provision of access to altered or added areas, or when a change in occupancy occurs, is often the most complex and ambiguous part of determining code compliance for existing buildings. "When in doubt, do it" should be the designer's motto regarding provision of accessibility in existing buildings. One code criteria for determining whether access can and should be provided is the definition of *technically infeasible*. This definition, contained in Code Section 3402, is repeated verbatim here for emphasis.

- Technically Infeasible. An alteration of a building or facility that has little likelihood of being accomplished because the existing structural conditions require the removal or alteration of a load-bearing member that is an essential part of the structural frame, or because other existing physical or site constraints prohibit modification or addition of elements, spaces or features that are in full and strict compliance with the minimum requirements for new construction and which are necessary to provide accessibility.

Application of this test occurs during renovation work in existing buildings as covered in this section. The test must be applied narrowly and consistently. Cost is a factor, but not the sole one. Complexity and the technical ability to accomplish the necessary alteration work, while clearly related to cost, is a better way to approach the application of the definition.

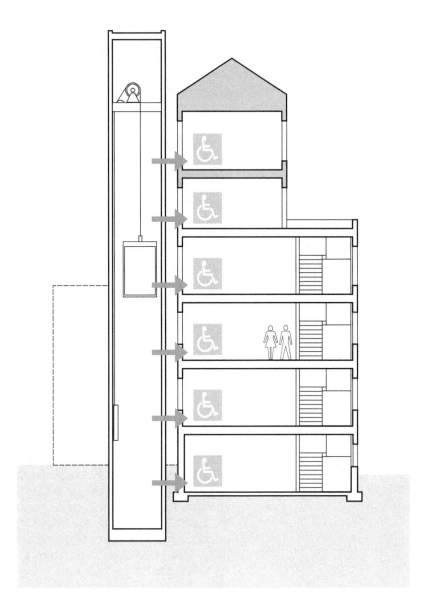

Per §3409.4, new construction provisions are to apply to a change of occupancy except where technically infeasible. This Code section applies this requirement to those portions of existing buildings that are altered concurrently with a change of occupancy. It might be argued that a change of occupancy that involved no alterations would not trigger accessibility requirements. However, the section goes on to require the following accessible features be provided. It should be noted that per §3409.4 and §3409.5 this list should be used for alterations and additions as well, except as modified by the alteration exceptions of 3409.6, as discussed below.

The following accessible features must be provided:

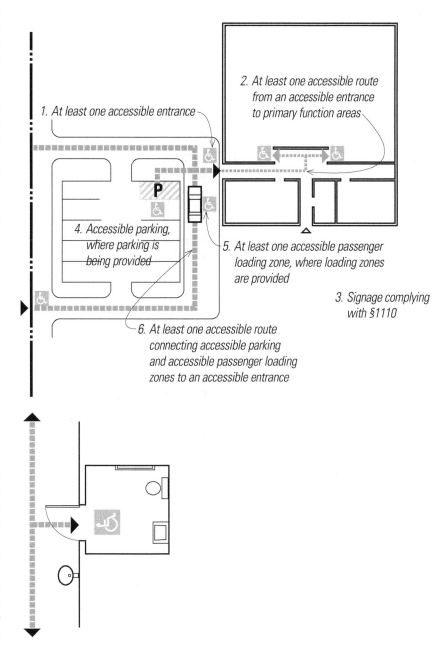

1. At least one accessible entrance

2. At least one accessible route from an accessible entrance to primary function areas

3. Signage complying with §1110

4. Accessible parking, where parking is being provided

5. At least one accessible passenger loading zone, where loading zones are provided

6. At least one accessible route connecting accessible parking and accessible passenger loading zones to an accessible entrance

Per §3409.7, when an alteration affects the accessibility to a primary function area, the route to that area is to be accessible and toilets and drinking fountains are to be provided on the path of travel. The cost of providing this accessible route is not required to exceed 20% of the cost of the alterations affecting the area or primary function. This calculation is best thought of in terms of base cost for the program area being equal to 100% with 20% of that number added to the base to determine the amount to be allocated to the accessible route.

For example:
- Base cost for alterations to primary function area: $600,000
- Cost allocated to providing accessible route at 20% of base cost: $120,000
- Total budget for program area and accessible route: $720,000

- Accessibility costs
- Base alteration costs

For alterations, which are defined as not adding height or area to an existing building, the following scoping provisions apply per §3409.8:

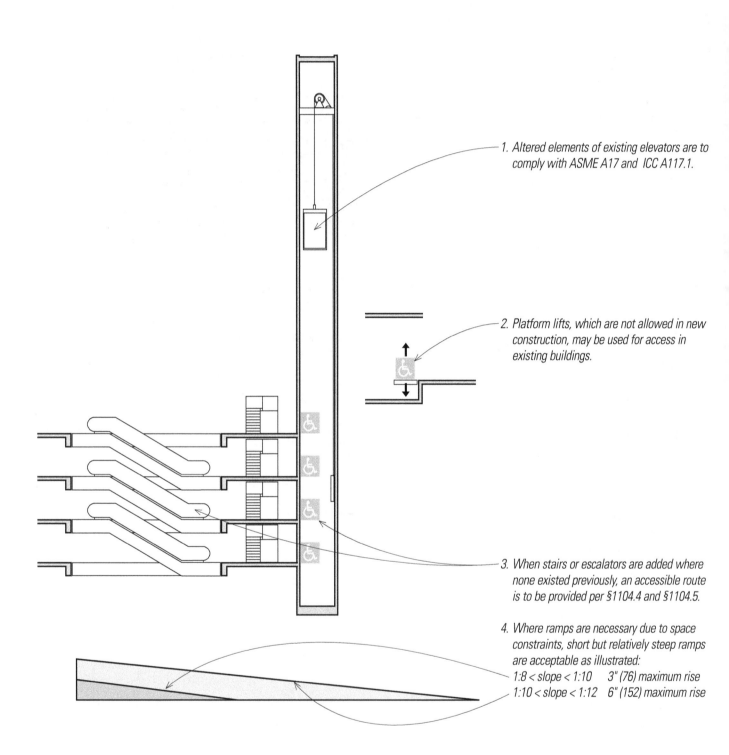

1. Altered elements of existing elevators are to comply with ASME A17 and ICC A117.1.

2. Platform lifts, which are not allowed in new construction, may be used for access in existing buildings.

3. When stairs or escalators are added where none existed previously, an accessible route is to be provided per §1104.4 and §1104.5.

4. Where ramps are necessary due to space constraints, short but relatively steep ramps are acceptable as illustrated:
   1:8 < slope < 1:10   3" (76) maximum rise
   1:10 < slope < 1:12   6" (152) maximum rise

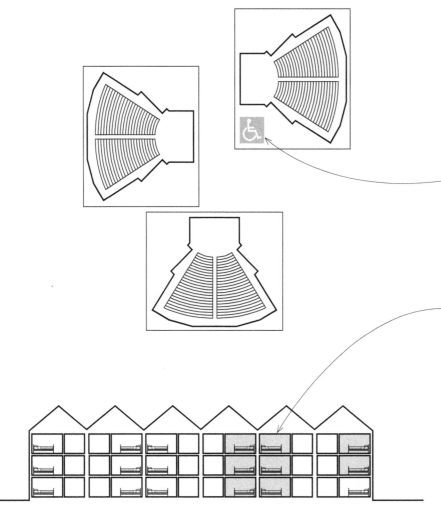

5. Performance areas need not all be accessible if it is technically infeasible to do so, but at least one of each type of performance area must be made accessible. This may be an issue at multiscreen cinemas that are the same types of spaces but offer different movies.

6. For dwelling or sleeping rooms in I-1, 2 or 3, and R-1, 2 or 4 occupancies, the requirements for accessible rooms from §1107 for Type A units and §907 for accessible alarms apply only to the quantity of rooms being altered or added. By using the word "quantity" in regard to rooms rather than applying the requirement to the rooms actually being altered it seems that the code intends the percentages to be used in determining the number of Type A units be applied to the number of spaces being altered, not the overall number of spaces in the existing buildings.

7. When it is technically infeasible to provide accessible toilet rooms or bathing facilities in existing rooms, an accessible unisex toilet or bathing facility is permitted. Such facilities must be located on the same floor and in the same area as the existing facilities.

8. When it is technically infeasible to provide accessible dressing, fitting or locker rooms at the same location as similar types of rooms, it is acceptable to provide an accessible room at the same level. If separate-sex rooms are provided then separate-sex accessible rooms are to be provided as well.

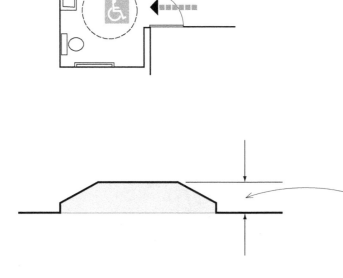

9. Thresholds at doorways are to be a maximum 3/4" (19.1 mm) high and must have beveled edges on each side.

COMPLIANCE ALTERNATIVES

§3410 allows evaluation of existing buildings with the intent to increase or maintain the level of public health, safety and welfare. This is to allow buildings to undergo repair, alteration, addition or change of occupancy without requiring full compliance with the new construction provisions of the code. These evaluations are an alternate path for Code approval from the other provisions in Chapter 34. The evaluation process is set forth in this section and is to be used in its entirety if this methodology for demonstrating code compliance is elected.

We will not go into these methods in detail here, but the designer should be aware of this alternate methodology. This system gives points for compliance with safety parameters for fire safety, means of egress and general safety. It is possible to receive negative points for noncompliant systems that are to remain as well as getting positive points for mitigating deficiencies. There are minimum point totals for each of the three areas of concern. The point totals are based upon the occupancy of the building with different weighted point requirements for fire safety, means of egress and general safety based upon the relative importance of each criteria for each occupancy. Buildings that achieve the necessary point totals for safety parameters pass and are permitted. Table 3410.7 is a summary sheet for use in applying the compliance alternative method.

It would seem that this system would prove most useful for instances of change in use or occupancy where renovations could be tailored to meet the safety parameters necessary for the new occupancy without spending money in areas not directly affecting the fire and life safety of the buildings.

Bibliography

2006 International Building Code. International Code Council, Inc., 2006.

2006 IBC Handbook, Fire- and Life-Safety Provisions. International Conference of Building Officials (ICBO), 2006.

Building Construction Illustrated, third edition. Francis D.K. Ching and Cassandra Adams. John Wiley and Sons, Inc., 2001.

Design Guide to the 1997 Uniform Building Code. Richard Conrad and Steven R Winkel. John Wiley and Sons, Inc., 1998.

From Model Codes to the IBC: A Transitional Guide. Rolf Jensen and Associates. R.S. Means Company, Inc., 2001.

Illustrated 2000 Building Code Handbook. Terry L. Patterson. McGraw-Hill, 2001.

International Energy Conservation Code. International Code Council, Inc., 1999.

NFPA 101, Life Safety Code Handbook, Tenth Edition, Edited by Ron Coté, P. E. and Gregory Harrington, P. E. National Fire Protection Association, 2006.

A Visual Dictionary of Architecture. Francis D.K. Ching. John Wiley and Sons, Inc. 1995.

Index